MATHEMATICAL ASTRONOMY MORSELS

JEAN MEEUS

Published By

Willmann-Bell, Inc.
Publishers and Booksellers Serving
Astronomers Worldwide Since 1973
P.O. Box 35025
Richmond, Virginia 23235

Published by Willmann-Bell, Inc.
P.O. Box 35025, Richmond, Virginia 23235

Printed in the United States of America

Library of Congress Cataloging-in-Publication Data
Meeus, Jean
   Mathematical astronomy mosrsel / Jean Meeus.
      p.   cm.
   Includes index.
    ISBN 0-943396-51-4
    1. Astronomy--Mathematics. I. Title
QB43.2.M44   1997   96-534883
521--dc21   CIP

97 98 99 00 01 02 03 04 05 9 8 7 6 5 4 3 2

FOREWORD

Every time two full moons occur in the same month, pundits in the media take note. They explain that, according to folklore, the rare second one is called a *blue moon*—whence the saying, "once in a blue moon." However, they have got it backwards, says Philip Hiscock of the Folklore and Language Archive, Memorial University of Newfoundland. The fabled blue moon, meaning any rare or unusual occurrence, dates back at least 150 years in the English language. The link to lunar phases in a calendar month is more recent. Hiscock traces it from a 1937 Maine almanac to an obscure children's book published in 1985, followed the next year by a question card in the game of Trivial Pursuit!

Meanwhile, hardly a week goes by in the Internet's astronomy areas without a total novice dropping in to ask, "What about the alignment of planets coming in May of the year 2000? Are we in danger of a tidal wave or a massive earthquake?" A replay seems brewing of the public alarm over the supposed syzygy of 1982 March 10, widely touted as something that occurs once every 179 years.

Pop-culture obsessions like these do not originate in the magazines for amateur astronomers, let alone in scientific journals or textbooks. But they say something important about our society. Here we have naive but sincere people, whose astronomical curiosity has been stirred for perhaps the first time in their lives. Their interest will soon wane unless a teacher, commentator, or writer they respect can step in with a meaningful response. Even worse, their end-of-the-world fears may escalate.

Jean Meeus's latest book explores the frequency of blue moons, planetary groupings, and a great deal more, as only this master of astronomical calculations could. He predicted the May 2000 alignment in an article for *Sky & Telescope* magazine in December 1961, but without spreading the doomsday concern, of course. He has brought together these and other tidbits from his voluminous writings, spanning nearly half a century, on every sort of celestial configuration, cycle, and curiosity. His wide following in America can now enjoy these penetrating analyses, many of which originally appeared only in Europe. The collection is much more than mere anthology. Each conclusion has been checked, and virtually every numerical result calculated afresh, with all the rigor we have come to expect.

This Belgian astronomer is particularly attracted to the rarest of all celestial occurrences — things almost impossible to find by paging through almanacs or scrolling through time with a computer's planetarium program. For example, he investigates how often a bright star or planet is occulted by the moon *during* a total lunar eclipse. He looks at how many times per century Jupiter can appear "without a visible moon," all the Galilean satellites being either in front of the disk, behind it, or in eclipse. He goes on to examine another elusive event, one that the English amateur Horace Dall was lucky enough to photograph with his 15-inch reflector on 1956 April 21: the shadow of not one, not two, but three satellites crossing Jupiter's disk at once! This book lists the occasions when we, too, can hope to witness something similar during our lifetimes.

The detection of patterns and cycles is a theme pursued throughout. Most readers have probably heard about the Saros in connection with solar eclipses, or the eight-year cycle of Venus risings that is a cornerstone of the Maya calendar. But here we find evidence for the half Saros (about 3293 days), for which the author proposes the name Sar. A mysterious 586-year period also emerges among the lunar eclipses. It is so long a span that a few writers have fallen into a statistical trap. Using data for the entire 20th and 21st centuries, they have concluded that total lunar eclipses are more common than partial ones. Wrong! As Jean Meeus demonstrates with his beautiful diagram of eclipse clumps (page 104), the opposite is true over the long haul.

Many celestial cycles are fleeting, destined to fade away after a few iterations as others overlap them or start up afresh. It is a fallacy to think that you can recreate planetary motions for many years by spinning back or fast-forwarding a planetarium projector. Only someone with a profound grasp of astronomical motions and relationships could have produced an authoritative book like this.

Some readers will see here an antidote to the claims of astrology. Others will gain a deep insight into the misuse of statistics, especially in such areas as the sunspot cycle and its relation to weather on Earth. But all of us can acquire plenty of ammunition to settle bets at star parties, test computer programs, and amaze our friends (or an astronomy professor) with some little-known surprises about the sky and calendar.

So why exactly does Christmas fall more often on a Tuesday than on a Monday? How many centuries will elapse before 10 successive Easters occur in April? What is the reason that total solar eclipses are more common for observers in the Northern Hemisphere than in the Southern? Turn these pages, and you'll find out!

Roger W. Sinnott
Sky & Telescope magazine

Preface

This book contains about sixty chapters on various aspects of mathematical astronomy. These chapters are independent and they may be read in any order.

The pieces in this collection were written at different times and for several different journals, principally for *Heelal*, the monthly journal of the Belgian Dutch-language astronomical society 'Vereniging voor Sterrenkunde' (VVS). The texts have been translated and updated.

No glossary is given of the terms frequently used: this work is not a textbook on astronomy, and the reader is supposed to possess the appropriate astronomical background.

I would like to thank the publishers of *Heelal*, *Ciel et Terre* (Société Royale Belge d'Astronomie), *Hemel en Dampkring* (Nederlandse Vereniging voor Weer- en Sterrenkunde), *l'Astronomie* (Société Astronomique de France), the *Journal* of the British Astronomical Association, and the *Journal* of the Royal Astronomical Society of Canada, for permission to reprint my texts in the present volume.

Special appreciation is owed to Edwin Goffin, the Belgian specialist on minor planet orbits, who provided the material used in Chapters 33 and 34.

Jean Meeus

Table of Contents

Planetary Motions

Planetary Phenomena

On the Celestial Sphere

Statistics, etc.

Varia

Note on Time Reckoning

Calendar Dates

For the calendar dates, the Julian Calendar is used till 1582 October 4, while the Gregorian Calendar is used from 1582 October 15 onwards.

There is a disagreement between astronomers and historians about how to count the years preceding the year 1. In this book, the 'B.C.' years are counted astronomically. Thus, the year before the year $+1$ is called the year zero, and the year preceding the latter is the year -1. The year which the historians call 585 B.C. is actually the year -584.

In general, we have written the dates as is usual in astronomy, namely in the order year − month − day of month. For instance, 1994 Aug. 6 means the 6th day of August, 1994. This 'scientific' form for calendar dates reads from the largest to the smallest unit of time.

The dates are based either on the Universal Time or on the Dynamical Time, according to the case (see below). For example, 1984 November 22 indicates an event which took place between 0^h00^m and 24^h00^m Universal Time on 1984 November 22.

Dynamical Time and Universal Time

The Universal Time (UT), or Greenwich Civil Time, is based on the rotation of the Earth. The UT is necessary for civil life and for the astronomical calculations where local hour angles are involved.

However, the Earth's rotation is generally slowing down and, moreover, this occurs with unpredictable irregularities. For this reason, the UT is not a uniform time.

Astronomers, however, need a uniform time scale for their accurate calculations (celestial mechanics, orbits, ephemerides). From 1960 to 1983, in the great astronomical almanacs such as the *Astronomical Ephemeris*, use was made of the uniform time scale called the *Ephemeris Time* (ET) which was defined by the laws of dynamics and was based on the planetary motions. In 1984, the ET was replaced by the *Dynamical Time*, which is defined by atomic clocks. In practice, the Dynamical Time is a prolongation of the Ephemeris Time.

In this book, Dynamical Time will be indicated by TD.

TABLE   A

ΔT = TD − UT (in seconds) for the beginning of some years
Subtract the value given by this Table in order to convert TD to UT

Year	ΔT	Year	ΔT	Year	ΔT	Year	ΔT	Year	ΔT
1620	+124	1700	+ 9	1780	+17	1860	+ 7.9	1940	+24.3
1630	85	1710	10	1790	17	1870	+ 1.6	1950	29.1
1640	62	1720	11	1800	13.7	1880	− 5.4	1960	33.1
1650	48	1730	11	1810	12.5	1890	− 5.9	1970	40.2
1660	+37	1740	+12	1820	+12.0	1900	− 2.7	1980	+50.5
1670	26	1750	13	1830	7.5	1910	+10.5	1985	54.3
1680	16	1760	15	1840	5.7	1920	+21.2	1990	56.9
1690	10	1770	16	1850	7.1	1930	+24.0	1995	60.8

TABLE   B

Approximate values of ΔT

Year	Minutes	Year	Minutes	Year	Minutes
0	177	1000	35	2075	4
100	158	1100	27	2200	8
200	140	1200	20	2300	13
300	123	1300	14	2400	19
400	107	1400	9	2500	26
500	93	1500	5	2600	34
600	79	1600	2	2700	43
700	66	1700	0	2800	53
800	55	1800	0	2900	64
900	45	1980	1	3000	76

Table A gives the value of ΔT for the *beginning* of some years. Except for the two last values, they are taken from the *Astronomical Almanac* for 1988 (Washington, D.C.), pages K8 and K9. For epochs before the year 1620 or in the future, an *approximate* value of ΔT, in seconds, can be deduced from the following relation due to Morrison and Stephenson:

$$\Delta T = -15 + 0.00325 \, (\text{year} - 1810)^2$$

This leads to Table B. It should be noted that for the very past and future, the tabulated values of ΔT may be in error by many minutes, as the fluctuations due to the variable rotation of the Earth are unknown for those remote epochs.

Tabulated times in TD may be converted into Universal Time by *subtracting* from them the quantity ΔT, since we have UT = TD − ΔT.

THE MOON

1. The instantaneous lunar orbit

In 1991 Michelle Chapront-Touzé and Jean Chapront, two astronomers at the Bureau des Longitudes, Paris, France, published their *Lunar Tables and Programs from 4000 B.C. to A.D. 8000* (Willmann-Bell, ed.).

These tables enable the calculation, for any instant over the years -4000 to $+8000$, of

— the geocentric position of the Moon (longitude, latitude, distance);
— the osculating elements of the lunar orbit.

The osculating elements are those of the 'instantaneous' orbit of the Moon, that is, the elements of the elliptic orbit of a fictitious Moon whose position and velocity for the given instant would be the same as those of the real Moon.

The motion of the Moon is disturbed mainly by the gravitational attraction of the Sun, and the osculating elements of the orbit are complicated functions of time. As examples, we give in the Figures 1.*a* to 1.*d* the variation of four osculating elements of the Moon's orbit from 1996 January 1 to 1999 January 1.

The first drawing shows the variation of the instantaneous orbital eccentricity. This eccentricity is a maximum when the major axis of the lunar orbit is directed toward the Sun, as in January 1996, July 1996, February 1997, etc., which occurs at mean intervals of 205.9 days. This period is somewhat longer than six months by reason of the motion of the longitude of the Moon's perigee in the direct sense ($+0.11140$ degree per day). However, it appears from Figure 1.*a* that there are large secondary oscillations. While the mean value of the eccentricity is 0.055, its 'instantaneous' value can vary between the extremes 0.026 and 0.077.

Our second figure shows the variation of the inclination of the instantaneous lunar orbit on the ecliptic — the plane of the Earth's orbit around the Sun. This inclination is a maximum when, as in March and September 1997, the line of nodes of the lunar orbit is directed toward the Sun; this occurs at mean intervals of 173.3 days. This period is somewhat shorter than six months by reason of the retrograde motion of the longitude of the Moon's ascending node (-0.05295 degree per day). Here too we see that there are secondary oscillations of short period, and that their effect is larger when the value of the inclination is least.

The retrograde motion of the lunar nodes has a period of 18.6 years, more exactly of 6798.38 days with respect to the moving equinox, or 6793.48 days with respect to the stars. However, the motion of the nodes is not regular. The principal inequality has a period of 173.3 days, the same as that of the inclination. The variation of the longitude of the Moon's ascending node, referred to the mean equinox of the date, is illustrated in Figure 1.*c* for the years 1996 to 1998.

The line of nodes is almost stationary when it is directed toward the Sun. This coincides with the maximum value of the orbital inclination, and it is near these epochs that solar and lunar eclipses take place.

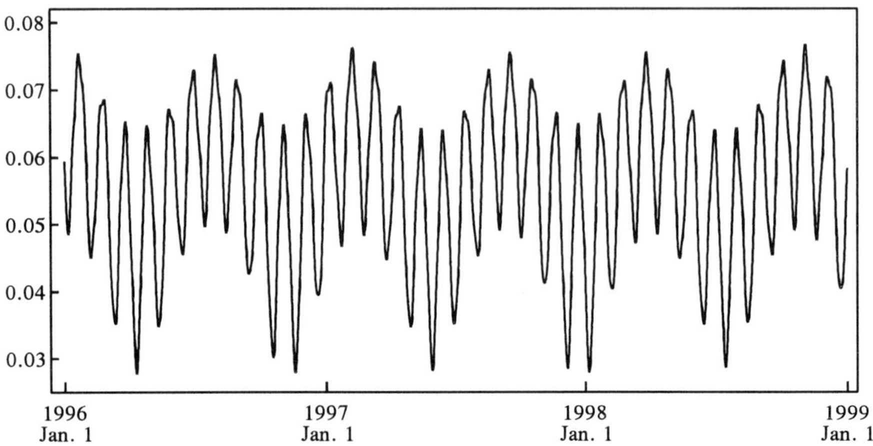

Fig. 1.a : *The instantaneous eccentricity of the lunar orbit, 1996 to 1998*

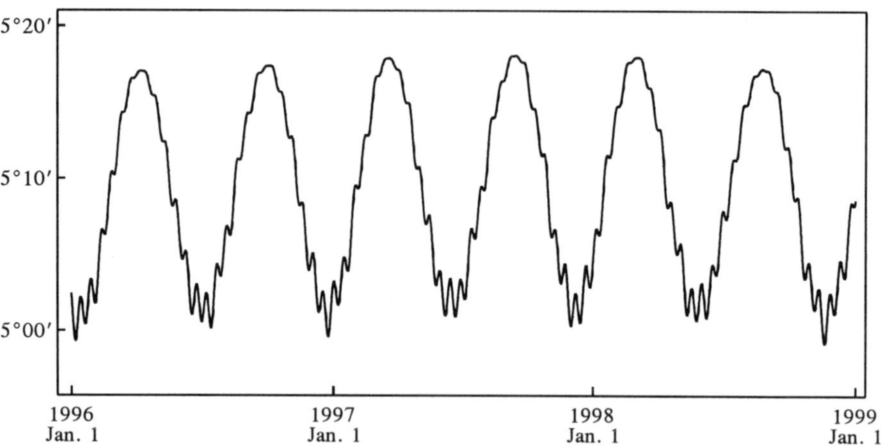

Fig. 1.b : *The instantaneous inclination of the lunar orbit on the ecliptic,*
1996 to 1998

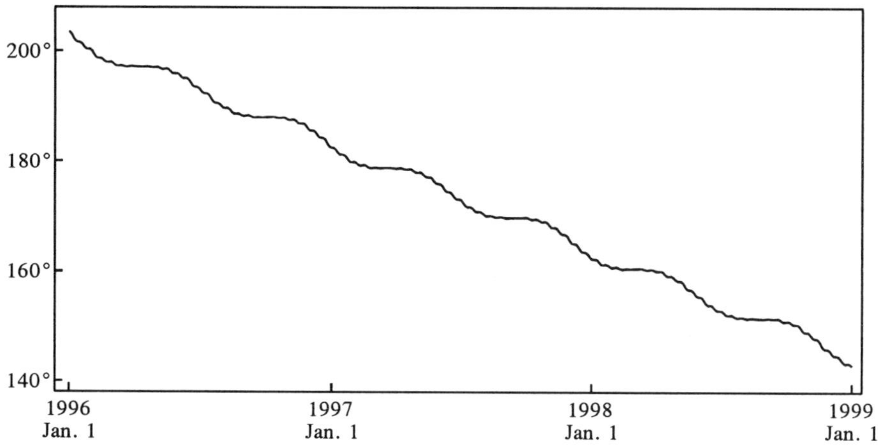

Fig. 1.c : *The longitude of the ascending node of the lunar orbit,*
1996 to 1998

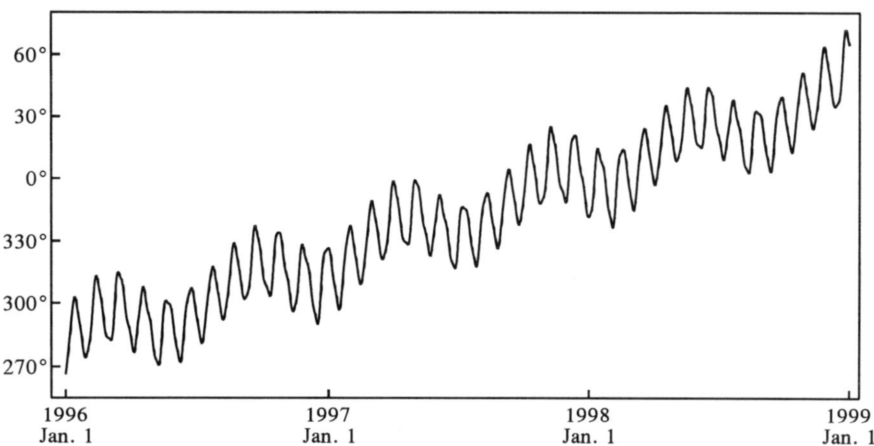

Fig. 1.d : *The longitude of the perigee of the lunar orbit,*
1996 to 1998

While the line of nodes is retrograding, the major axis of the lunar orbit moves in the direct sense, that is, in the same direction as the revolution of the Moon around the Earth. One complete revolution of the lunar perigee takes 8.85 years; more exactly, the period is 3231.50 days with respect to the moving equinox (tropical period), or 3232.61 days with respect to the stars (sidereal period).

The longitude of the perigee of the instantaneous lunar orbit undergoes important periodic inequalities, as is shown in Figure 1.*d*. The difference in longitude between this perigee and that of the mean lunar orbit can be as large as 30 degrees.

It should be noted that the longitude of the perigee is measured in two different planes: it is equal to the longitude of the ascending node, measured along the ecliptic from the vernal equinox to this node, increased by the arc from the ascending node to the perigee measured in the orbital plane.

2. *The extreme values of the distance of the Moon to the Earth*

In astronomy textbooks we read that the mean distance a from the Earth to the Moon is 384 400 kilometers, and that the eccentricity of the lunar orbit is $e = 0.0549$. From these values one can deduce that the minimum (perigee) and the maximum (apogee) distances between the centers of the two bodies are $a(1-e) = 363 296$ km and $a(1+e) = 405 504$ km, respectively.

However, these are not the smallest and largest possible distances between the Earth and the Moon. The Moon's motion is strongly perturbed by the attraction of the Sun, and in a lesser way by that of the planets and by the flattened shape of the Earth. Every 206 days — a little more than half a year — the major axis of the lunar orbit is directed toward the Sun. As we have seen in Chapter 1, near these epochs the eccentricity of the lunar orbit reaches a maximum, and the perigee distance of the Moon is much smaller than normal and the apogee distance is larger. When, however, the major axis of the lunar orbit is perpendicular to the direction of the Sun, the eccentricity reaches a minimum; at these epochs, perigee and apogee distances are less extreme. See Figure 2.*a*.

Figure 2.*b* shows the variation of the Earth-Moon distance over a two-year period. This plot reveals the 206-day cycle of maxima and minima mentioned above. The remarkable fact is that the variation of the perigee distance is much larger than that of the apogee. Using the ELP 2000-82 and ELP 2000-85 lunar theories of Michelle Chapront-Touzé and Jean Chapront, we calculated the perigee and apogee distances of the Moon for the years 1960 to 2040. During this 81-year period, the perigee distances of the Moon vary between 356 445 km (on 2034 November 25) and 370 354 km (on 1988 December 16), which gives a spread of 13 909 kilometers.

On the other hand, during this same period the apogee distance varies from 404 064 km (on 1976 July 19) to 406 712 km (on 1984 March 2), a variation of only 2648 kilometers.

What are the smallest and the largest possible values for the distance between the centers of the Earth and the Moon? To answer this question, we made a calculation for the much longer period of A.D. 1500 to 2500. During these ten centuries, 49 perigee distances are less than 356 500 km, while 33 apogees are larger than 406 700 km. During the same period, fourteen times the Moon approaches the Earth to less than 356 425 kilometers, and the same number of times the distance grows to larger than 406 710 km. These extreme cases are listed in Table 2.A.

It thus appears that, during the time interval of ten centuries considered, the extreme distances between the centers of Earth and Moon are

356 371 kilometers on 2257 January 1
406 720 kilometers on 2266 January 7

We further see from the table that the extreme perigees and apogees take place only during the winter in the northern hemisphere, the period of the year when the Earth is closest to the Sun. For instance, all 14 closest perigees mentioned in the table occur between December 6 and February 9. It is evident that the Earth's variable distance from the Sun somewhat affects the Earth-Moon distance.

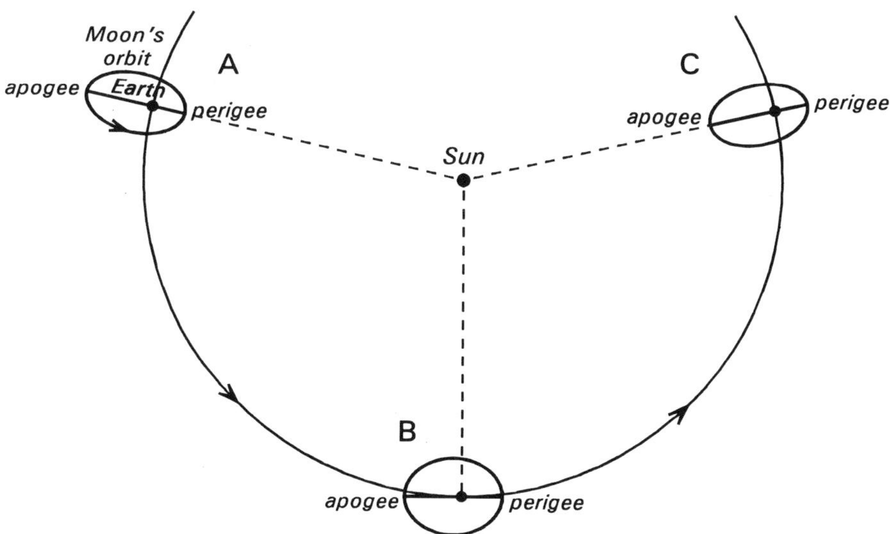

Fig. 2.a : When the major axis of the Moon's orbit is aligned with the Earth - Sun line (A), the orbital eccentricity exceeds its mean value. About 103 days later, in B, the two lines are at right angle and the eccentricity reaches a minimum. A new maximum is reached again after another 103 days (C). Sizes and distances are not to scale!

But that's not yet all. In the table we recognize the well-known periodicity of 18 years + 11 days, the *Saros*! However, here this famous period has nothing to do with solar or lunar eclipses. See, for example, the extreme perigees of 1893 − 1912 − 1930, or the extreme apogees of 2107 − 2125 − 2143.

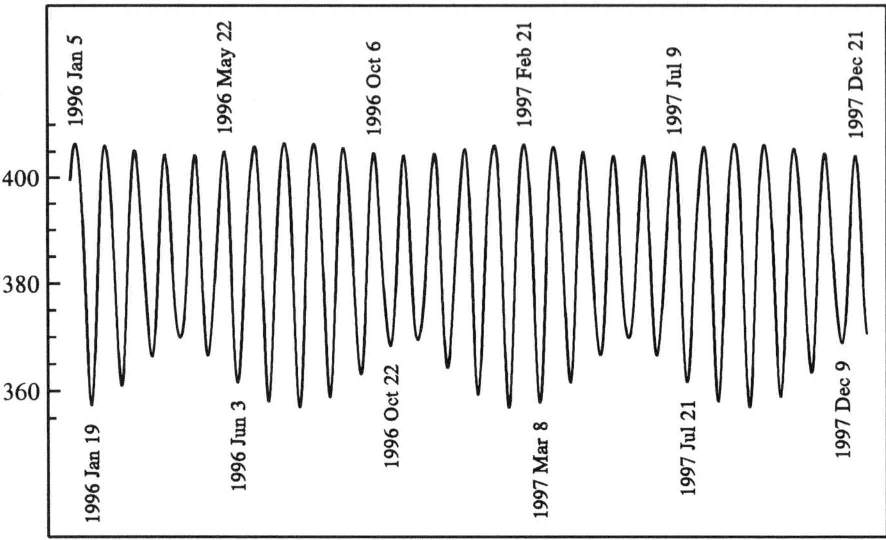

Fig. 2.b : *The distance between the centers of Earth and Moon, 1996-1997.*
Verticale scale : thousands of kilometers.

TABLE 2.A : *Extreme perigees and apogees, A.D. 1500 to 2500*

perigee < 356 425 km		apogee > 406 710 km	
1548 Dec 15	356 407 km	1921 Jan 9	406 710 km
1566 Dec 26	356 399	1984 Mar 2	406 712
1771 Jan 30	356 422	2107 Jan 23	406 716
1893 Dec 23	356 396	2125 Feb 3	406 720
1912 Jan 4	356 375	2143 Feb 14	406 713
1930 Jan 15	356 397	2247 Dec 27	406 715
2052 Dec 6	356 421	2266 Jan 7	406 720
2116 Jan 29	356 403	2284 Jan 18	406 714
2134 Feb 9	356 416	2388 Nov 29	406 715
2238 Dec 22	356 406	2406 Dec 11	406 718
2257 Jan 1	356 371	2424 Dec 21	406 712
2275 Jan 12	356 378	2452 Jan 21	406 710
2461 Jan 26	356 408	2470 Feb 1	406 714
2479 Feb 7	356 404	2488 Feb 12	406 711

3. *The distribution of the Moon's perigee and apogee distances*

The following text was first published in the Belgian journal Heelal *of September 1984, where the distribution of the Moon's perigee and apogee distances was given for the period 1924-2005. We now have performed a new calculation for the years 1960-2040, based on the lunar theory ELP of Michelle Chapront-Touzé and Jean Chapront; the climatological data for Uccle have been updated.*

In a given population the mean height of the adult men is 163 centimeters. It is evident that the height of most men of this group will not differ much from this mean value: for instance, there will be very few dwarfs of 120 cm or giants of 205 cm. The lengths are so-called 'normally' distributed; the distribution of the lengths present a single maximum near the middle: the mean length is the most frequent one (Figure 3.*a*).

Other examples of distributions with one maximum can be found, for instance, in meteorology. For the month of October, the mean air temperature at the Royal Meteorological Institute at Uccle (near Brussels, Belgium) is 10.5 °C. For the years 1901-1994 the October means are distributed as shown in Table 3.A and illustrated in Figure 3.*b*. Another example: the annual amount of precipitation (rain, snow, etc.) at Uccle; see Table 3.B and Figure 3.*c*.

However, other types of distributions are possible. As we have seen in Chapter 2, the mean perigee and apogee distances of the Moon are 363296 and 405504 kilometers, respectively. And, as we know, the motion of the Moon is strongly perturbed mainly due to the gravitational attraction of the Sun. For this reason, the perigee as well as the apogee distances will vary greatly. But it is *not* true that the *mean* perigee (or apogee) distances are the most frequent. Prof. Dr. E. Hantzsche, of Berlin, Germany, drew my attention on the fact that each of these distributions has *two* maxima.

For the years 1960-2040 the distributions of the perigee and apogee distances of the Moon are given in Table 3.C, and illustrated in Figures 3.*d* and 3.*e*. It appears that the *extreme* (smallest and largest) perigee and apogee distances are the most frequent. At first sight this is a most peculiar phenomenon. The explanation was given by C. Steyaert [now president of the Belgian astronomical society 'Vereniging voor Sterrenkunde'].

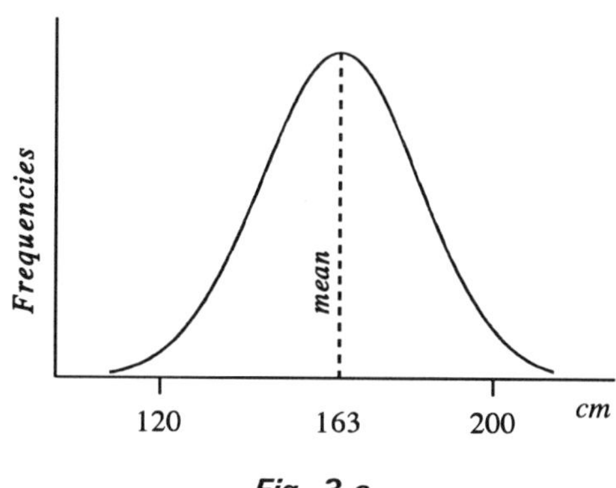

Fig. 3.a

TABLE 3.A

Distribution of the mean
air temperature in October
at Uccle, 1901 to 1994

Temperature in °C	Number of cases
6.0 — 6.9	2
7.0 — 7.9	4
8.0 — 8.9	9
9.0 — 9.9	21
10.0 — 10.9	23
11.0 — 11.9	22
12.0 — 12.9	12
13.0 — 13.9	0
14.0 — 14.9	1

TABLE 3.B

Distribution of the annual
amount of precipitation
at Uccle, 1901 to 1994

Amount in millimeters	Number of cases
400 — 499	1
500 — 599	4
600 — 699	10
700 — 799	27
800 — 899	25
900 — 999	20
1000 — 1099	7

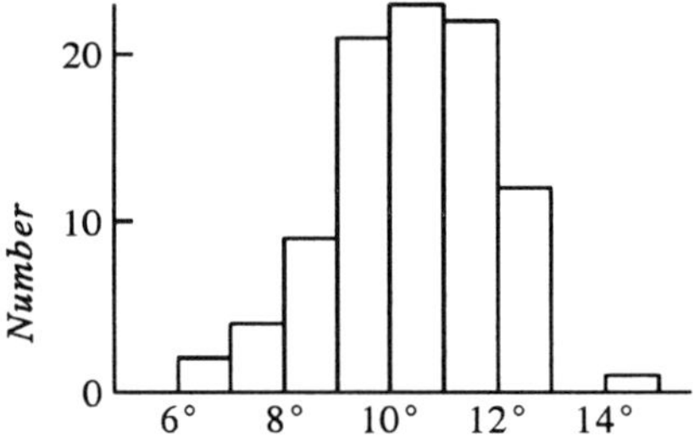

Fig. 3.b *: Distribution of the*
mean air temperature in October
at Uccle, Belgium, 1901 to 1994

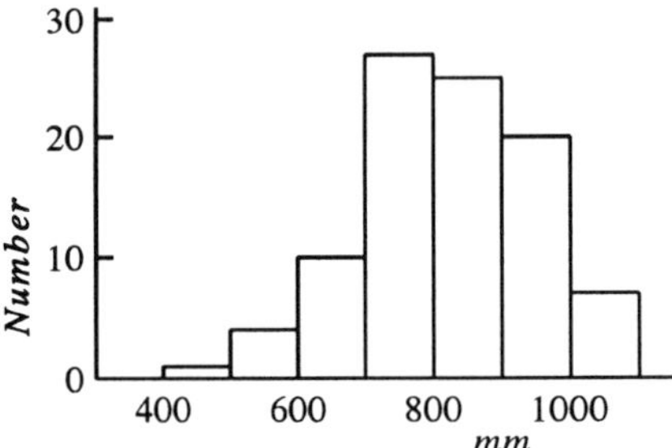

Fig. 3.c *: Distribution of the*
annual amount of precipitation
at Uccle, 1901 to 1994

It is, in fact, quite normal that both distributions have two peaks, and it is unnecessary to invoke complicated considerations of celestial mechanics.

The perigee distance $q = a (1 - e)$ of the *orbit* of the Moon can be regarded as an orbital element. So we can write

$$q = q_0 + \text{sum of perturbation terms}$$

where q_0 is the mean value of the perigee distance. Each perturbation term is a periodic function (the sine of a well-defined argument, and a well-defined amplitude).

TABLE 3.C

Distribution of the perigee and apogee distances of the Moon, 1960 to 2040

Perigee distance	Number of cases		Apogee distance	Number of cases
356 000			404 000	
	1			8
356 500			404 100	
	82			35
357 000			404 200	
	98			59
357 500			404 300	
	62			70
358 000			404 400	
	55			51
358 500			404 500	
	40			47
359 000			404 600	
	40			36
359 500			404 700	
	37			33
360 000			404 800	
	36			34
360 500			404 900	
	31			29
361 000			405 000	
	32			33
361 500			405 100	
	33			33
362 000			405 200	
	25			28
362 500			405 300	
	27			29
363 000			405 400	
	28			27
363 500			405 500	
	28			28
364 000			405 600	
	27			34
364 500			405 700	
	25			33
365 000			405 800	
	30			36
365 500			405 900	
	25			30
366 000			406 000	
	25			46
366 500			406 100	
	28			44
367 000			406 200	
	28			64
367 500			406 300	
	33			73
368 000			406 400	
	31			53
368 500			406 500	
	33			49
369 000			406 600	
	43			29
369 500			406 700	
	63			3
370 000			406 800	
	27			
370 500				

The distances are those between the centers of Earth and Moon, in kilometers.

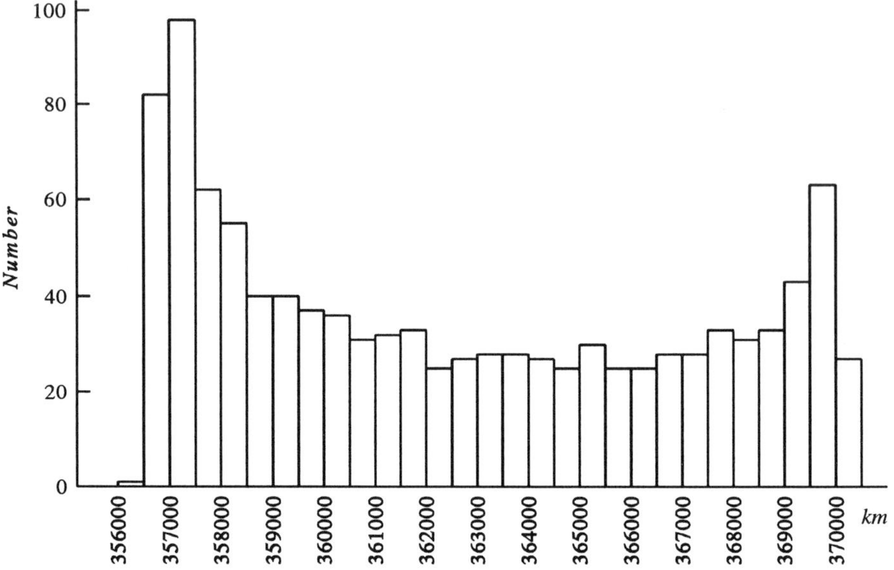

Fig. 3.d : *Distribution of the perigee distances of the Moon, 1960 to 2040*

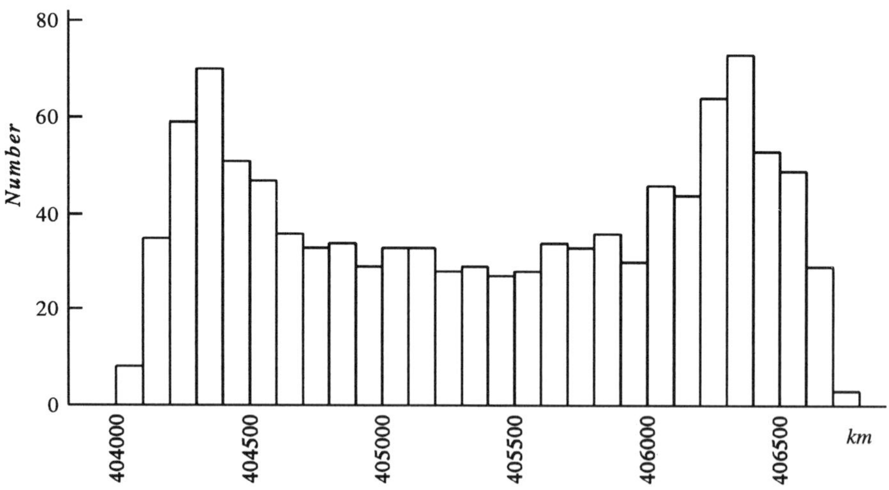

Fig. 3.e : *Distribution of the apogee distances of the Moon, 1960 to 2040*

First, suppose that there is but one single perturbation term, thus

$$q = q_0 + c \sin (a + bT)$$

where T is the time, and a, b, c are constants. In such a case, the values of q near the extremes $q_0 + c$ and $q_0 - c$ will occur more frequently than those near the mean value q_0, because the sine function varies much slower near its extreme values ($+1$ and -1) than near its mean value 0. For the same reason, the satellites of Jupiter are seen more often in the vicinity of their greatest elongations (where their motion with respect to Jupiter is slow) than close to the planet (where their apparent motion is fastest).

When more perturbation terms are added, the obtained distribution can become asymmetric, but the 'depression' in the middle will remain.

Moreover, it appears from Figures 3.*d* and 3.*e* that in both cases the maxima have different heights: the *smallest* perigee distances are more probable than the largest, and the *largest* apogee distances are more probable than the smallest. The Moon seems to like the extremes!

4. *What is the mean value of the Earth - Moon distance ?*

The following text was first published in Heelal *of September 1985.*

S u m m a r y . — Different values can be given for the 'mean' distance between the centers of Earth and Moon, depending on the definition adopted for this mean value.

Nowadays, the motion of the Moon around the Earth is very accurately known. For any given instant — at least, not too far in the past or in the future — it is possible to calculate the Moon's position (longitude and latitude, or right ascension and declination) to within a fraction of an arcsecond, and the distance to the Earth to an accuracy of a few meters. (By means of the laser technique, the Moon's distance can be *measured* with an accuracy of about 30 centimeters, but the *calculation* — the prediction — of the distance is another problem!).

Then, one might think that the *mean* distance too between Earth and Moon is very accurately known. And indeed it is. Yet, if one asks a specialist for the exact value of this mean distance, he will reply — perhaps at one's astonishment: 'What do you precisely mean by the mean distance?'.

General remark: In what follows, we ever speak about the distance between the *centers* of Earth and Moon.

Mathematical intermezzo

Suppose that the value of the variable quantity A is given by the formula

$$A = 60 + 2 \cos t + 0.3 \cos 2t \tag{1}$$

that is, the constant 60 plus two periodic terms. What is the 'mean' value of A? For this we could adopt, by definition, the constant term in formula (1), namely **60**.

The extreme values of A are 62.3 (for $t = 0°$) and 58.3 (for $t = 180°$). If we define the mean value of A as the arithmetical mean of these two extremes, then we find **60.3**.

But let us now suppose that we don't know formula (1), and that we have only the following expression for the calculation of the *inverse* of A :

$$
\begin{aligned}
1/A = \; & 0.016\,676\,073\,03 \\
& - \; 0.000\,553\,255\,76 \; \cos t \\
& - \; 0.000\,074\,203\,29 \; \cos 2t \\
& + \; 0.000\,002\,617\,46 \; \cos 3t \\
& + \; 0.000\,000\,141\,96 \; \cos 4t \\
& - \; 0.000\,000\,008\,85 \; \cos 5t \\
& - \; 0.000\,000\,000\,19 \; \cos 6t
\end{aligned}
\tag{2}
$$

Then A will be obtained by taking the inverse of the result. By doing the calculation for some value of t, the reader can verify that the formulae (1) and (2) are equivalent. Take, for instance, $t = 33°$; both formulae should give the same result for A.

In formula (2) the constant term is $0.016\,676\,073\,03$. We could adopt as the 'mean' value of A the inverse of that constant. That inverse is $1/0.016\,676\,073\,03$, or **59.966156**, not 60.

The reader may ask why a formula should be used to calculate the *inverse* of the quantity A. Wait a moment: see further what is said about the Moon's parallax.

Now we have found *three* different means for A, namely 60, 60.3 and 59.966156. What is the best value? It's just a matter of definition! But, as you see, we first should exactly know what we are speaking about...

Back to the Moon

1. In the lunar theory which E. W. Brown constructed in the early years of the 20th century, and which later has been corrected and improved a few times, the horizontal equatorial parallax π of the Moon, expressed in degrees, is given by a long series of periodic terms, of which the most important are

$$\begin{aligned}
\pi \; = \;\; & 0.950\,7245 \\
+ \; & 0.051\,8128 \; \cos M' \\
+ \; & 0.009\,5303 \; \cos (2D - M') \\
+ \; & 0.007\,8422 \; \cos 2D \\
+ \; & 0.000\,8571 \; \cos (2D + M') \\
+ \; & \ldots\ldots
\end{aligned} \tag{3}$$

where M' is the mean anomaly of the Moon, and D is the Moon's mean elongation from the Sun, but this is irrelevant here. So, Brown did not provide a formula for the direct calculation of the Earth-Moon distance; instead, he gave a formula for the parallax π. Once π is known, the Earth-Moon distance can be obtained from

$$\frac{6378.14 \text{ kilometers}}{\sin \pi} \tag{4}$$

where the value in the numerator is the equatorial radius of the Earth. So, the distance is inversely proportional to $\sin \pi$ and, because π is a small angle, it is also almost inversely proportional to π itself. (As you see, here does the *inverse* of the distance come into play!).

Well, the constant term in formula (3) is 0.950 7245 degrees and, if we enter this value in formula (4), we find for the 'mean' distance between Earth and Moon the value **384 399** kilometers.

2. As we mentioned in Chapter 1, a completely new theory for the motion of the Moon has later been constructed by Chapront-Touzé and Chapront at the Bureau des Longitudes of Paris, France. This theory does not give the parallax of the Moon, but it gives directly its distance to the Earth in kilometers. Here too we have to do with a long series of periodic terms, of which the most important are :

$$\begin{aligned}
\textit{distance in kilometers} \; = \;\; & 385\,000.5584 \\
- \; & 20\,905.3550 \; \cos M' \\
- \; & 3\,699.1109 \; \cos (2D - M') \\
- \; & 2\,955.9676 \; \cos 2D \\
- \; & 569.9251 \; \cos 2M' \\
+ \; & 246.1585 \; \cos (2D - 2M') \\
+ \; & \ldots\ldots\ldots
\end{aligned} \tag{5}$$

By definition, we could adopt the constant term of this series as the 'mean' distance of the Earth to the Moon. Rounded to the nearest integer kilometer, this is **385 001** km. This value differs from the preceding one, just as the 'mean' value of A based on series (2) differs somewhat from that based on series (1).

3. As we have seen in Chapter 2, during the period A.D. 1500 to 2500 the extreme distances between Earth and Moon are 356 371 and 406 720 kilometers. The 'mean' lies close to the arithmetical mean of these two values : **381 546** kilometers.

It is dangerous, however, to take this as the definition for the mean distance. In the case of the curve of Figure 4.*a* it is evident that the arithmetical mean of the heights of the extreme points *A* and *B* is not a good choice for the mean height.

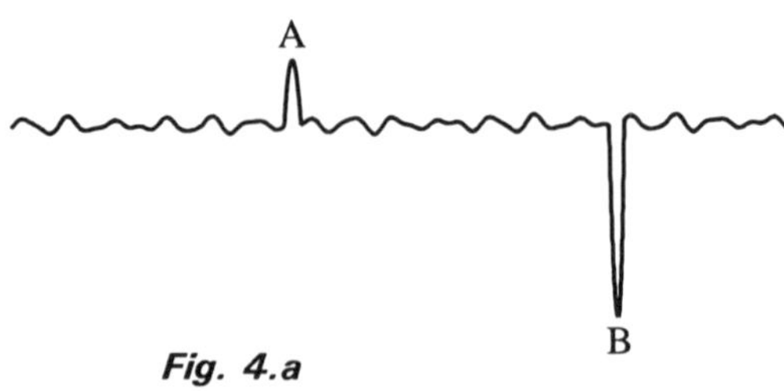

Fig. 4.a

4. Perhaps it would be better to consider, instead of a 'mean distance', the semimajor axis a of the elliptic orbit of the Moon. Although the Moon's motion is strongly perturbed by the attraction of the Sun, we can use the mean length of the sidereal revolution P of the Moon, which is very accurately known. The formula to be used is

$$a^{3/2} = \frac{k P \sqrt{\Sigma m}}{2 \pi} \tag{6}$$

where

$$k = \text{the Gaussian gravitational constant} = 0.017\,202\,098\,95,$$
$$P = 27.321\,6615 \text{ days},$$
$$\Sigma m = \text{sum of the masses of Earth and Moon} = 1/328\,900.5, \text{ the mass}$$
of the Sun being taken as unity,
$$\pi = 3.141592653\ldots.$$

Then the semimajor axis a of the lunar orbit is expressed in astronomical units. We obtain $a = 0.002\,571\,8814$ AU. Multiplying this value by $149\,597\,870$ we find the required result in kilometers: **384\,748** km. This is the value given on page XXXVIII of the *Connaissance des Temps* for 1984.

5. But this is not yet the end of the story. The gravitational attraction of the Sun not only strongly perturbs the motion of the Moon (whence the many periodic terms in the expressions for the Moon's longitude, latitude and distance), but it also has as consequence that formula (6) is not 'exact'. Due to the presence of the Sun, the sidereal revolution period P of the Moon is *not* equal to what might be deduced from the value of a by means of formula (6). According to A. Danjon (*Astronomie Générale*, Paris, page 275 of the edition of 1959), all is going on as if the presence of the Sun would *decrease* the attraction Earth-Moon by the factor $F = 1.002723$.

The true (mean) semimajor axis of the lunar orbit can then be found by means of formula (6) on the condition that Σm be divided by the above-mentioned quantity F. One then finds $a = $ **384\,399** kilometers. This is (quite accidentally?) exactly the value we found earlier on the basis of the constant term in the expression for the Moon's parallax!

6. However, we also could consider the mean value of the distance *in time*. This is *not* the same as the semimajor axis of the orbit.

The semimajor axis of the elliptic orbit of the Earth around the Sun is equal to a = 149 597 870 kilometers. Very often, principally in popular books, this value is called the 'mean distance of the Earth to he Sun'. However, a is not the average distance *with respect to the time*, as is readily visible from an extreme case: the periodic comet Halley. For this comet, a approximately equals 18 astronomical units. But the body remains a very long time near its far aphelion, at about 35 AU from the Sun, where it moves very slowly. The comet remains only a short time close to the Sun, near the perihelion of its orbit, because there its velocity is very large. So it is quite evident that for this comet the mean distance to the Sun (the mean *in the time*) must be much larger than a = 18 AU.

It can be shown (see for instance A. Danjon, *Astronomie Générale*, page 173 of the 1959 edition) that in the case of a purely elliptical motion (Keplerian orbit) the average distance (with respect to the time) to the central body is equal to

$$a\ (1+\frac{e^2}{2})$$

where e is the orbital eccentricity. For the orbit of the Earth, using the value e = 0.016 708 62 (epoch 2000.0), one finds an average distance of 149 618 752 km.

If we perform a similar calculation in the case of the Moon, using the value a = 384 399 km found above, and for the eccentricity the mean value e = 0.05490, we obtain 384 978 kilometers.

However, the Moon doesn't follow an unperturbed elliptic orbit. This you can see from formula (5). There, the terms in cos M' and in cos $2M'$ are in fact *not* perturbations by the Sun; they are the periodic terms resulting from the Moon describing its elliptic orbit: the so-called equation of the center — see, for instance, the formula on page 222 of my *Astronomical Algorithms* (Willmann-Bell, ed.; 1991). There are also smaller terms in $3M'$, $4M'$, . . . But all other periodic terms are true perturbations due to the attraction of the Sun.

For this reason, the actual average distance *in time* will differ somewhat from the value of 384 978 km just found. That mean distance in time is easily found: it just is the constant term in expression (5); rounded to the nearest integer, this is **385 001** kilometers. To find out that it's really this value, we could imagine that we take the average value of each term over a very long period. Then the mean value (in time) of each cosine is *zero* (the argument of each cosine is a linear function of time). In such a case, on the average only the very first, constant term will remain.

But, as you see, all that is not so easy. It's just a question of definition...

5. *Extreme declinations of the Moon*

The regression of the lunar nodes on the ecliptic has a period of 6798 days, or 18.61 years. Hence, at intervals of 18.61 years the (mean) ascending node of the Moon's orbit coincides with the vernal equinox, and its longitude Ω is zero. Here are the dates in the period A.D. 1900 to 2100 when $\Omega = 0°$:

1913 May 27	2006 June 19
1932 January 6	2025 January 29
1950 August 17	2043 September 10
1969 March 29	2062 April 22
1987 November 8	2080 December 1
	2099 July 13

Near these epochs, the inclination of the Moon's orbit on the Earth's equator is $23°26' + 5°09' = 28°35'$, where $23°26'$ is the obliquity of the ecliptic (near A.D. 2000), and $5°09'$ is the inclination of the Moon's orbit on the ecliptic.

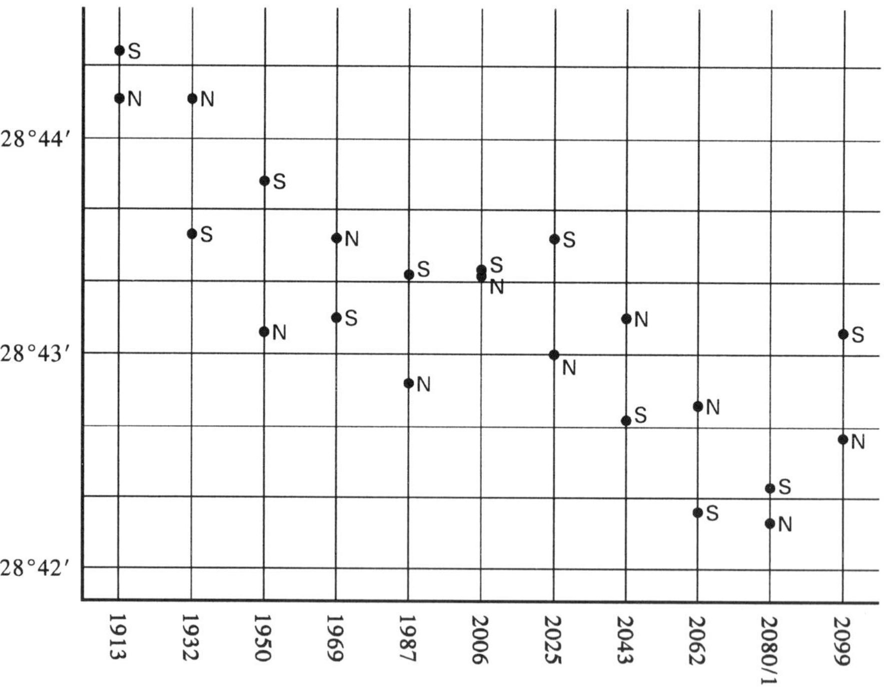

Fig. 5.a : *Extreme values, north (N) and south (S),*
of the geocentric declination of the Moon, 1900 to 2100.

Actually, as we have seen in Chapter 1, the latter inclination is itself variable, being affected by an inequality whose period is 173.31 days and whose amplitude is 8'. (There are other, smaller periodic terms). The inclination reaches its greatest value (5°17') when the line of the nodes is directed to the Sun, that is at each conjunction of the Sun with one of the nodes. When Ω is approximately zero, these conjunctions take place near the times of the equinoxes (about March 21 and September 23). Hence, the extreme value of the geocentric declination of the Moon's center is not 28°35', but 23°26' + 5°17' = 28°43'. This occurs near the times when $\Omega = 0°$, and when the longitude of the Moon is about 90° or 270°.

The table below contains the greatest values, north and south, of the geocentric declination of the Moon's center between the years 1900 − 2100. The times are given in Dynamical Time (TD). The difference between Dynamical Time and Universal Time was zero in A.D. 1902, and is expected to be 4 minutes near A.D. 2075 — see page 8. The declinations given in the table are *apparent*, that is, the effect of the nutation has been taken into account.

The extreme values given in the table are plotted in Figure 5.*a*, from which it can be seen that the value of the greatest declination of the Moon, reached at mean intervals of 18.61 years, is generally decreasing with time. This decrease is due to the secular diminution of the obliquity of the ecliptic. At present, this diminution is 8.7 arcseconds per 18.61 years. The mean obliquity was 23°27'08" in 1900, and will be 23°25'35" in 2100.

It should be noted that before 200 B.C. the obliquity of the ecliptic was larger than 23°43', so the Moon's geocentric declination could then exceed 29°.

TABLE 5.A

Extreme geocentric declinations, north and south, of the Moon,
1900 to 2100

North Declination			South Declination		
Date	*TD*	*Declination*	*Date*	*TD*	*Declination*
	h m	° ' "		h m	° ' "
1913 Mar. 16	7 17	+28 44 11	1913 Mar. 28	22 09	−28 44 24
1932 Mar. 15	21 33	+28 44 11	1932 Mar. 28	11 59	−28 43 33
1950 Oct. 3	10 25	+28 43 06	1950 Sep. 19	4 21	−28 43 48
1969 Mar. 25	14 24	+28 43 32	1969 Mar. 11	23 46	−28 43 10
1987 Sep. 15	17 12	+28 42 52	1987 Sep. 29	23 50	−28 43 22
2006 Sep. 15	1 28	+28 43 22	2006 Mar. 22	16 55	−28 43 23
2025 Mar. 7	15 44	+28 43 00	2025 Mar. 22	6 39	−28 43 32
2043 Sep. 25	14 10	+28 43 10	2043 Sep. 12	11 31	−28 42 42
2062 Mar. 18	9 55	+28 42 46	2062 Mar. 31	23 07	−28 42 16
2081 Mar. 18	10 30	+28 42 13	2080 Sep. 21	2 15	−28 42 23
2099 Sep. 8	22 53	+28 42 37	2099 Sep. 21	11 04	−28 43 06

We further see that in 1969, although $\Omega = 0°$ occurred on March 29, very close to the date of the equinox, this produced no particularly high peak. On 2025 March 22, the absolute value of the declination will be the same as that of 1969 March 25, although $\Omega = 0°$ will occur seven weeks before the equinox, and although the obliquity of the ecliptic will be less by $26''$ than in 1969.

In A.D. 2006 and in 2080-2081, the greatest northern declination and the greatest southern one will take place six months apart. For the nine other cases listed here, they occur with an interval of one-half sidereal month.

In what precedes, we considered only the *extreme* declinations of the Moon, which are reached every 18.6 years. But, of course, the Moon has a maximum northern and southern declination every month. As it may be instructive to look at these monthly maxima, we give in Table 5.B their values for a part of the years 1987 (when the ascending node of the lunar orbit was near the vernal equinox), for 1997 (when the *descending* node was near that equinox), and for 1992 (halfway between those epochs). The effect of the 173-day periodic variation in the inclination of the lunar orbit, described in Chapter 1, is readily visible from these values.

TABLE 5.B

*Maximum northern and southern geocentric declinations of the Moon
in the years 1987, 1992 and 1997*

1987		1992		1997	
Apr. 5	$+28°38'$	Jan. 17	$+24°52'$	Jan. 20	$+18°22'$
Apr. 18	$-28°37'$	Jan. 30	$-24°50'$	Feb. 4	$-18°17'$
May 2	$+28°34'$	Feb. 13	$+24°45'$	Feb. 17	$+18°13'$
May 16	$-28°30'$	Feb. 26	$-24°39'$	Mar. 3	$-18°10'$
May 29	$+28°27'$	Mar. 11	$+24°30'$	Mar. 16	$+18°08'$
June 12	$-28°25'$	Mar. 24	$-24°24'$	Mar. 31	$-18°09'$
June 25	$+28°24'$	Apr. 8	$+24°15'$	Apr. 12	$+18°11'$
July 10	$-28°26'$	Apr. 21	$-24°10'$	Apr. 27	$-18°16'$
July 23	$+28°28'$	May 5	$+24°06'$	May 10	$+18°20'$
Aug. 6	$-28°33'$	May 18	$-24°04'$	May 24	$-18°24'$
Aug. 19	$+28°37'$	June 1	$+24°03'$	June 6	$+18°27'$
Sep. 2	$-28°41'$	June 14	$-24°03'$	June 20	$-18°28'$
Sep. 15	$+28°43'$	June 29	$+24°04'$	July 3	$+18°27'$
Sep. 29	$-28°43'$	July 11	$-24°03'$	July 18	$-18°25'$
Oct. 13	$+28°42'$	July 26	$+24°01'$	July 31	$+18°22'$
Oct. 27	$-28°38'$	Aug. 8	$-23°58'$	Aug. 14	$-18°18'$
Nov. 9	$+28°35'$	Aug. 22	$+23°51'$	Aug. 27	$+18°16'$
Nov. 23	$-28°30'$	Sep. 4	$-23°45'$	Sep. 11	$-18°15'$
Dec. 6	$+28°27'$	Sep. 19	$+23°35'$	Sep. 23	$+18°15'$
Dec. 20	$-28°26'$	Oct. 1	$-23°29'$	Oct. 8	$-18°18'$

6. The librations of the Moon

It is a well-known fact that the Moon always keeps the same side toward the Earth, so that one half of the Moon's surface is never seen at all. In fact, this is not completely exact: the globe of the Moon has a periodic oscillation, which is called *libration* (Latin: *librare* = to swing). Accordingly, from the Earth about 59 per cent of the lunar surface can be observed altogether. But let us begin by stating the three empirical laws describing the rotation of the Moon; they were formulated by Jean-Dominique Cassini in 1721.

First law: The Moon rotates in the direct sense (that is, in the same sense as the Earth), with a constant angular velocity (with respect to the stars), and the sidereal period of rotation is equal to the mean sidereal period of revolution of the Moon around the Earth, which is 27.32166 days.

Second law: The inclination I of the mean plane of the lunar equator to the plane of the ecliptic is constant. According to F. Hayn (1907), this inclination is $1°32'06''$. Later, Hayn derived the improved value $I = 1°32'20''$. Watts (1955) found $1°33'50''$, but the value presently adopted by the International Astronomical Union is $I = 1°32'32''.7$.

Third law: The rotation axis of the Moon, the axis of the ecliptic (that is, the perpendicular to this plane) and the axis of the lunar orbit (the perpendicular to its plane) are always situated in a same plane, and in that order. In other words: the intersection of the equatorial plane of the lunar globe with the ecliptic is parallel to the line of nodes of the Moon's orbit; the *descending* node of the lunar equator coincides with the *ascending* node of the orbit.

If you find this rather complicated, look at Figure 6.*a*. In this drawing the diameters of Earth and Moon, and their relative distance are not to scale! The plane of the figure is perpendicular to the plane of the orbit of the Earth (the ecliptic) and also to the plane of the orbit of the Moon. *MR* is the rotation axis of the Moon, *ME* is perpendicular to the plane of the ecliptic, and *MB* is perpendicular to the plane of the Moon's orbit. The second law of Cassini states that the angle *RME* is

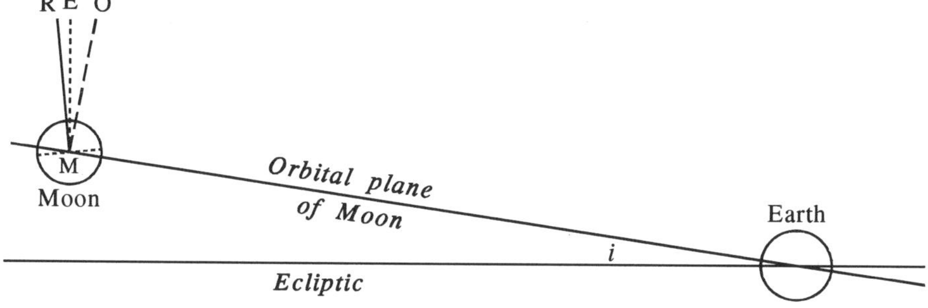

Fig. 6.a

constant (1°32′33″). The third law states that *MR*, *ME* and *MB* are situated in the same plane, and in such a way that *MR* and *MB* are at opposite sides with respect to *ME*.

The angle *EMB* is equal to the inclination of the plane of the lunar orbit to the ecliptic (*i* in Figure 6.*a*), and the mean value of this angle is 5°09′. Consequently, the (mean) inclination of the rotational axis of the lunar globe to the orbital plane of the Moon is equal to

$$RMB \ = \ RME + EMB \ = \ 1°32' + 5°09' \ = \ 6°41'.$$

Let us now go back to the libration of the Moon. One distinguishes a libration in longitude and a libration in latitude.

Libration in latitude

Because the lunar equator makes an angle (mean value 6°41′) with the plane of the lunar orbit, the north and south lunar poles will be alternatively inclined toward the Earth during a revolution of the Moon (27 days). A similar situation

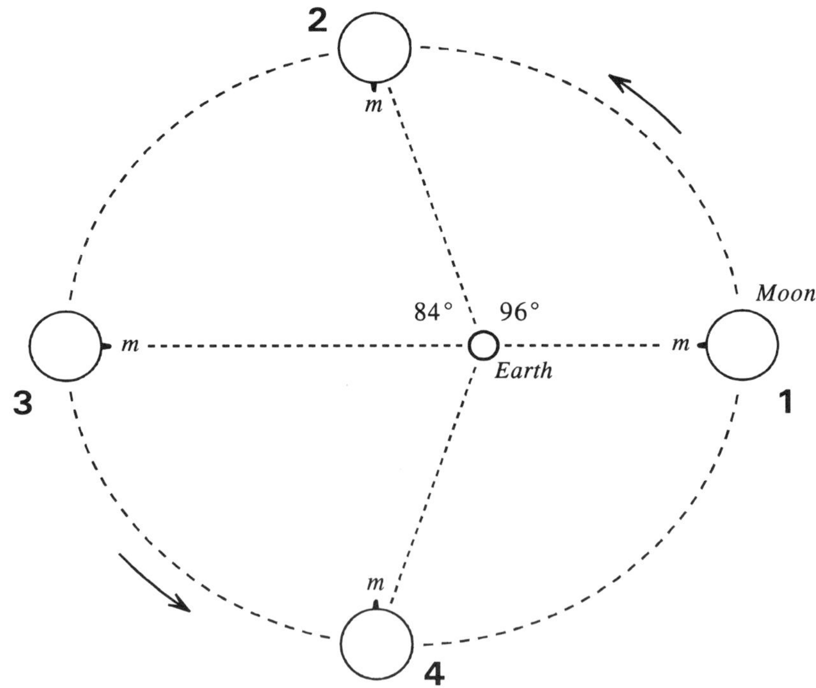

Fig. 6.b : *Explanation of the libration in longitude*

gives rise to the seasons on Earth, the Sun illuminating alternatively the North Pole and the South Pole in the course of a year. In the case of Figure 6.*a*, the Moon is situated north of the ecliptic, approximately halfway between the ascending and the descending node of its orbit, so the lunar southern pole is turned toward the Earth. The lunar equator (dotted in the drawing) is perpendicular to the rotation axis *MR*.

Libration in longitude

This libration arises from the fact that the lunar globe rotates uniformly (constant angular speed), while the velocity of the Moon in its orbit around the Earth is variable. Near the perigee, where the distance to the Earth is smallest, the orbital velocity is larger than near the apogee where the distance to the Earth is a maximum.

Figure 6.*b* shows the motion of the Moon around the Earth. In this figure too the diameters of Earth and Moon and the distances are not to scale. In position **1** the Moon is at perigee. Let *m* be the center of the lunar disk at that instant (as seen from the Earth). You may imagine that at point *m* there is a big lunar mountain directed toward the Earth.

After a quarter of a revolution (that is $27.32/4 = 6.83$ days) the lunar globe has rotated over exactly 90 degrees. The Moon then has reached position **2**. This position is not 90, but about 96 degrees further in the orbit than position **1**, and point *m* now seems displaced toward the east (to the 'left'). At the western lunar limb we see a little beyond the edge of the 'mean' visible half of the Moon.

Half a revolution after perigee the Moon reaches apogee, and it has travelled over 180 degrees as seen from the Earth — position **3** in Figure 6.*b*. The Moon's globe has now rotated over exactly 180° too, and consequently point *m*, as seen from the Earth, is again exactly at the center of the lunar disk.

A quarter of a revolution after apogee, point *m* has rotated over another 90 degrees with respect to the axis of the lunar globe. But the Moon has travelled over only 84 degrees with respect to the Earth, because in that part of its orbit the speed is smaller than its mean value. The Moon is now in position **4**, and we see somewhat beyond the mean eastern limb of its disk; point *m* is now seen somewhat to the west ('right') of the center of the disk.

After another quarter of a revolution, the Moon has returned in perigee with an increased velocity, and as seen from the Earth point *m* is once more exactly at the center of the visible half of the lunar globe.

The above-mentioned angle of 96 degrees is a mean value: actually, this angle can vary between 95° and 98°, because besides travelling on an elliptic orbit the Moon is perturbed by the attraction of the Sun.

In what precedes, the observer was imagined to be located at the center of the Earth. But an actual observer is situated at the Earth's surface. Therefore, the direction at which he looks at the Moon makes an angle with the line connecting the centers of Earth and Moon (Figure 6.*c*). For this observer, the center *o* of the lunar

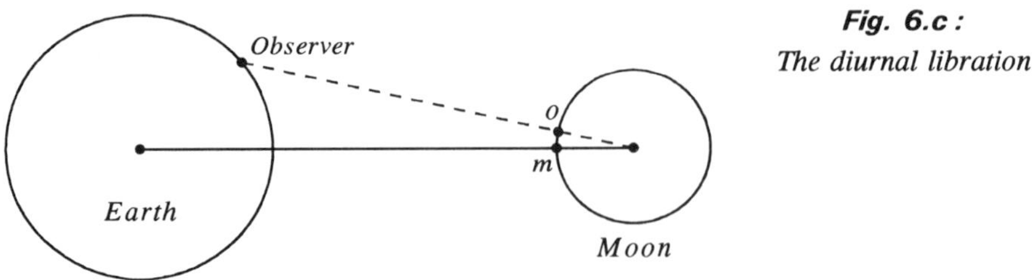

Fig. 6.c :
The diurnal libration

disk is not the same point of the lunar globe as the center *m* for a geocentric observer. This topocentric effect, which can reach 1°02′, varies in the course of the day, because the observer turns around along with the Earth's surface (rotation of the Earth). For this reason, the effect is called the *diurnal* libration.

Physical libration

The librations in longitude and latitude described above, together with the diurnal libration, are not real oscillations of the lunar globe: they just are apparent librations for an observer on the Earth. For this reason they are called *optical librations*. However, there does exist a real oscillation of the lunar globe. This *physical libration* is an irregularity of the rotation of the Moon with respect to its mean rotation rate. It results from the fact that the diameter of the lunar globe which is turned toward the Earth is a little longer than the mean diameter. This very small 'protuberance' is attracted by the Earth, causing an oscillation of the Moon's globe. The physical libration is much smaller than the optical libration, never exceeding 2′ in longitude and 3′ in latitude.

Curves

Many astronomical almanacs give, for each day of the year at 0^h UT, the values of the librations in longitude (*l*) and in latitude (*b*). Of course, these are the geocentric values — as seen from the center of the Earth — and they include the physical librations. As an example, we give in Table 6.A the values of *l* and *b* in September 1995. It appears that the libration in longitude reached a greatest positive value (+5°.78) on September 10, and a greatest negative value (−4°.75) on September 23. The extreme values of the libration in latitude were −6°.63 on September 5, and +6°.81 on September 19.

The maximum value of the libration in longitude is very variable by reason of perturbations in the motion of the Moon due to the gravitational attraction of the Sun. Sometimes the value of *l* can be somewhat larger than 8°, as for instance on 1971 October 10 when *l* reached a maximum of +8°.14. But at other times the maximum of *l* barely exceeds 4½ degrees ; on 1974 October 31, the greatest negative value of *l* was only −4°.53.

The maximum values of the libration in latitude vary between narrower limits, namely between 6.49 and 6.89 degrees.

TABLE 6.A

*Geocentric librations of
the Moon in longitude
(l) and in latitude (b)
at 0h UT
in September 1995*

Date	l	b
	°	°
1	−3.39	−2.49
2	−2.61	−3.91
3	−1.66	−5.11
4	−0.54	−6.00
5	+0.70	−6.52
6	+1.98	−6.62
7	+3.20	−6.31
8	+4.28	−5.60
9	+5.11	−4.56
10	+5.62	−3.28
11	+5.78	−1.84
12	+5.56	−0.33
13	+5.01	+1.15
14	+4.16	+2.55
15	+3.08	+3.82
16	+1.85	+4.90
17	+0.55	+5.76
18	−0.73	+6.38
19	−1.93	+6.73
20	−2.97	+6.79
21	−3.80	+6.54
22	−4.38	+5.98
23	−4.69	+5.11
24	−4.73	+3.95
25	−4.50	+2.56
26	−4.03	+0.99
27	−3.37	−0.65
28	−2.55	−2.27
29	−1.61	−3.77
30	−0.59	−5.03

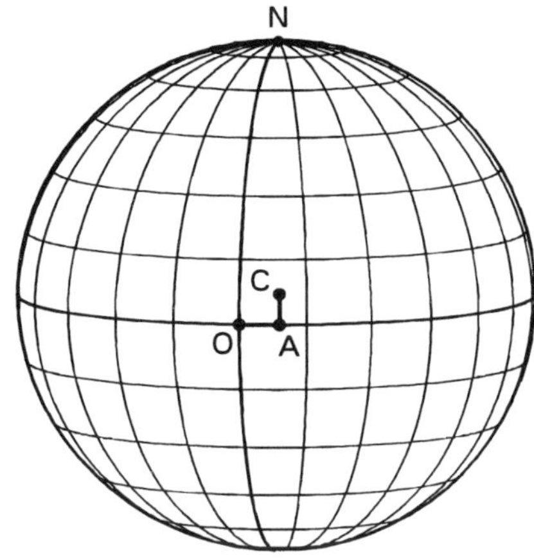

Fig. 6.d

*The circle represents the disk of the Moon as
seen from the Earth. C is the center of the disk,
and O is the 'zero-point' on the lunar globe, that
is, the point at selenographic longitude 0° and
latitude 0°. The length of the arc OA is the
libration in longitude (l), while AC is the libration in latitude (b).*

When we plot the points (*l*, *b*) for
successive days in a graph, we obtain curves
such as those in Figure 6.*e*. These curves show
the apparent motion of the zero-point of the
lunar globe around its mean position (the center
of the lunar disk) by reason of the librations in
longitude and in latitude, for a geocentric
observer. The 'zero-point' is the point at
selenographic longitude = 0°, and latitude =
0°; it is the point of the lunar globe which
lies, on the average, at the center of the disk.

The curves so obtained are distorted
'ellipses', which are more or less flattened, and
whose shapes continuously vary with time. The
cross at the center of each drawing represents
the center of the Moon's disk. The width of
each drawing corresponds to 20 degrees on the
lunar surface. North is up. The numbers indicate days of the month, each time at 0^h UT.

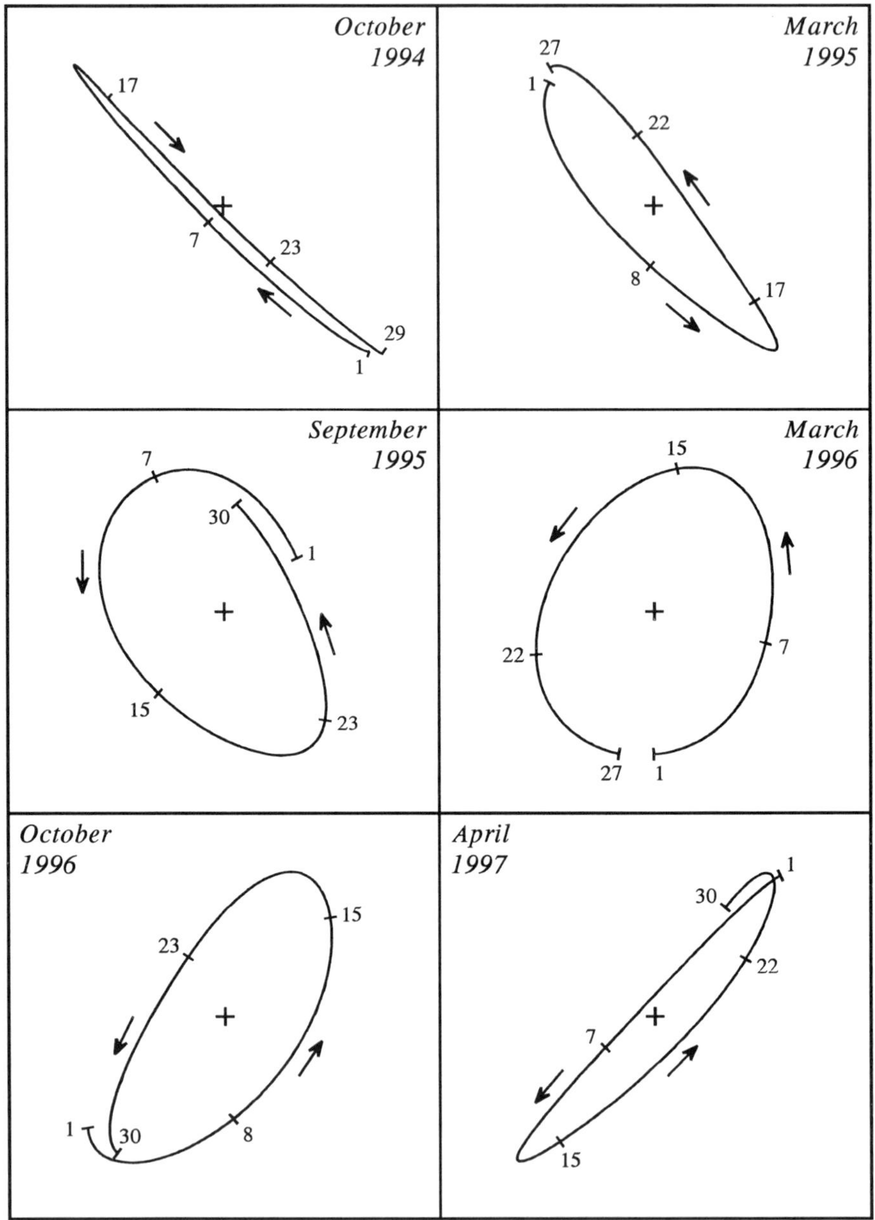

Fig. 6.e (part I): *Curves showing the motion of the 'zero-point' of the lunar globe around its mean position, due to the librations in longitude and in latitude.*

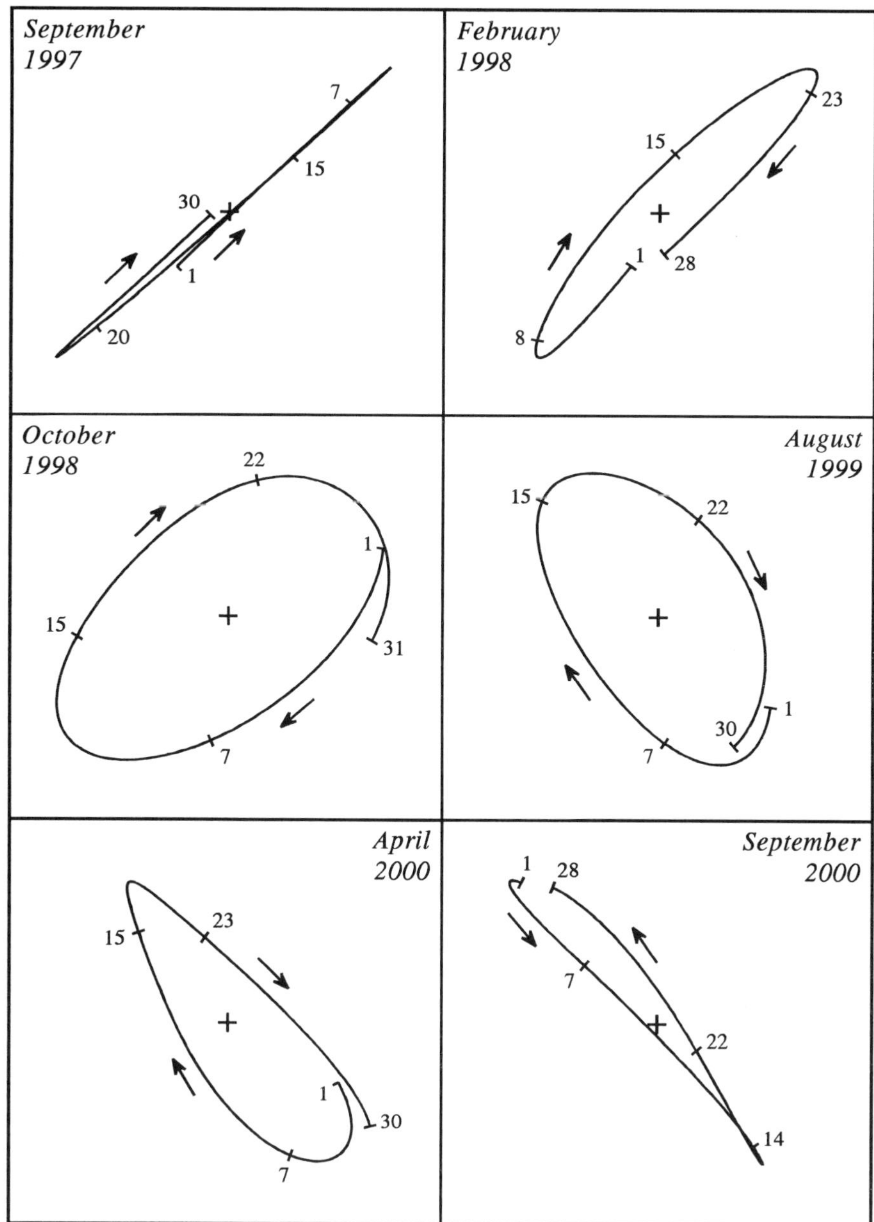

Fig. 6.e (part II) : *Curves showing the motion of the 'zero-point' of the lunar globe around its mean position, due to the librations in longitude and in latitude.*

When, as in September 1994 and in September 1997, the major axis of the lunar orbit coincides with the line of nodes of this orbit, the 'ellipse' reduces to approximately a straight line. Indeed, when the Moon is in its perigee or its apogee, then — as explained above — the libration in *longitude* is zero. And when the Moon is at the ascending or at the descending node, then the libration in *latitude* is zero. In September 1994, the longitude of the perigee was equal to that of the *ascending* node; then the Moon passed almost simultaneously through the perigee and the ascending node ; and, of course, half a revolution later it passed almost simultaneously through the apogee and the descending node. Halfway between, the librations in longitude and in latitude reached a maximum almost simultaneously too. In September 1997, the lunar perigee coincides with the *descending* node of the orbit.

On the other hand, when the major axis of the lunar orbit is perpendicular to the line of nodes, as in March 1996, then the 'ellipses' are the most open and they don't differ much from circles. See the curves for March 1996 and October 1998.

The curves repeat — besides the perturbations — after a period of almost exactly six years. After six years the major axis of the lunar orbit has performed one complete revolution with respect to the line of nodes. The exact value of the period is 2190.35 days, or 5.9969 Julian years. Neglecting secular terms in the second and higher powers of time, the epochs when the (mean) major axis of the lunar orbit coincides with the (mean) line of nodes are given by

$$\text{Julian Day} \;=\; 2451\,798.66 \;+\; 1095.175\,k$$

where k is an integer. If k is even, then the perigee coincides with the ascending node, and the apogee with the descending node. If k is odd, the perigee coincides with the descending node, and the apogee with the ascending node. The values $k = -2$ to $+2$ correspond to the dates 1994 September 12, 1997 September 11, 2000 September 11, 2003 September 11, and 2006 September 10, respectively. Note that, in the formula given above, non-integer values for k would give meaningless results !

Finally, we see from the graphs that the rotation of the point (l, b) around the center of the Moon's disk reverses every three years. In 1995 and 1996 this motion was performed counterclockwise, but in 1998 and 1999 the motion is clockwise, etc.

In fact, the gradual evolution of the shape of the curves is due to the fact that the periods of the librations in longitude and in latitude are inequal. Their values are, respectively, 27.55455 days (the *anomalistic* or perigee-to-perigee revolution period of the Moon) and 27.21222 days (the *draconic* or node-to-node revolution period).

7. *Months with five lunar phases*

In 1996 there was a Full Moon on July 1, and again that same month on July 30. So, in July 1996 there were *five* lunar phases: Full Moon (FM), Last Quarter (LQ), New Moon (NM), First Quarter (FQ), and then a second Full Moon.

It is possible to have five lunar phases in a month whenever a lunar phase occurs on the first or second day. Since about 29½ days separate a lunar phase from the next similar phase, all of the months in our Gregorian calendar except February can have five lunar phases. In February this is never possible because that month contains at most 29 days.

At this point, we need an exact definition of the lunar phases. In practical life — for a person who looks at the Moon, or for a nightly walker — the Moon can be said to be 'full' for two or even four consecutive days. Astronomically, however, the lunar phases have no 'duration'; they occur at well-defined instants. By definition, the times of New Moon, First Quarter, Full Moon, and Last Quarter are the times at which the excess of the apparent longitude of the Moon over the apparent longitude of the Sun is exactly 0, 90, 180, and 270 degrees, respectively. Here, 'apparent' means that in the calculation of the longitude of the Moon the effect of light-time is included; for the Sun, the so-called aberration is taken into account. (The effect of the nutation is irrelevant here, since this correction is exactly the same for the longitude of the Sun as for that of the Moon).

How often are there two lunar phases of the same type in a calendar month? Does there exist a periodicity? We note that over successive months a given lunar phase gradually moves toward the beginning of the month, because a calendar month is a little longer than a lunation. Only after a February month is there a small jump backwards, because February is always shorter than a lunation. [A *lunation* is the (mean) interval between two similar phases, for instance between two successive New Moons]. For example, Full Moon took place on the following dates:

1994	Sep.	19	1995	May	14	1996	Jan.	5
1994	Oct.	19	1995	June	13	1996	Feb.	4
1994	Nov.	18	1995	July	12	1996	Mar.	**5**
1994	Dec.	18	1995	Aug.	10	1996	Apr.	4
1995	Jan.	16	1995	Sep.	9	1996	May	3
1995	Feb.	15	1995	Oct.	8	1996	June	1
1995	Mar.	**17**	1995	Nov.	7	1996	July	1
1995	Apr.	15	1995	Dec.	7	1996	July	30

Finally, in July 1996 Full Moon occurred so early in the month, that a second similar phase took place before that month's end.

How long does it take for a given lunar phase to slide through all the days of the month? In the Gregorian calendar the mean length of the month is

$$M = \frac{365.2425}{12} = 30.436875 \text{ days}$$

while the mean length of a lunation is $L = 29.530589$ days. Hence, the lunation is shorter than a mean month by $M - L = 0.906286$ days.

After $30.436875 / 0.906286$ or 33½ lunations the lunar phase has moved through all the days of the month, from the end to the beginning of the month. So this occurs at mean intervals of 32½ months. And because there are *four* different lunar phases, it takes place every eight months — more exactly, the mean frequency is one case every 8.15 months.

Consequently, at mean intervals of a little more than eight months one of the lunar phases occurs in the early days of the month, and a second similar phase at the end. But here too the shorter February month sometimes can perturb the trend.

In 1995, for instance, we had a New Moon on January 1 and another on January 31, but no New Moon in February (that month, there were only *three* lunar phases). Then in March there were again two New Moons. Similar 'double' events occur in January / March of the years 1997, 1999, 2012, 2014, etc.

The table below lists all the months with five lunar phases from A.D. 1990 to 2020. The phase which occurs twice is mentioned.

TABLE 7.A

Months with five lunar phases, 1990 to 2020

FQ	1990 May		LQ	1999 Oct.	FQ	2012 Jan. 2012 Mar.	
FM	1990 Dec.		NM	2000 July			
LQ	1991 Oct.		FQ	2001 Apr.	FM	2012 Aug.	
NM	1992 June		FM	2001 Nov.	LQ	2013 May	
FQ	1993 Jan. 1993 Mar.		LQ	2002 Aug.	NM	2014 Jan. 2014 Mar.	
			NM	2003 May			
FM	1993 Sep.		FQ	2003 Nov.	FQ	2014 Oct.	
LQ	1994 June		FM	2004 July	FM	2015 July	
NM	1995 Jan. 1995 Mar.		LQ	2005 May	LQ	2016 Mar.	
			NM	2005 Dec.	NM	2016 Oct.	
FQ	1995 Oct.		FQ	2006 Aug.	FQ	2017 July	
FM	1996 July		FM	2007 June	FM	2018 Jan. 2018 Mar.	
LQ	1997 Jan. 1997 Mar.		LQ	2007 Dec.			
			NM	2008 Aug.	LQ	2018 Oct.	
NM	1997 Oct.		FQ	2009 May	NM	2019 Aug.	
FQ	1998 July		FM	2009 Dec.	FQ	2020 Apr.	
FM	1999 Jan. 1999 Mar.		LQ	2010 Oct.	FM	2020 Oct.	
			NM	2011 July			

When there are two Full Moons in a month, "the second full Moon of the month is called a Blue Moon. Nobody knows why." (E. C. Krupp, *Sky and Telescope*, September 1993, p. 59).

In Table 7.A we find Meton's 19-year cycle several times, for instance:

— two Full Moons in December 1990 and in December 2009;
— two Last Quarters in October 1991 and in October 2010.

However, in many cases after 19 years there is no repetition in exactly the same month. The Metonic cycle is only approximately exact, and the number of bissextile days (4 or 5) in a period of 19 years can give rise to an extra jump of one day. For instance, there were two Full Moons in *September* 1993, but 19 years later there will be two Full Moons in *August* 2012. The reason is that the Full Moon of 1993 September 1 will repeat, after 19 years (235 lunations), on 2012 August 31, not on September 1.

February is the only month which can have only three lunar phases. Table 7.B gives the complete list of these cases during the period 1800 — 2100. For each case, the missing phase is indicated. For instance, there was no New Moon in February 1995. The sign *B* refers to a bissextile (leap) year.

TABLE 7.B

February months with only three lunar phases, 1800 to 2100
(the missing phase is indicated)

1805	NM	1870	NM		1925	FQ		1993	FQ	*B*	2052	NM

1805	NM	1870	NM		1925	FQ		1993	FQ	*B* 2052	NM
1807	LQ	1879	FQ		1934	FM		1995	NM	2054	LQ
1809	FM	1883	LQ		1938	NM		1997	LQ	2058	FQ
1811	FQ	1885	FM	*B*	1940	LQ		1999	FM	2067	FM
1826	LQ	1889	NM		1955	FQ	*B*	2012	FQ	2071	NM
1830	FQ	1900	NM		1957	NM		2014	NM	2077	FQ
1843	NM	1902	LQ		1959	LQ		2018	FM	2081	LQ
1845	LQ	1911	NM		1961	FM		2031	FQ	2090	NM
1847	FM	1915	FM		1970	LQ		2033	NM	*B* 2092	LQ
1849	FQ	1919	NM		1974	FQ		2035	LQ	2094	FM
1866	FM	1921	LQ		1978	LQ		2037	FM		

It is important to note that all the dates mentioned to this point are based on the *Universal Time*. Different results are obtained for dates based on another standard time. Examples are given in Table 7.C. So, if one uses the Pacific Standard Time there are two Full Moons in June, not in July 1996; and under the same conditions there will be only three lunar phases instead of four in February 2016.

Of course, things may be completely different in other calendars. For example, in the Moslem calendar the months begin 1 to 3 days after New Moon, so in this calendar each month has always exactly four lunar phases, in the order FQ, FM, LQ, and NM.

TABLE 7.C

Examples of change of date when another zone time is used

Lunar phase	Universal Time		Pacific Standard Time	
	1996 *h m*		**1996** *h m*	
Full Moon	June 1	20 47	June 1	12 47
Last Quarter	June 8	.11 05	June 8	3 05
New Moon	June 16	1 36	June 15	17 36
First Quarter	June 24	5 23	June 23	21 23
Full Moon	July 1	3 58	June 30	19 58
Last Quarter	July 7	18 55	July 7	10 55
New Moon	July 15	16 15	July 15	8 15
First Quarter	July 23	17 49	July 23	9 49
Full Moon	July 30	10 35	July 30	2 35
	2016 *h m*		**2016** *h m*	
Last Quarter	Jan. 2	5 30	Jan. 1	21 30
New Moon	Jan. 10	1 30	Jan. 9	17 30
First Quarter	Jan. 16	23 26	Jan. 16	15 26
Full Moon	Jan. 24	1 46	Jan. 23	17 46
Last Quarter	Feb. 1	3 28	Jan. 31	19 28
New Moon	Feb. 8	14 39	Feb. 8	6 39
First Quarter	Feb. 15	7 46	Feb. 14	23 46
Full Moon	Feb. 22	18 20	Feb. 22	10 20
Last Quarter	Mar. 1	23 10	Mar. 1	15 10

ECLIPSES AND OCCULTATIONS

8. *The number of eclipses in a year*

Eclipse types

Let us first review the different types of solar eclipses, and the abbreviations we shall use in this and the next chapters to designate them.

For a given place on the Earth's surface where a *solar eclipse* is visible, the event is either partial, total, or annular. But when we consider a solar eclipse *for the Earth generally*, we can distinguish six types:

P   a *partial* solar eclipse, when only a part of the penumbral cone of the Moon passes over the Earth. Observers in the region of visibility can see only a partial eclipse;

T   a *total* eclipse: a central eclipse at which the inner (umbral) cone passes over the Earth. [At a *central* eclipse the axis of the lunar shadow passes over the surface of the Earth. At such an eclipse there exists a so-called 'central line'; for observers situated on this line, the center of the Moon's disk passes exactly over the center of the solar disk];

A   an *annular* eclipse: a central eclipse at which the *extension* of the umbral cone passes over the Earth;

A-T   an *annular-total* eclipse: a central eclipse which is total for a part of the path and annular for the rest;

(T)   *non-central total* eclipse: only a part of the umbral cone passes over the Earth's surface (in the polar regions); the axis of the umbra does not touch the Earth, and hence the eclipse is not central;

(A)   *non-central annular* eclipse: as the preceding case, but now only a part of the *extension* of the umbral cone passes over the Earth's surface.

The most frequent types are P, T, and A. When a solar eclipse is not central, in most cases it is of type P.

It is important to note that a total (or annular) eclipse is seen as total (or annular) only from a rather narrow path on the Earth's surface. To the north and to the south of this path, there is a much larger region from which only a partial eclipse is visible.

There are three types of *lunar eclipses*:

t   a *total* eclipse: the Moon passes completely into the Earth's umbral cone;

p   a *partial* eclipse, when the Moon passes only partly through the umbra;

pen   a *penumbral* eclipse, when the Moon passes through the outer (penumbral) cone, but does not touch the umbra.

Penumbral lunar eclipses are not detectable visually unless their magnitude is greater than about 0.7. Small penumbral eclipses are undistinguishable, but the theory and statistics of eclipses would be incomplete without them; hence, for reason of completeness, they will be taken into consideration in this book.

Note on the lunar eclipses

Observation has shown that the atmosphere of the Earth has the effect of increasing the apparent radius of the Earth's shadow by about one fiftieth. Following an old tradition, most astronomical almanacs increase by 1/50 the geometric radii of both the umbra and the penumbra. However, since 1951 the French almanacs *Connaissance des Temps* and *Annuaire du Bureau des Longitudes* use another theory for the computation of the radii. A. Danjon correctly pointed out [*L'Astronomie*, vol. 65, 51-53 (February 1951)] that the only reasonable way to take into account the presence of an opaque atmospheric layer around the Earth is to increase the Earth's radius, which can be performed by increasing proportionally the Moon's parallax. In that case, the radii of the umbra and the penumbra will undergo the same absolute correction, and not the same relative correction as the traditional rule states. Thus, even if the rule of 1/50 were true for the umbra, it cannot be correct for the penumbra. As a consequence, the magnitudes of lunar eclipses, as calculated by using the traditional (1/50) rule, are too large (as compared to the 'French' values) by about 0.005 for umbral eclipses, but by about 0.026 for penumbral eclipses.

For instance, the magnitude of the partial lunar eclipse of 1977 April 4 was 0.198 according to the *Astronomical Ephemeris*, but only 0.193 according to the *Annuaire du Bureau des Longitudes*. For the magnitude (in the penumbra) of the penumbral eclipse of 1973 July 15, the same publications gave 0.130 and 0.104, respectively.

In this book, statistics about lunar eclipses are based on the *Canon of Lunar Eclipses* by Jean Meeus and Hermann Mucke (Astronomisches Büro, Vienna, Austria; 1983), in which the magnitudes of lunar eclipses have been calculated with the 'French' rule, used by the *Connaissance des Temps*. This should be taken into account when our statements are compared with the data from other sources.

As a consequence, small penumbral 'eclipses', found by using the classical rule of 1/50 for the enlargement of the penumbra, do in fact not exist. This was the case for the 'eclipse' of 1951 February 21, for which Bao-Lin Liu and Alan D. Fiala (*Canon of Lunar Eclipses, 1500 B.C. to A.D. 3000*; Willmann-Bell, ed.; 1992) give a magnitude (in the penumbra) of 0.007. A similar case will take place at the Full Moons of 2016 August 18 and 2042 October 28.

Similarly, very small partial eclipses in the umbra (according to the rule of 1/50) are in fact penumbral eclipses where the Moon comes very close to the edge of the umbra. This was the case for the penumbral eclipse of 1988 March 3; according to Liu and Fiala, this eclipse was an umbral one (type P) with a magnitude of 0.002. A similar case will occur on 2042 September 29. Actually, there is no sensible difference between a very small partial eclipse with a magnitude of 0.002 in the umbra, and a penumbral eclipse with an umbral magnitude of, say −0.003 (negative!).

Finally, eclipses of type P which are nearly total in the umbra are found to be just total (type T) when the rule of 1/50 is used for the calculation. This will be the case for the eclipse of 2015 April 4, for which Liu and Fiala find a magnitude of 1.003, while Meeus and Mucke give 0.998.

The number of eclipses in a year

Let us now return to the actual subject of this chapter. But first two important remarks should be made. When we speak of the number of eclipses in a year, what we mean are eclipses *for the Earth generally*. Not all of these eclipses are visible from any given place on the Earth's surface.

Secondly, by 'year' we mean a calendar year, that is, a time lapse beginning on the first day of January and ending on December 31. When we ask how often there are five solar eclipses in a year, then in a certain sense this is a rather 'false' problem because what we call the first day of the year is just a convention. Nothing special distinguishes January 1 from any other date. Let us give two examples.

As we shall see, the last times there were five solar eclipses in a calendar year were in A.D. 1805 and 1935, and the next occurrence will happen in 2206. However, five solar eclipses in a period of 365 days (a 'year') take place more often than that, for instance :

1916	July	30	A	2134	April 24	P
1916	Dec.	24	P	2134	May 23	P
1917	Jan.	23	P	2134	Oct. 17	P
1917	June	19	P	2134	Nov. 16	P
1917	July	19	P	2135	April 13	A

It is possible to have three total eclipses of the Moon in a calendar year. The last such occurrence was in A.D. 1982, but it will not take place again before the year 2485. However, often there are three total lunar eclipses in a period of 365 days, so for instance :

1989	Feb.	20	1992	Dec.	9	2000 Jan. 21
1989	Aug.	17	1993	June	4	2000 July 16
1990	Feb.	9	1993	Nov.	29	2001 Jan. 9

So, let us now consider the number of eclipses in a *calendar* year, in the Julian/Gregorian calendar.

The least possible number of eclipses in a year is *four*, and in that case there are two eclipses of the Sun and two of the Moon. Of the latter, one or both can be penumbral eclipses. Here are some examples :

1989 Feb. 20	*t*		1990 Jan. 26	A		1970 Feb. 21	*p*
1989 Mar. 7	P		1990 Feb. 9	*t*		1970 Mar. 7	T
1989 Aug. 17	*t*		1990 July 22	T		1970 Aug. 17	*p*
1989 Aug. 31	P		1990 Aug. 6	*p*		1970 Aug. 31	A
1995 Apr. 15	*p*		1987 Mar. 29	A-T			
1995 Apr. 29	A		1987 Apr. 14	*pen*			
1995 Oct. 8	*pen*		1987 Sep. 23	A			
1995 Oct. 24	T		1987 Oct. 7	*pen*			

The greatest possible number of eclipses per year is *seven*, and this is possible in four different ways :

5 eclipses of the Sun + 2 eclipses of the Moon, as in 1805, 1935, 2206, 2709;
4 eclipses of the Sun + 3 eclipses of the Moon, as in 1917, 1982, 2094, 2159;
3 eclipses of the Sun + 4 eclipses of the Moon, as in 1908, 1973, 2038, 2103;
2 eclipses of the Sun + 5 eclipses of the Moon, as in 1879 and 2132.

Here is a detailed example for each of these four cases :

1935 Jan. 5	partial solar eclipse	1982 Jan. 9	total lunar eclipse
1935 Jan. 19	total lunar eclipse	1982 Jan. 25	partial solar eclipse
1935 Feb. 3	partial solar eclipse	1982 June 21	partial solar eclipse
1935 June 30	partial solar eclipse	1982 July 6	total lunar eclipse
1935 July 16	total lunar eclipse	1982 July 20	partial solar eclipse
1935 July 30	partial solar eclipse	1982 Dec. 15	partial solar eclipse
1935 Dec. 25	annular solar eclipse	1982 Dec. 30	total lunar eclipse

1973 Jan. 4	annular solar eclipse	1879 Jan. 8	penumb. lunar eclipse
1973 Jan. 18	penumb. lunar eclipse	1879 Jan. 22	annular solar eclipse
1973 June 15	penumb. lunar eclipse	1879 Feb. 7	penumb. lunar eclipse
1973 June 30	total solar eclipse	1879 July 3	penumb. lunar eclipse
1973 July 15	penumb. lunar eclipse	1879 July 19	annular solar eclipse
1973 Dec. 10	partial lunar eclipse	1879 Aug. 2	penumb. lunar eclipse
1973 Dec. 24	annular solar eclipse	1879 Dec. 28	partial lunar eclipse

One of the seven eclipses can be a very small one. The partial solar eclipse of 1935 January 5 was a grazing case, having a maximum magnitude of only 0.001 — see Figure 8.*a*. The magnitude of the small penumbral lunar eclipse of 1879 January 8 was only 8 %.

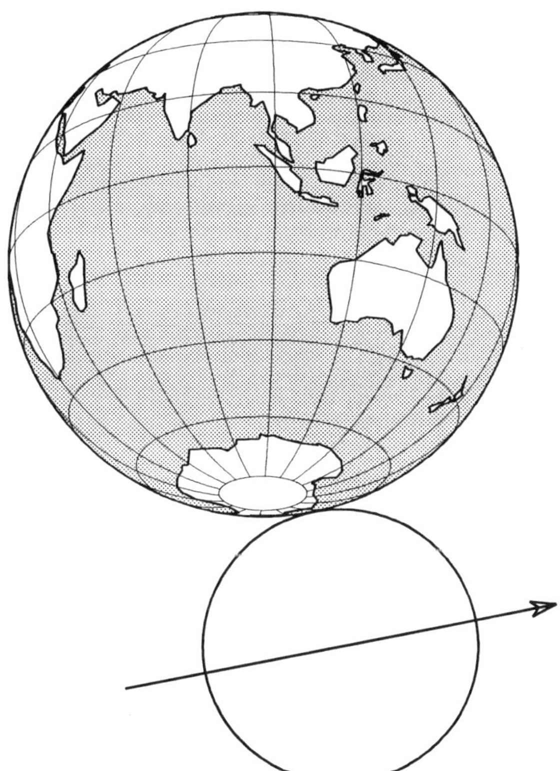

Fig. 8.a

The 'grazing' solar eclipse of 1935 January 5. The lower circumference represents the edge of the lunar penumbra in the fundamental plane — the plane through the center of the Earth and perpendicular to the axis of the shadow of the Moon. The position of the penumbra is that for the time of maximum eclipse, 5 h 35 m Universal Time. The arrow indicates the direction of motion. The penumbra hardly touched the Earth. Maximum 'eclipse' took place at 110° west longitude, 65° south latitude, near Antarctica. Maximum magnitude was only 0.001. It is likely that nobody saw this imperceptible eclipse.

In a calendar year, there are at most five solar eclipses. This maximum possible number is rarely reached. Between the years −600 and +3400 there are only 14 such 'rich' years, namely: −568, −503, −438, −373, 1255, 1805, 1935, 2206, 2709, 2774, 2839, 2904, 3295, and 3360. Note their irregular distribution: there are three cases from −568 to −438, and three from 2709 to 2839, but *none* between the years −373 and +1255. In all those 14 cases, four of the five eclipses are partial (type P); the remaining eclipse is either of type A (as in 1935 and 2206) or of type T (as in 2774).

The least number of solar eclipses in a calendar year is two. Both can be partial, as in 1996 and in 2004.

The maximum number of lunar eclipses in a calendar year is five. Between 1600 and 2500 there are five lunar eclipses in the following years: 1676, 1694, 1749, 1879, 2132, 2262, and 2400. In such cases, often four of the five eclipses are of type *pen*. However, in 1694 and 1749 only three eclipses were of that type; the other two were partial eclipses in the umbra (one of them very small).

The least number of lunar eclipses in a year is two. Both can be penumbral eclipses, as in 1966 and in 2016.

In a calendar year, the number of *umbral* lunar eclipses can be zero (as in 1998), one (as 1995), two (1997), or three (1898, 1982). During the period A.D. 1500 to 2499, the distribution of these cases is as follows:

191 years without a lunar eclipse in the umbra,
105 years with 1 lunar eclipse in the umbra,
676 years with 2 lunar eclipses in the umbra,
 28 years with 3 lunar eclipses in the umbra.

So, in about 2 cases out of 3, there are exactly two umbral lunar eclipses in the year. Both of these eclipses can be total (as in 1996), both can be partial (as in 1970), or one can be total and the other partial (as in 1992 and 2001).

The mean frequency of three umbral lunar eclipses in the year is once every 36 years. From A.D. 1500 to 2499, this happens in the following years:

1544	1656	1787	1917	2113	2243	2373
1563	1703	1833	1982	2159	2289	2420
1591	1722	1852	2028	2178	2308	2438
1610	1768	1898	2094	2224	2354	2485

In a calendar year there may be three *total* lunar eclipses in the umbra. Between the years 0 and 3200, this occurs 17 times, namely in the following years:

307 —	372 —	437
828 —	893 —	958
1414 —	1479 —	1544
1917 —	1982	
2485 —	2550 —	2615
3006 —	3071 —	3136

Note the repetitions after 65 years: $307 + 65 = 372$, $1917 + 65 = 1982$, etc. Each 'series' except the fourth contains three cases. Sixty-five years after 1982, in A.D. 2047, the repetition misses by a little: the third total eclipse occurs a few hours *after* the end of December 31!

The period of 65 years + about two days is equal to 804 lunations, or 2 Saros periods + 1 Inex. (The Inex will be defined in the next chapter). After 65 years, the eclipses are repeated with an accuracy not as good as that of the well-known Saros. So we have, for example, the following lunar eclipses in the umbra:

1896 August 23	magnitude	= 0.73
1961 August 26	magnitude	= 0.99
2026 August 28	magnitude	= 0.93
2091 August 29	magnitude	= 1.23
2156 August 30	magnitude	= 1.41
2221 September 2	magnitude	= 1.40

When there are three total lunar eclipses in a calendar year, the first one occurs early in January, the second at the end of June or the first few days of July, and the third eclipse at the end of December. See on page 46 the dates in 1982.

All the solar eclipses in one calendar year can be of type P, so for example in 1996 (two eclipses), 2018 (three eclipses), and 2000 (four eclipses). Hence, in those years no total or annular eclipse can be seen.

It is possible to have two total solar eclipses during one calendar year, and this is the maximum possible number. The next case will occur in 2057. It is *not* possible to have three total solar eclipses in one year, even if we include eclipses of the types A-T and (T).

It is possible to have two annular solar eclipses during a calendar year, for example in 1951 and in 1973. The maximum number of 'pure' annular solar eclipses in a calendar year is two. By *pure* we mean annular eclipses which are not of the type A-T (annular-total eclipses).

However, if one includes A-T eclipses, the maximum possible number of 'annular' eclipses in a year is three. In such cases, the year contains either one pure annular and two annular-totals, or two pure annulars and one annular-total. Between the years −2000 and +3400, this occurs eleven times, namely in the years −1944, −484, −400, −139, 1144, 1228, 1339, 1405, 1489, 1666, and 3192. (Note the large gaps between −1944 and −484, between −139 and 1144, and between 1666 and 3192).

9. *Solar eclipses : some periodicities*

It is well-known that solar and lunar eclipses are repeated, with slightly altered characteristics, after a period of 223 lunations, or 6585 $^1/_3$ days, that is 18 years and approximately 11 days : the *Saros*. Because the length of the Saros is 6585 days + approximately 8 hours, the region of visibility of a solar eclipse is displaced, after one such period, by approximately 120 degrees in geographical longitude. Compare, for example, the map of the total solar eclipse of 1994 November 3 with that of the total eclipse of 2012 November 13 (Figure 9.*a)*. The path of totality of the first eclipse crossed South America and the Atlantic Ocean, while that of the second eclipse passes mainly over the Pacific Ocean. — The dashed lines on the maps are 'isomagnitudes' : for the places on these lines, the magnitude of the partial eclipse is 0.2, 0.4, 0.6 or 0.8. The other curves on the maps will be explained in Chapter 11.

In this chapter we shall consider solar eclipses only. But of course the several periodicities which we will mention (Inex, Hepton, etc.) are valid for eclipses of the Moon too. The reader can find examples in published lists of lunar eclipses.

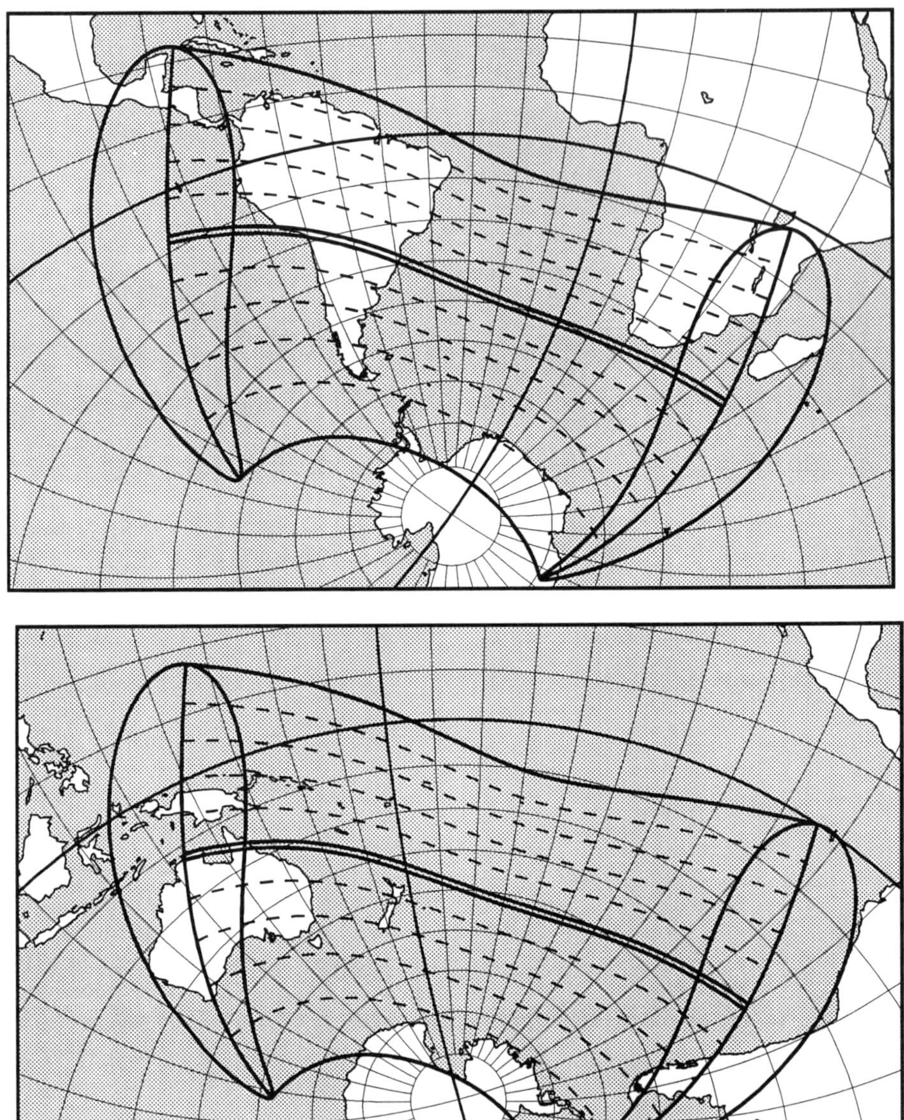

Fig. 9.a : *The region of visibility of two solar eclipses. Top : the total eclipse of 1994 November 3. Bottom : one Saros later, the total eclipse of 2012 November 13. In each case, a total eclipse is visible only inside of the narrow central track.*

Eclipses taking place at intervals of one Saros belong to the same *Saros series*. Examples: the eclipses of 1973 June 30, 1991 July 11, and 2009 July 22. Saros series do not live for ever. They are born and then they die.

Saros series contain from 69 to 86 eclipses, so their durations range from 1226 to 1532 years. A Saros series begins and ends with several eclipses of type P. In between the beginning and ending P-eclipses, central eclipses take place.

Let us give some examples. A very long Saros series (86 eclipses) began on −1378 August 14 and ended on +155 February 19. It started as a small partial eclipse visible from the southern hemisphere, and ended with a small partial eclipse in the northern hemisphere. This series consisted of 24 partial eclipses (type P), followed by 40 annular eclipses (A), and finally by another 22 eclipses of type P. So, this series did not contain any total solar eclipse.

A new Saros series will start on 2011 July 1 with a partial eclipse visible from the southern hemisphere. It will end on 3237 July 14 with a partial event in the northern hemisphere, and will contain only 69 eclipses: 8 of type P, 52 of type A, and finally 9 of type P. So, in this series too no total eclipse will occur.

The famous eclipse of 1991 July 11, which was total in Mexico, belonged to a Saros series which began with a small partial eclipse in the southern hemisphere. After the eighth eclipse, the events were central: 6 eclipses of type A, then 6 of type A-T, then 44 of type T. The last seven eclipses of the series will be of type P again, now visible from the northern hemisphere, the last one taking place on 2622 July 30. The eclipses of this series are summarized in the following list. The value following the 'P' is the maximum magnitude of the eclipse for an observer on the Earth's surface. The last column gives the least distance γ of the axis of the lunar shadow to the center of the Earth, the equatorial radius of the Earth being taken as unity. The quantity γ is positive or negative, depending on the axis of the shadow passing to the north or to the south of the Earth's center. When the absolute value of γ is less than 0.997, the eclipse is central. (The limit is 0.997, not 1, by reason of the flattening of the Earth's globe.)

No. of eclipse in the series	Date	Type of eclipse	γ
1	1360 June 14	P 0.049	−1.52
8	1486 Aug. 29	P 0.985	−1.00
9	1504 Sept. 8	A	−0.95
14	1594 Nov. 12	A	−0.78
15	1612 Nov. 22	A-T	−0.77
20	1703 Jan. 17	A-T	−0.73
21	1721 Jan. 27	T	−0.73
36	1991 July 11	T	−0.00
64	2496 May 13	T	+0.96
65	2514 May 25	P 0.951	+1.03
71	2622 July 30	P 0.104	+1.49

The three Saros series described above all started in the southern hemisphere and ended in the northern one. This is always so when the eclipses take place near the *descending* node of the Moon's orbit. (All the eclipses of the same Saros series occur near the same node).

When, however, the eclipses occur near the *ascending* node of the lunar orbit, they gradually shift southward in the series. This is the case for the series containing the European total eclipse of 1999 August 11. This Saros series began on 1639 January 4 with a small partial eclipse in the northern hemisphere, and will end with a small partial eclipse in the southern hemisphere on 3009 April 17.

The Saros has been described in many books on astronomy, and we will not say more about it here. There are other eclipse periodicities than the Saros, however, and we shall describe some of them briefly.

In his excellent work *Periodicity and Variation of Solar (and Lunar) Eclipses* (Haarlem, Netherlands, 1955), Prof.Dr. G. van den Bergh († 1966) investigated the period of 358 lunations, or 29 years minus approximately 20 days, which he called the *Inex*. Inex series have a much longer lifetime than Saros series, because after one Inex the shift of the Moon with respect to the node of its orbit is much smaller than for the Saros.

The Inex period is equal to 388½ draconic (node-to-node) revolutions of the Moon. The fraction ½ has as consequence that, unlike eclipses in a Saros series, those in an Inex series take place alternately at the one and at the other node. Hence, an eclipse visible in the northern hemisphere of the Earth will be followed, after one Inex, by an eclipse in the southern hemisphere; 1 Inex later there will again be an eclipse in the northern hemisphere, and so on during many centuries. Let us give an example:

			node
1845 May 6	annular	Arctic Ocean	descending
1874 April 16	total	Antarctica	ascending
1903 Mar. 29	annular	Siberia	descending
1932 Mar. 7	annular	Antarctica	ascending
1961 Feb. 15	total	Southern Europe, Russia	descending
1990 Jan. 26	annular	Antarctica	ascending

and so on . . . . .

It is not possible to mention here all the properties of the Inex. As compared with the Saros, the Inex has several drawbacks; it has, however, some advantages too. Prof. G. van den Bergh showed that the time interval between *any* two solar (or lunar) eclipses can be represented by the formula $t = mS + nI$, where S is the length of the Saros, I that of the Inex, and m and n are integer numbers (positive, negative, or zero). He has arranged all 8000 solar eclipses listed in Oppolzer's famous *Canon der Finsternisse* (Vienna, 1887) in a large *Saros - Inex Panorama*. In this panorama, the eclipses are grouped column by column according to Saros intervals, and line by line according to Inex intervals. So, one step downward in the panorama means one Saros later, and one step to the right means one Inex later. Such a panorama combines the advantages of Saros and Inex. And so the Inex really is a 'natural' period just as the Saros.

Table 9.A lists some periods after which eclipses repeat to a more or less good approximation. Most names in the first column are due to G. van den Bergh. An asterisk indicates that there is a change of node from one eclipse to the other in the series.

TABLE 9.A

Some eclipse periodicities

Name of the period	Combination S = Saros I = Inex	Number of synodic revolutions (lunations)	Period in days	Period in years
Semester *	$5I - 8S$	6	177	0.49
Hepton *	$5S - 3I$	41	1 211	3.32
Octon	$2I - 3S$	47	1 388	3.80
Tritos *	$I - S$	135	3 987	10.92
Saros	S	223	6 585	18.03
Meton's Cycle	$10I - 15S$	235	6 940	19.00
Inex *	I	358	10 572	28.94
Exeligmos	$3S$	669	19 756	54.1
Tetradia *	$19I + 2S$	7 248	214 038	586.0
Heliotrope	$58I + 6S$	22 102	652 685	1787
Megalosaros	$58I + 7S$	22 325	659 270	1805
Accuratissima	$58I + 9S$	22 771	672 441	1841
Horologia	$110I + 7S$	40 941	1 209 012	3310

The *Semester* is a period of 6 lunations — a little less than six months, whence its name. In a series of Semesters, just as in the case of the Inex, there is a change of node from one eclipse to the next. A series begins, often as a partial eclipse, near one of the poles of the Earth; the next eclipse in the series takes place in the other hemisphere, generally closer to the equator, and so on. The middle eclipses of the series are visible from the equatorial regions. Here is an example:

1990 January 26	annular	south	$\gamma = -0.95$
1990 July 22	total	north	$+0.76$
1991 January 15	annular	equatorial	-0.27
1991 July 11	total	equatorial	-0.00
1992 January 4	annular	equatorial	$+0.41$
1992 June 30	total	south	-0.75
1992 December 24	partial	north	$+1.07$

A series of Semesters generally contains 7 or 8 eclipses. A series of *Heptons* (eclipses at intervals of 41 lunations, or 3.32 years) has a longer lifetime. In such a series too the eclipses 'alternate' : they jump from one hemisphere to the other. The two groups (the one shifting northward, the other southward) cross each other in the equatorial regions. Here is an example of a complete series of Heptons. Note the 'crossing' between the eclipses of 1991 and 1994 (see the sign of γ).

1971 August 20	partial	south	$\gamma =$	-1.27
1974 December 13	partial	north		$+1.08$
1978 April 7	partial	south		-1.11
1981 July 31	total	north		$+0.58$
1984 November 22	total	equat./south		-0.31
1988 March 18	total	equat./north		$+0.42$
1991 July 11	total	equatorial		-0.00
1994 November 3	total	south		-0.35
1998 February 26	total	equatorial		$+0.24$
2001 June 21	total	south		-0.57
2004 October 14	partial	north		$+1.03$
2008 February 7	annular	south		-0.96
2011 June 1	partial	north		$+1.21$

In this series, seven successive eclipses are total. Indeed, the Hepton preserves, better than the Semester, the total or annular character of the eclipse, because it contains almost an integer number of anomalistic (perigee-to-perigee) revolutions of the Moon : 1 Hepton = 43.**94** anomalistic revolutions, while 1 Semester is equal to 6.**43** anomalistic revolutions.

Let us now consider the *Tritos*. With this period of eleven years minus one month, we obtain eclipse series of still longer lifetime than the Hepton, because the displacement of the Moon with respect to the node is smaller (-2.51 degrees after 1 Hepton, but only $+0.52$ after 1 Tritos). While a Hepton series generally contains 13 or 14 eclipses and has a lifetime of 40 or 43 years, a Tritos series contains about sixty eclipses and lasts for seven centuries. As an example, here is the beginning of a Tritos series :

Date	*Type*	γ	*Date*	*Type*	γ
1931 Sep. 12	P	$+1.51$	2029 Dec. 5	P	-1.06
1942 Aug. 12	P	-1.52	2040 Nov. 4	P	$+1.10$
1953 July 11	P	$+1.44$	2051 Oct. 4	P	-1.21
1964 June 10	P	-1.14	2062 Sep. 3	P	$+1.02$
1975 May 11	P	$+1.06$	2073 Aug. 3	T	-0.88
1986 Apr. 9	P	-1.08	2084 July 3	A	$+0.82$
1997 Mar. 9	T	$+0.92$	2095 June 2	T	-0.64
2008 Feb. 7	A	-0.96	2106 May 3	T	$+0.47$
2019 Jan. 6	P	$+1.14$	etc..		

As you see, in a Tritos series too the successive eclipses alternate from one hemisphere to the other (see the sign of γ). Further, it is interesting to note that the small partial eclipse of 1931 September 12 was both the last eclipse of a Saros series (which began on 651 July 23) and the first one of a Tritos series.

The *Metonic Cycle* is well-known. After 19 years, the lunar phases are repeated on nearly the same calendar dates. Meton's cycle is an interesting periodicity to find rapidly the approximate lunar phase in the near past or future. For instance, 190 years (ten cycles) after the total solar eclipse of 1991 July 11 (New Moon!) we have the New Moon of 2181 July 11. However, there will be *no* eclipse on this date. Indeed, eclipse series with a period of 19 years contain only 4 or 5 events. Therefore, Meton's cycle is not very useful for the prediction of eclipses. Here is an example of a 19-year series occurring at the ascending node of the lunar orbit; this series contains only five eclipses :

1923 August 12	(no eclipse)	
1942 August 12	partial eclipse	Antarctica
1961 August 11	annular eclipse	southern Atlantic
1980 August 10	annular eclipse	Pacific Ocean, South America
1999 August 11	total eclipse	Europe
2018 August 11	partial eclipse	Arctic regions, Siberia
2037 August 11	(no eclipse)	

The *Octon* is merely one fifth of the Metonic Cycle : one Metonic Cycle = 5 Octons. As an example, the following eclipses, separated from each other by 1 Octon, can be inserted into the above-mentioned 19-year series :

1980 August 10	annular eclipse	Pacific Ocean, South America
1984 May 30	annular eclipse	North America, Atlantic Ocean
1988 March 18	total eclipse	Borneo, Pacific Ocean
1992 January 4	annular eclipse	Pacific Ocean
1995 October 24	total eclipse	Asia
1999 August 11	total eclipse	Europe

Finally, the *Tetradia* rules the occurrence of the lunar eclipse *tetrads*, which will be one of the subjects of Chapter 16. The period of 586 years will be mentioned several times in the next chapters.

10. Curious and interesting facts about solar eclipses

1. We mentioned in Chapter 8 that an annular-total (A-T) eclipse is a central eclipse which is total for a part of the path and annular for the rest. More precisely, along the central line such an eclipse starts as annular, then becomes total as the curvature of the Earth brings its surface closer to the Moon so that it intercepts the tip of the umbra; later the eclipse returns to annular before the end of the path (Figure 10.*a*).

The A-T eclipse of 2013 November 3 will be a special case: here the central eclipse will start as an annular one, change 15 seconds later to a total eclipse, *and remain total up to its end*. This is due to the fact that at this eclipse the tip of the Moon's umbral cone will lie close to the fundamental plane; because at eclipse time the Moon is approaching the Earth (on its way from apogee to perigee), the tip of the umbra will actually *cross* the fundamental plane in the course of the eclipse.

The previous similar case took place on 1854 November 20, and the next one after 2013 will occur on 2172 October 17.

2. Two consecutive solar eclipses can never both be T-eclipses. However, it is possible to observe two total solar eclipses in less than half a year's time, but then at least one of them will be an A-T eclipse seen near noontime. Here are two examples:

1912 April 17	A-T		2067 Dec. 6	A-T	
1912 Oct. 10	T		2068 May 31	T	

However, two consecutive *central* eclipses can both be total ones. For example: the T-eclipses of 1999 August 11 and 2001 June 21, which are separated by the four eclipses of A.D. 2000, all partial events.

3. Two consecutive solar eclipses may both be A-T eclipses, for example the eclipses of 1908 December 23 and 1909 June 17, those of 1986 October 3 and 1987 March 29, and those of 2049 November 25 and 2050 May 20.

There are eleven such A-T *couples* between A.D. 1800 and 2500. During this period of seven centuries there are 44 eclipses of the type A-T, and it is remarkable that *half of them* are grouped in couples!

4. In the period −599 to A.D. 3400, there are 9493 solar eclipses for the Earth generally. This leads to an average of 237 eclipses per century. During those 40 centuries, the numbers of eclipses of each type are as follows: 3344 of type P, 3071 of type R, 2508 of type T, 493 of type A-T, 58 of type (A), and 19 of type (T). The distribution of the eclipses over the centuries is given in Table 10.A.

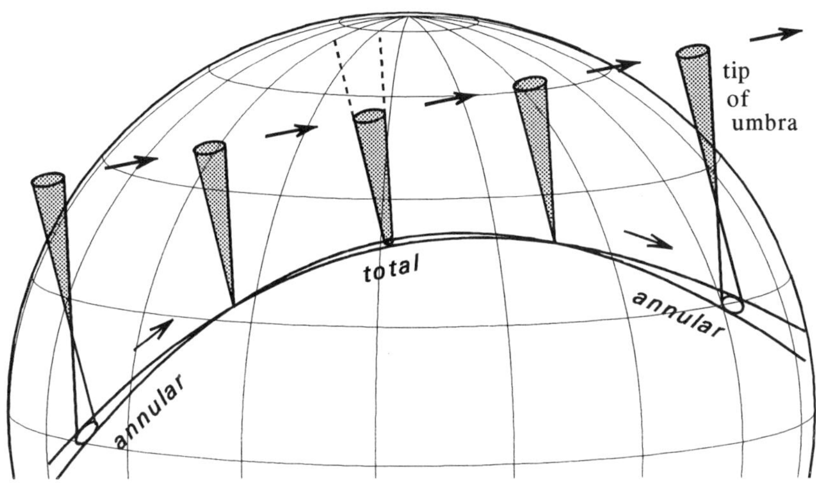

Fig. 10.a : *Some solar eclipses start as annular, then become total, and finally return to annular because the roundness of the Earth reaches up and intercepts the lunar shadow near the middle of the path. These events are called annular-total eclipses.*

The values given in the table have been plotted in Figure 10.*b*, and some remarkable facts come readily into view. There exists a cycle with a length of a little less than six centuries, giving alternatively 'rich' and 'poor' periods. So, for example, the 20th and the 21st centuries (the years 1901 to 2100) are poor periods, with only 228 and 224 eclipses, and only 145 and 144 central eclipses, respectively. Compare this with the period 1701 − 1800, when there were 251 eclipses, 159 of them being central.

It thus appears that the number of solar eclipses generally, and that of the central eclipses, vary in *antiphase* with the number of the lunar eclipse tetrads, which we shall discuss in Chapter 16. When there are *many* tetrads, there are *few* solar eclipses.

The lower curve of Figure 10.*b* reveals a more puzzling fact. Here it appears that for what concerns the A-T eclipses too there are rich and poor periods, but here the length of the cycle seems to be 17 or 18 centuries. We know of no explanation for that curious fact, and we did not investigate the subject further.

5.   The distribution of the *non-central* total or annular eclipses too is very irregular. There are five such eclipses during the 20th century, two of which took place during the same year :

1928 May 19	(T)		1957 April 30	(A)
1950 March 18	(A)		1957 October 23	(T)
			1967 November 2	(T)

TABLE 10.A

Number of solar eclipses per periods of 100 years

Years	all ecl.	central ecl.	A-T ecl.	Years	all ecl.	central ecl.	A-T ecl.
−599 to −500	255	159	8	1401 to 1500	222	141	19
−499 to −400	241	155	17	1501 to 1600	227	150	19
−399 to −300	225	141	23	1601 to 1700	248	158	24
−299 to −200	226	140	24	1701 to 1800	251	159	19
−199 to −100	237	154	21	1801 to 1900	242	155	15
− 99 to 0	251	158	17	1901 to 2000	228	145	6
+ 1 to 100	248	157	25	2001 to 2100	224	144	7
101 to 200	237	154	16	2101 to 2200	235	152	4
201 to 300	227	144	5	2201 to 2300	248	156	3
301 to 400	222	146	7	2301 to 2400	248	160	8
401 to 500	233	152	2	2401 to 2500	237	153	1
501 to 600	251	157	6	2501 to 2600	225	140	6
601 to 700	251	160	4	2601 to 2700	227	147	5
701 to 800	233	154	2	2701 to 2800	242	158	3
801 to 900	222	140	6	2801 to 2900	254	158	9
901 to 1000	227	149	3	2901 to 3000	248	155	11
1001 to 1100	241	157	6	3001 to 3100	230	150	17
1101 to 1200	250	158	15	3101 to 3200	224	139	25
1201 to 1300	246	158	18	3201 to 3300	231	152	22
1301 to 1400	229	150	24	3301 to 3400	250	157	21

Between A.D. 3100 and 3200 there will be seven such eclipses. On the other hand, there were no (A) or (T) eclipses at all between the years 994 and 1277, between 1656 and 1928, and there will be none between 2141 and 2459.

6.   In Chapter 8 we mentioned the very small partial solar eclipse of 1935 January 5. Between A.D. 1000 and 3000 the following P-eclipses have a maximum magnitude of 0.005 or less :

1175 Oct. 16	0.002		2883 Aug. 23	0.001
1639 Jan. 4	0.001		2893 Dec. 29	0.003
1935 Jan. 5	0.001		2904 June 5	0.004
2098 Oct. 24	0.005		2995 Aug. 17	0.004

For these shallow eclipses too we note an irregular distribution in the course of time: there is none between A.D. 2098 and 2883, and then four cases take place in the course of the next 113 years.

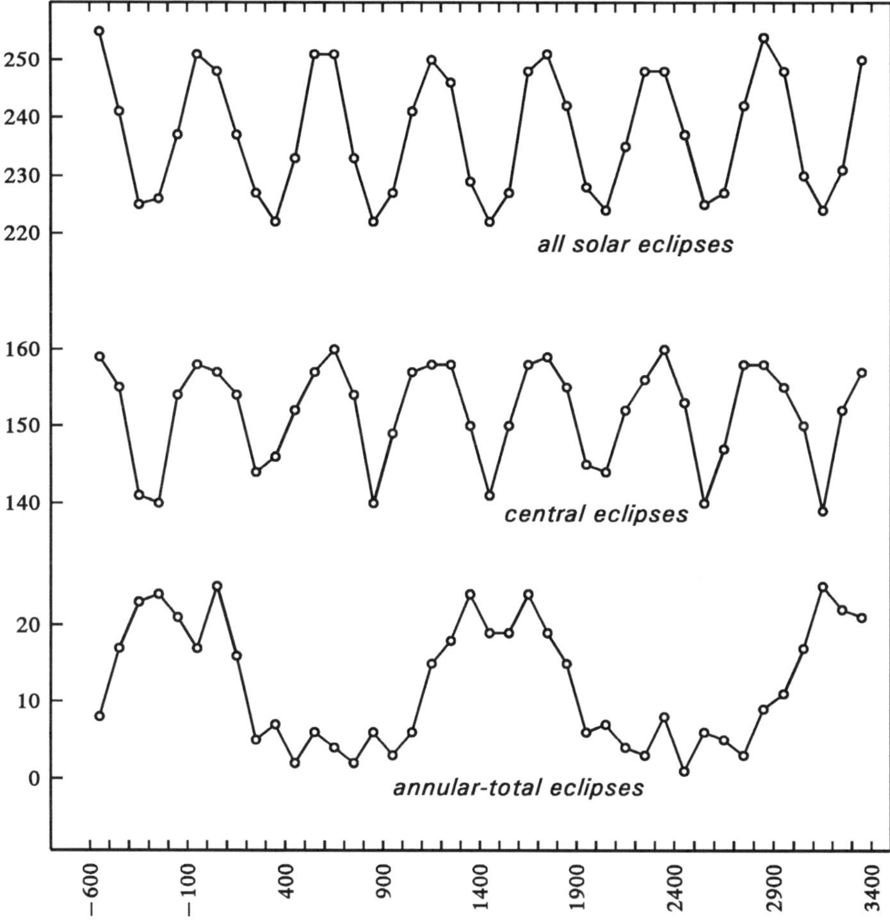

Fig. 10.b : *The number of solar eclipses per periods of 100 years*

7.   Two successive New Moons can both give rise to a solar eclipse. In almost all cases these eclipses are both of type P, visible from opposite (northern and southern) hemispheres. Here are two examples :

1982 June 21	southern hemisphere	(southern Atlantic, South Africa)
1982 July 20	northern hemisphere	(Arctic, Scandinavia, W. Europe)
2011 June 1	northern hemisphere	(Siberia, Arctic, Canada, Greenl.)
2011 July 1	southern hemisphere	(southern Indian Ocean)

It is extremely rare that one of the two eclipses is central. Between the years −599 and +3400 it occurs only five times :

−575 May 19	T	$\gamma = +0.9708$		2195 July 7	P	$M = 0.036$
−575 June 18	P	$M = 0.017$		2195 Aug. 5	T	$\gamma = -0.9845$
−434 April 21	T	$\gamma = -0.9831$		2912 July 6	T	$\gamma = -0.9854$
−434 May 20	P	$M = 0.030$		2192 Aug. 4	P	$M = 0.032$
1248 May 24	T	$\gamma = +0.9799$				
1248 June 22	P	$M = 0.022$				

We see that in these cases the central eclipse is always a total one taking place in high northern or southern latitudes (see the value of γ), while the partial eclipse is always a very small one. For the partial eclipses, the maximum magnitude M is given.

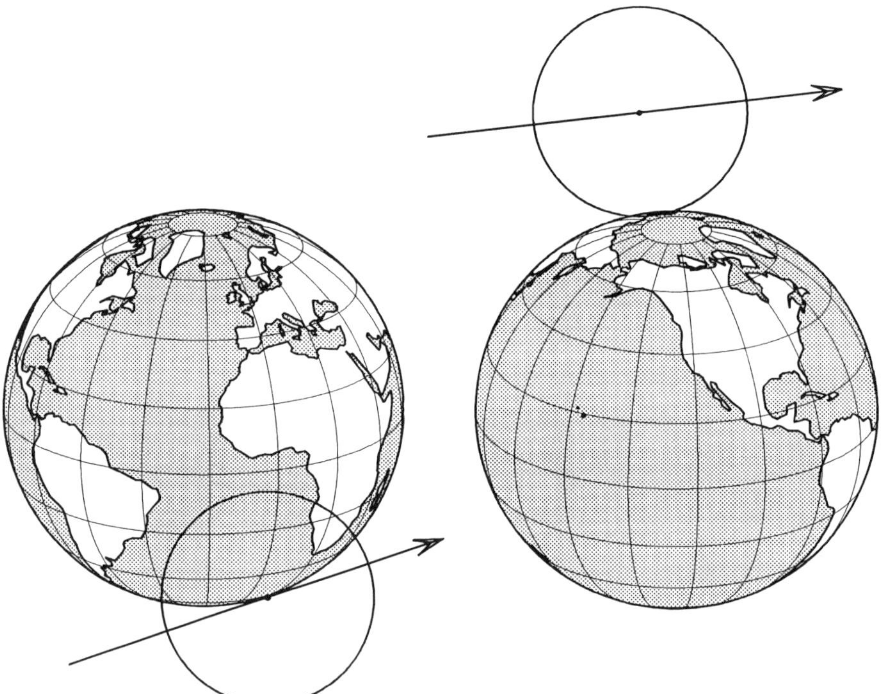

Fig. 10.c : *Two solar eclipses taking place at successive New Moons. In this special case, the first event (at left), on 1928 May 19, was a non-central total eclipse in the southern hemisphere; the second (right), on 1928 June 17, was a very small partial eclipse in the northern hemisphere. In each drawing, the circle having a diameter 53 % that of the Earth is the edge of the Moon's penumbra in the fundamental plane — the plane through the Earth's center perpendicular to the axis of the lunar shadow. The tiny dot at the center of the penumbra is the Moon's umbra, within which the eclipse is total.*

At four other times during the same period of 40 centuries, the axis of the shadow just misses the Earth, and the total eclipse is a non-central one :

-159 July 8	(T)	$\gamma = -1.0104$		1928 May 19	(T)	$\gamma = -1.0049$	
-159 Aug. 6	P	$M = 0.014$		1928 June 17	P	$M = 0.038$	
$-$ 26 May 10	P	$M = 0.027$		2459 May 3	P	$M = 0.022$	
$-$ 26 June 8	(T)	$\gamma = +1.0062$		2459 June 1	(T)	$\gamma = -1.0097$	

And look at the dates : in all nine cases mentioned above all eclipses take place during the months April — August!

8.   One might think that partial eclipses (type P) are visible only from the polar (northern or southern) regions of the Earth. This certainly is not the case.

The partial eclipse of 1974 December 13 ($\gamma = +1.0796$, maximum magnitude 0.827) was visible not only in the United States, but the region of visibility included Mexico, Cuba, and the northern part of South America.

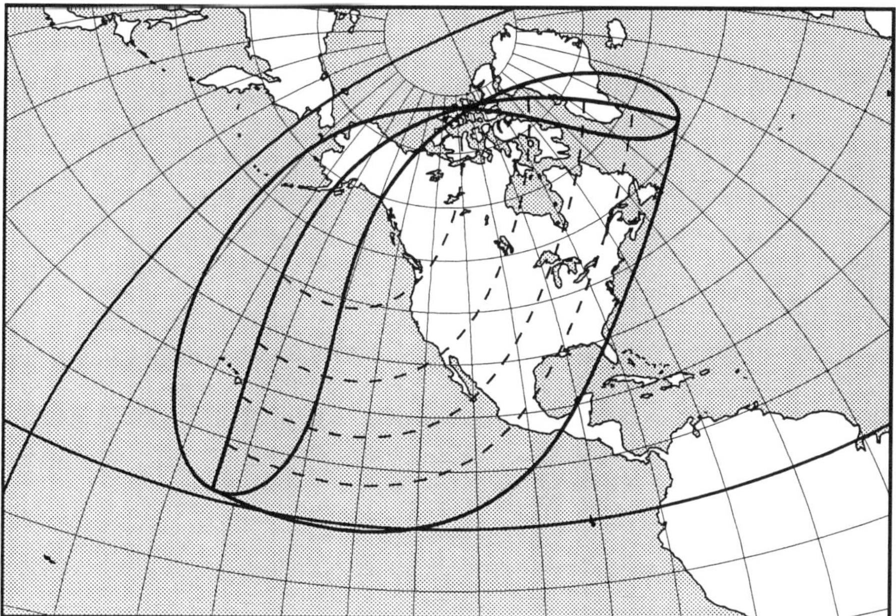

Fig. 10.d : *The region of visibility of the partial solar eclipse of 2083 February 16, for which $\gamma = +1.0168$, and the maximum magnitude is 0.944. The region of visibility will extend a little into the southern hemisphere. For example, at longitude 132°00' west and latitude 0°30' south, the maximum magnitude will be 0.014. In this map the equator and the meridian of 180° are drawn as thicker lines.*

In the case of the P-eclipse of 1996 October 12, the region of visibility includes the whole of Europe, and a large portion of Africa up to latitude 16° north.

It may come as a surprise that in some (rare) cases the region of visibility of an eclipse of type P extends right to the equator and even a little into the other hemisphere. An example is shown in Figure 10.*d* : the partial eclipse of 2083 February 16.

11. Regions of visibility of solar eclipses

Many astronomical almanacs give for every solar eclipse of the current year a map showing the region of visibility of the event. In this chapter we will describe the various forms such a region can take.

Let us first recall some classical concepts. In Figure 11.*a* the Sun, the Moon and the Earth are represented (*not* to scale!). *ABC* is the Moon's umbral cone; all points inside this cone see a total solar eclipse (or an annular one in the extension of this cone). *JABK* is the Moon's penumbral cone, inside which a partial eclipse is seen.

During a solar eclipse, the tip *C* of the lunar umbra is always in the vicinity of the Earth. Consequently, this umbra (or its extension) always can intercept only a small part of the Earth's surface. If, as in Figure 11.*a*, the umbra reaches the surface, then the eclipse is *total* there (in *C*). If the umbral cone ends in space above the Earth's surface, then only its extension can reach the surface, resulting in an *annular* eclipse. The penumbral cone of the Moon, however, can cover a much larger part of the Earth's surface.

In the theory of eclipses, one considers the plane through the center of the Earth and perpendicular to the axis of the lunar shadow — the dashed line in the drawing. That plane (not represented in the figure) is called the *fundamental plane*. The intersection of the umbral cone, or of its extension, with the fundamental plane is a relatively small circle, having a diameter that is at most 3 % that of the Earth. The intersection of the penumbra in the fundamental plane is a much larger circle, whose diameter (*JK* in the drawing) is between 53 % and 58 % that of the Earth. So, roughly speaking, in the vicinity of the Earth the diameter of the penumbra is equal to one half the Earth's. Consequently, the lunar penumbra never can cover the

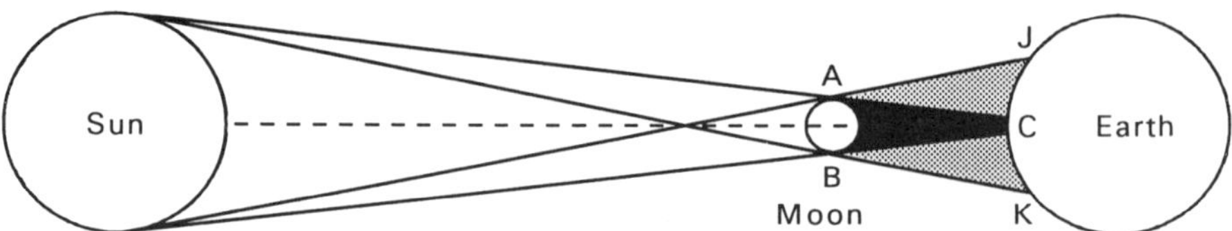

Fig. 11.a

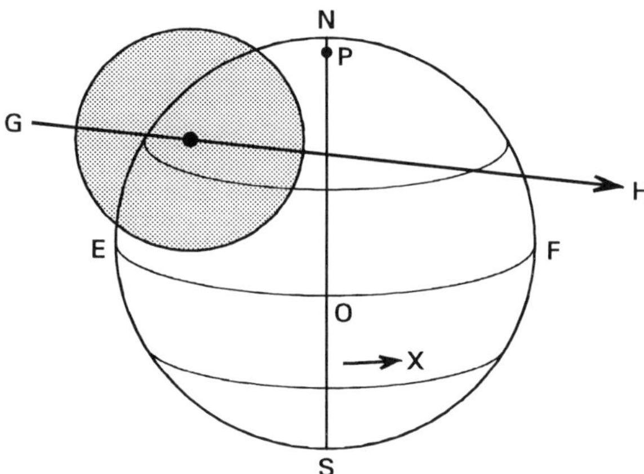

Fig. 11.b

In the course of a solar eclipse, the Moon's pen-umbra (the grey circle) and the umbra (the much smaller black circle) pass over the Earth. This view is as 'seen' by an observer on the Sun.

Earth completely. A given solar eclipse cannot be visible at both the North and the South Pole.

Let us now imagine that we are on the Sun and are looking at the Earth to see what happens there in the course of a solar eclipse. We then 'see' that in the period of a few hours the large penumbra of the Moon — the grey circle in Figure 11.*b* — and the much smaller umbra are passing over the Earth, from *G* to *H*. Meanwhile the Earth's globe is rotating around its axis (see the arrow *X*), but at a smaller linear speed than the lunar shadow. In Figure 11.*b*, *P* represents the Earth's North Pole. *EOF* is the equator. For all points of the arc *NES* it's sunrise: all those points are appearing 'from behind the horizon' — the Earth's limb. All points of the arc *NFS* have sunset. And for all points on the central meridian *NPOS* it is noon (true, or apparent, local noon).

In Figures 11.*b* and *c*, the North Pole *P* of the Earth is supposed to be visible, while the South Pole is behind the limb at *S*. Of course, the opposite can happen too.

Let us now examine the different possible cases. It should be noted that the subdivision we describe here concerns only the region of visibility in its entirety. An observer seeing a solar eclipse can have no idea of the shape of the region of visibility within which he is situated.

Type I

Let us first assume that the entire lunar penumbra passes over the Earth (Figure 11.*c*). In this case there exists a line of central eclipse. The eclipse is either total (T), annular (A), or annular-total (A-T).

The penumbra needs some time to pass over the edge *NGS*, and during this lapse of time a part of the Earth's surface comes into sight from behind the horizon. So, for all observers in this region the eclipse is already in progress at the time of sunrise.

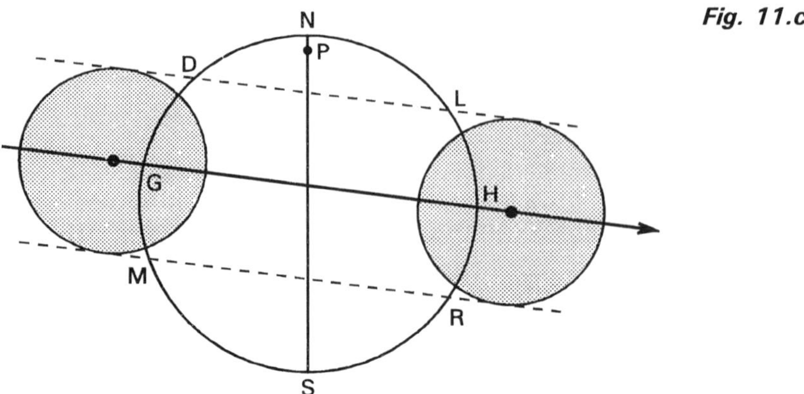

On a geographical map the region has an oval shape: $D'QM'U$ in Figure 11.*d*. For all points on the arc $D'UM'$ the eclipse is just beginning at sunrise; there the observers just see the whole eclipse, although the first contact between the disks of the Moon and the Sun takes place with the Sun on the horizon. For all points on the arc $D'QM'$ the eclipse *ends* at sunrise, so these observers just see nothing of the eclipse: last contact occurs at sunrise.

Moreover, there exists a line $D'G'M'$ (dashed in Figure 11.*d*) bisecting the oval and for all points of which the *maximum* of the eclipse occurs at sunrise. So, in the region $D'G'M'U$ more than half of the eclipse is visible: the Sun rises already eclipsed, but the maximum and the end of the phenomenon can be seen. In the area $D'QM'G'$ less than half of the eclipse is visible: here too the Sun is eclipsed when it rises, but first contact and maximum both take place before sunrise; only the last contact can be seen.

When the penumbra leaves the Earth, it takes some time to pass over the edge NHS (Figure 11.*c*), and during this lapse of time a part of the Earth's surface disappears from sight on this arc behind the Earth's limb. For all observers in this region the eclipse is still in progress at the time of sunset.

On a geographical map the region has the shape of an oval: $L'VR'T$ in Figure 11.*d*. For all points on the arc $L'VR'$ the eclipse is just ending at sunset, so these observers just see the whole eclipse, although last contact takes place with the Sun on the horizon. For all points on the arc $L'TR'$ the eclipse begins at sunset, so these observers just see nothing of the eclipse: first contact occurs at sunset.

Moreover, there exists a line $L'H'R'$ (dashed in the figure) for all points of which the maximum of the eclipse occurs at sunset. In the region $L'VR'H'$ more than half of the eclipse is visible: the beginning and the maximum can be seen, but the Sun is still eclipsed when it sets, and last contact takes place under the horizon. In the area $L'H'R'T$ less than half of the eclipse is visible: the Sun sets between first contact and maximum, so only the first contact can be observed there.

In the large region $D'L'VR'M'U$ the eclipse can be seen in its entirety, from first contact to last contact.

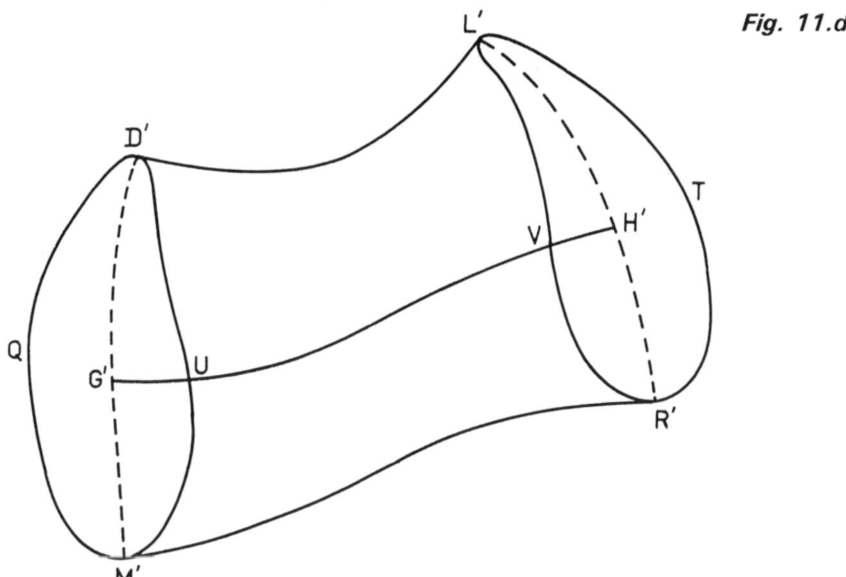

The center of the lunar shadow crosses the Earth from G to H (Figure 11.c), resulting in the *central line* on the Earth's surface: $G'H'$ in Figure 11.d. For all observers on this line the center of the lunar disk passes exactly over the center of the Sun's disk. The central line starts at sunrise in G', on the arc $D'G'M'$; it ends at sunset in H', on the arc $L'H'R'$.

In U the beginning of the partial phase (first contact) takes place at sunrise; about one hour later the observer enjoys a central (total or annular) eclipse, and last contact takes place with the Sun still higher in the sky. What happens for an observer in V?

Finally, there exists a northern limit $D'L'$ and a southern limit $M'R'$. These limits of the Moon's penumbra are the projection of DL and MR (Figure 11.c) on the Earth's surface; they delineate the region of visibility of the partial eclipse. For all points on the limit $D'L'$ the event is a *grazing* one: the lunar disk just touches the southern limb of the Sun; the magnitude of the eclipse is exactly 0.000 there. The same holds for an observer on the southern limit of partial eclipse $M'R'$, but here the Moon grazes the northern limb of the Sun.

North of the line $D'L'$ and south of $M'R'$ no eclipse takes place; there the Sun is above the horizon, but the Moon misses the Sun completely.

The limits $D'L'$ and $M'R'$ are tangent to the loops $D'QM'U$ and $L'VR'T$. In D' the grazing eclipse occurs at sunrise. What happens for observers situated at L', at M', at R'?

Figure 9.a (page 50) shows the regions of visibility of two solar eclipses of type I. On these maps the central line itself is not shown. Instead, both the northern and southern limits of the track of total eclipse are drawn.

Type II

It often happens that only a part of the lunar penumbra passes over the Earth, although there exists a line of central eclipse. Just as for eclipses of type I, the event is either total, annular, or annular-total.

In Figures 11.*e* and 11.*f* the event takes place in the northern part of the Earth. In this case, there exists a *southern* limit of partial eclipse (*M'R'*, the projection of *MR* on the Earth's surface), but there is no northern limit because the northern edge of the penumbra moves along the line *DL* which lies *outside* of the Earth.

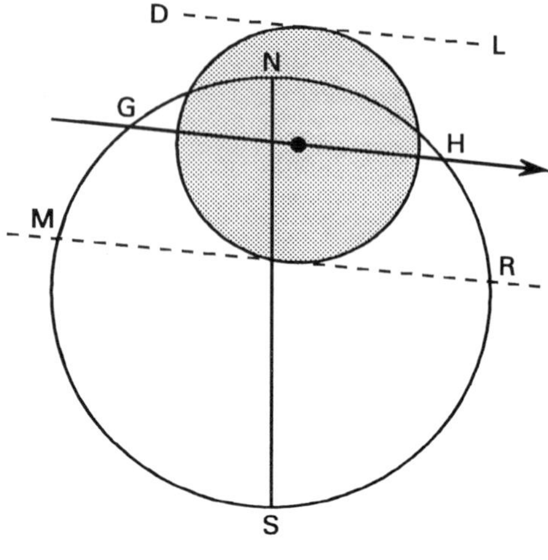

Fig. 11.e

Unlike to type I, the limiting curves at sunrise and sunset no longer form two separate, closed loops. Instead, the curves are connected into a distorted figure 8: the curve *M'QWVR'TWU* in Figure 11.*f*). The 'node' *W* lies in the polar regions.

Of course, the event can also take place in the southern hemisphere of the Earth. In this case, there exists a *northern* limit of partial eclipse, while the node *W* is situated south of it.

On the arc *WQM'* the eclipse ends at sunrise; on *WUM'* it begins at sunrise. For all places in the region *WQM'U* the Sun rises already eclipsed. On the arc *WVR'* the eclipse ends at sunset; on the arc *WTR'* the eclipse begins at sunset. So, for observers in the area *WTR'V* the Sun sets while the eclipse is still in progress.

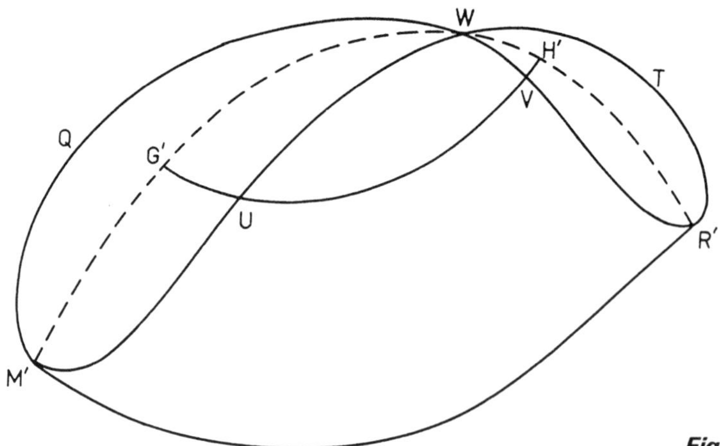

Fig. 11.f

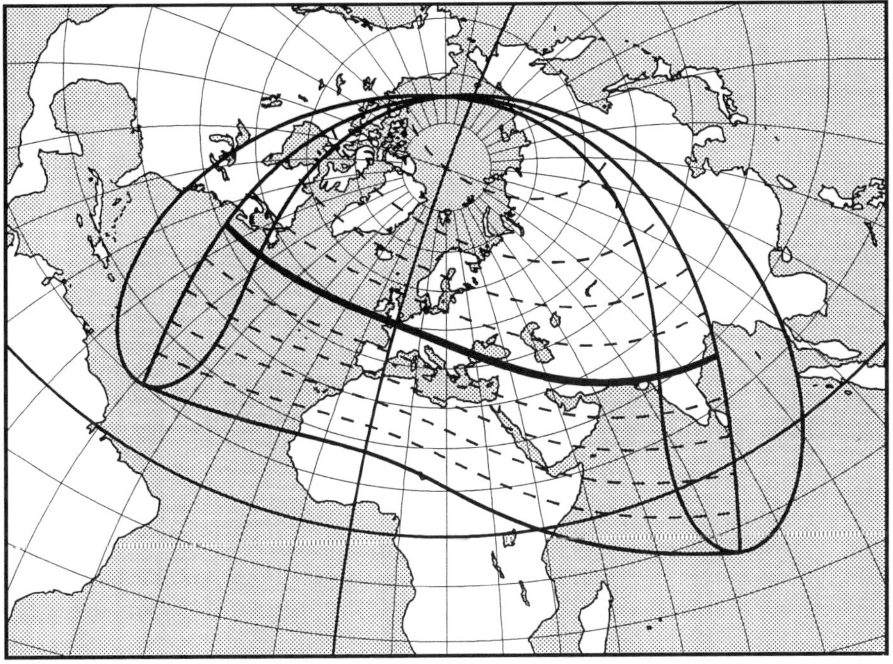

Fig. 11.g : *The total solar eclipse of 1999 August 11. The region of visibility of this eclipse is of type II. The path of total eclipse starts at sunrise in the Atlantic Ocean south of Nova Scotia, crosses the northern Atlantic Ocean, then Europe from southwestern England to Turkey, southwestern Asia, and ends at sunset shortly after leaving India. The dashed lines are the 'isomagnitudes' of 0.2, 0.4, 0.6, and 0.8.*

In the region $WUM'R'V$ the eclipse is visible in its entirety, from first to last contact. The curve 'maximum on the horizon' doesn't consist of two sections as for type I, but is one single arc $M'G'H'R'$ (the dashed line in Figure 11.f).

As for type I, we have the central line $G'UVH'$, which is the projection of GH on the Earth's surface. This central line begins at sunrise in G', and ends at sunset in H'. What happens near the node W will be discussed later.

An example of an eclipse of type II is illustrated in the map above.

Type III

This type occurs when less than half of the penumbral cone passes over the Earth and the umbral cone (or its extension) does not touch it, as in Figure 11.h. So, the eclipse is partial (type P). The region of visibility has the same appearance as for type II, except that there is no central line (Figure 11.i).

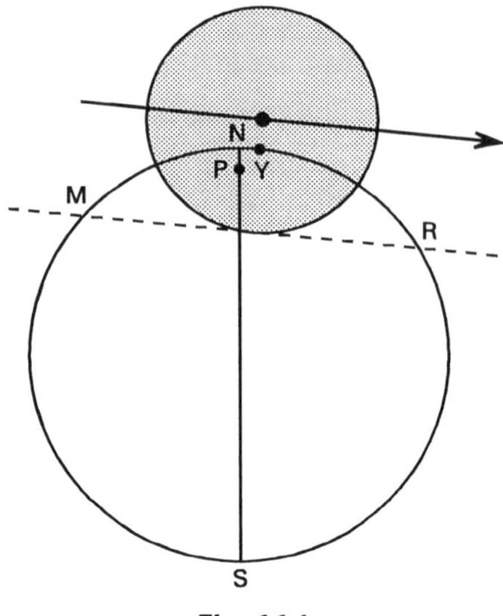

Fig. 11.h

The maximum magnitude on the Earth's surface occurs somewhere on the curve 'maximum on the horizon', not far from the node W, for instance in Y'. This is the point Y on the surface of the Earth which most closely approaches the axis of the shadow (Figure 11.h).

An example of type III is shown in Figure 10.d (page 61).

The three types (I, II, III) we have described are the most frequent. The next four are much rarer.

Type IV

The entire lunar penumbra passes over the Earth. However, unlike what happens in the case of type I, the northern (or southern) edge of the penumbra reaches *and* leaves the Earth on the same side of the central meridian NIS. Figure 11.j shows an event where the northern limit of the region of visibility begins *and* ends at sunrise, respectively in D and in L, on the arc $NLDGMS$.

In such a case, the northern limit on the Earth's surface is a rather short arc ($D'L'$ in Figure 11.k). North of the line DL there is a small region that remains outside of the penumbra; this small area lies completely on one side of the central meridian NIS. The points 'begin or end of the eclipse on the horizon' form (as for type I) two separate curves; but now one of these curves has the shape of a distorted 8 and thus has a node W.

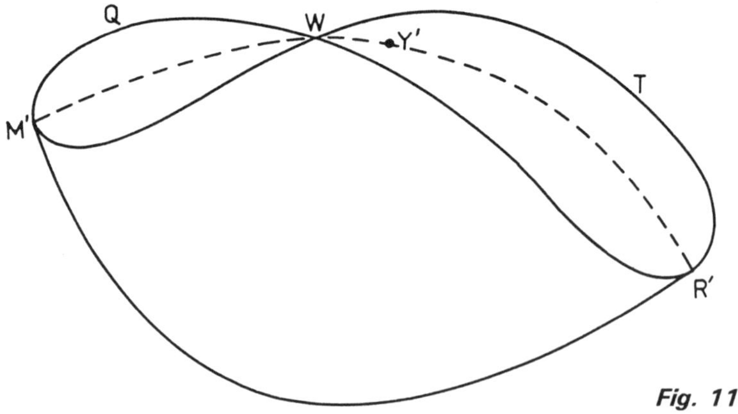

Fig. 11.i

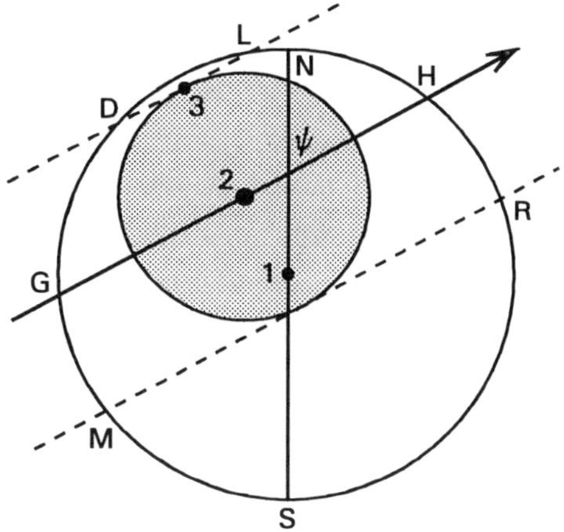

Fig. 11.j

As for type I, the eclipse begins at sunrise for all points of the arc $D'UM'$; it ends at sunrise on $D'QM'$. But for the second curve the situation is different. For all points of the arc $R'VW$ the eclipse ends at sunset, but on the small arc WaL' it ends at *sunrise*. On the arc $R'TW$ the eclipse begins at sunset, but on the small arc WbL' it begins at *sunrise*.

For all the points in the small loop $WaL'b$ the Sun *rises* eclipsed, but in the larger region $WVR'T$ the eclipse is in progress when the Sun is *setting*.

As for type I, there are a normal southern limit $M'R'$, a central line $G'H'$, and two curves 'maximum on the horizon'. On the curve $D'G'M'$ and on $L'W$ maximum eclipse occurs at sunrise; on the arc $WH'R'$ maximum takes place at sunset.

Of course, it is possible that the small loop $WaL'B$ occurs at the first curve, in the vicinity of D', or also at the southern part of one of these regions, near M' or near R'.

An example of an eclipse of type IV is shown in Figure 11.*l*.

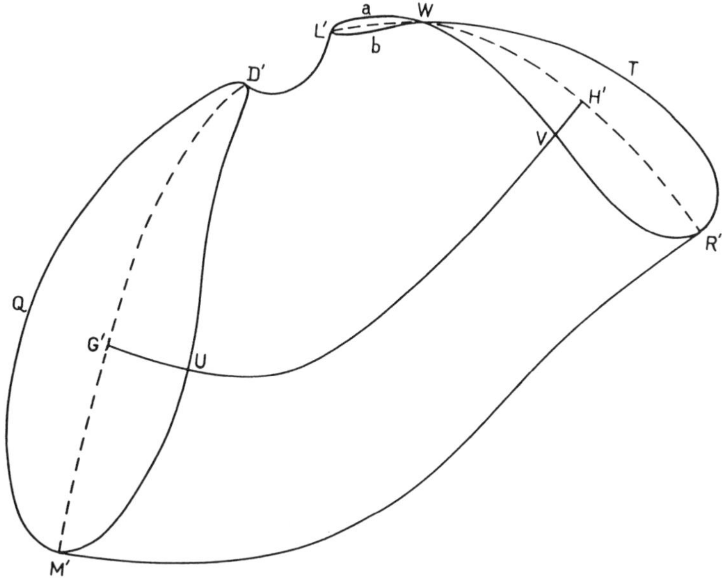

Fig. 11.k

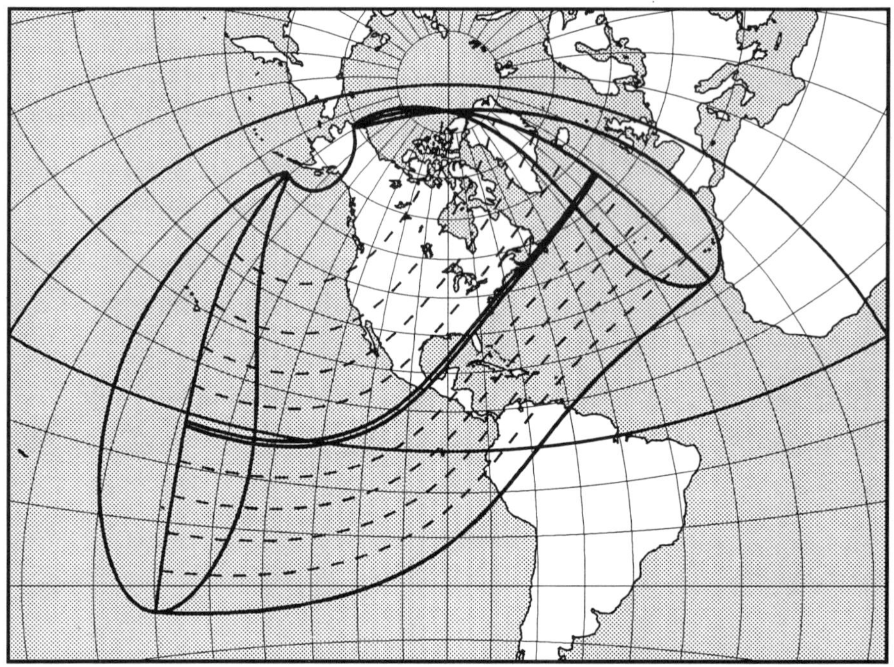

Fig. 11.l : *Map of the total solar eclipse of 1970 March 7. The path of total phase crossed Mexico, the eastern United States and eastern Canada. The eclipse was of type IV; see the small loop north of Alaska.*

During the evolution of the shape of the region of visibility from type II to type I for eclipses of the same Saros series, there suddenly occurs a 'break' into separate loops. This break occurs through type IV. For instance, the eclipse of 1941 September 21 was of type II; one Saros later, the eclipse of 1959 October 2 was of type IV; the next eclipses in that series, those of 1977 October 12 and 1995 October 24, were of type I.

In some Saros series, *several* consecutive eclipses are of type IV. This is the case, for example, for the eclipses of 1970 March 7, 1988 March 18, and 2006 March 29. But in some other Saros series, no eclipse of type IV takes place. For instance, the eclipse of 1965 May 30, which was of type I, was followed, after one Saros, by the eclipse of 1983 June 11, which was of type II.

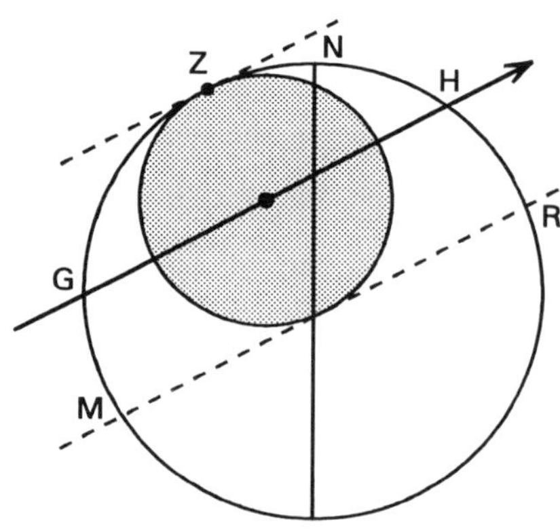

Fig. 11.m

It should be noted that the break from the single curve in the shape of a distorted 8 into two separate loops, or vice versa, does not occur at the node W of the figure 8, but at a short distance from it. For an observer in W, the times of beginning and end of the eclipse do not coincide; the event has some duration there, and the magnitude of the eclipse is not zero. The duration of the eclipse is zero at the point where the break occurs. This would be the case at an eclipse where the Moon's penumbra were exactly touching the Earth internally, such as point Z in Figure 11.m. At Z the magnitude of the eclipse is 0.000. At W the magnitude is small, but not zero. The 'small loop' occurs along the short arc ZN.

At the eclipse of 1952 February 25 (Figure 11.n) it was evident that a break was approaching. That eclipse was of type II and thus had a continuous rising-and-setting curve. But the region within which the eclipse was in progress at sunrise appeared strangled (near Iceland). One Saros later, the eclipse of 1970 March 7 was indeed of type IV — see Figure 11.l.

For an eclipse to be of type IV, the following two conditions must be satisfied simultaneously according to Figure 11.j:

a. the entire penumbra of the Moon should pass over the Earth;

b. the northernmost point N or the southernmost one S (according to the case) should be covered by the penumbra in the course of the eclipse.

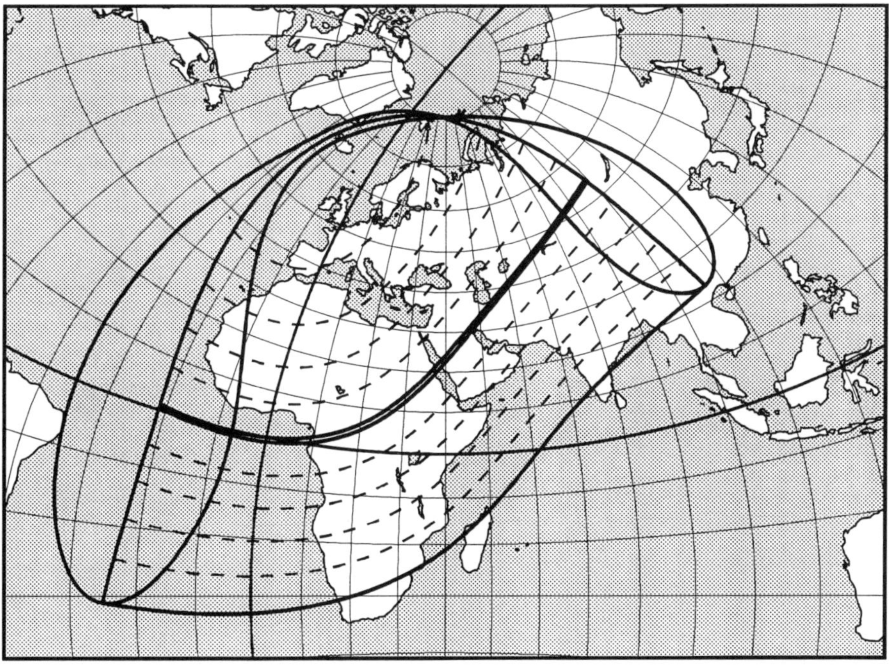

Fig. 11.n : *The region of visibility of the total solar eclipse of 1952 February 25.*

Let us now give the needed formulae for the interested reader. Use is made of some of the so-called Besselian elements of the given eclipse, which can be found in some astronomical almanacs, or in our *Elements of Solar Eclipses 1951-2200* (Willmann-Bell, ed.; 1989).

Let γ be the least distance of the axis of the lunar shadow cone to the center of the Earth; this is the distance *1-2* in Figure 11.*j*. The equatorial radius of the Earth is taken as unity. The quantity γ is positive or negative, depending on the axis of the shadow passing to the north or to the south of the center of the Earth. For each eclipse of the years 1951 to 2200, the quantity γ is given in our *Elements*. If γ is not known, it can easily be calculated: it is the minimum value of the square root of $x^2 + y^2$, where x and y are two of the Besselian elements.

Let L be the radius of the penumbra in the fundamental plane; this is the distance *2-3* in Figure 11.*j*. Here too the equatorial radius of the Earth is taken as unity.

According to condition *a*, the quantity $|\gamma| + L$ must be smaller than 0.997. (The value 0.997 is used instead of 1, to take the flattening of the Earth's globe into account). According to condition *b*, the projection of the segment *1-N* on *1-3* should be smaller than the distance *1-3*. Let ψ be the angle between *NS* and *GH*, as shown in Figure 11.*j*. Then $0.997 \sin \psi$ must be smaller than the distance *1-3*.

Accordingly, the conditions for the occurrence of type IV are

$$0.997 \sin \psi \ < \ |\gamma| + L \ < \ 0.997$$

The angle ψ can be calculated from $\tan \psi = x'/y'$, where x' and y' are the variations of x and y in an arbitrary unit of time, for instance per hour. We have $\sin \psi > 0$, and the angle ψ always lies between $60°$ and $120°$. On the other hand, L always lies between 0.530 and 0.575. From this it follows that type IV only *can* occur when the absolute value of γ is between 0.29 and 0.47.

During the period A.D. 1950 to 2030, there are 17 eclipses of type IV:

1959	Oct.	2	$\gamma =$	+0.4206	T	2006	Mar. 29	$\gamma =$ +0.3841	T

1959 Oct. 2	$\gamma = $ +0.4206	T		2006 Mar. 29	$\gamma = $ +0.3841	T
1965 Nov. 23	+0.3904	A		2006 Sep. 22	−0.4064	A
1970 Mar. 7	+0.4471	T		2010 Jan. 15	+0.4000	A
1973 Dec. 24	+0.4169	A		2017 Aug. 21	+0.4365	T
1977 Apr. 18	−0.3992	A		2023 Apr. 20	−0.3953	AT
1988 Mar. 18	+0.4186	T		2023 Oct. 14	+0.3752	A
1992 Jan. 4	+0.4089	A		2024 Oct. 2	−0.3510	A
2005 Apr. 8	−0.3475	AT		2028 Jan. 26	+0.3900	A
2005 Oct. 3	+0.3304	A				

It's remarkable that four *consecutive* solar eclipses, those of the years 2005 and 2006, are all of type IV.

When $|\gamma| + L$ is only a little larger than $0.997 \sin \psi$, then the small loop (*WaL'b* in Figure 11.*k*) is so small that it is not visible on the map. Near that place,

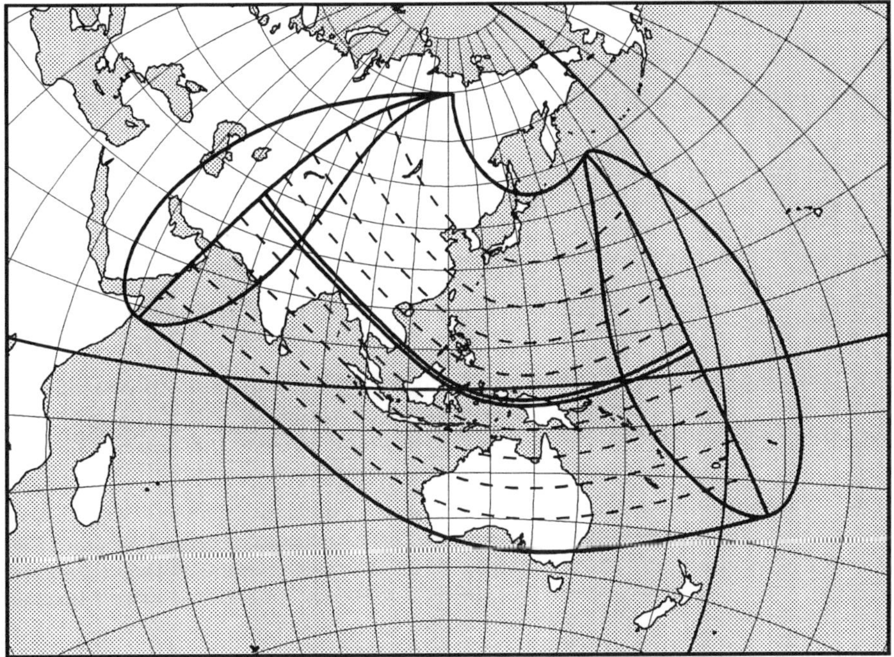

Fig. 11.o : *Map showing the region of visibility of the annular solar eclipse of 1965 November 23. Note the 'cusp' in northern Siberia.*

the curve 'eclipse begins or ends on the horizon' presents a cusp, as seen for instance in Figure 11.*o*.

The node *W* (Figure 11.*k*) lies in the northern or in the southern polar regions according to the sign (positive or negative) of γ. It occurs at the beginning of the eclipse when y' and γ have opposite signs, at the end of the eclipse if y' and γ have the same sign.

Type V

This case occurs when the *central line* behaves as the line *DL* in the previous case. The situation is as depicted in Figure 11.*p*. Here the central line begins *and* ends at sunrise: *G* and *H* are both on the arc *NMS*. But of course both *G* and *H* can also lie on the arc *NRS*; in this case the central line begins and ends at sunset.

The shape of the eclipse region is as depicted in Figure 11.*q*. This differs little from type II (Figure 11.*f*). But now the node *W* no longer is situated between *G'* and *H'*. The latter two points now both are either between *M'* and *W*, or between *W* and *R'*.

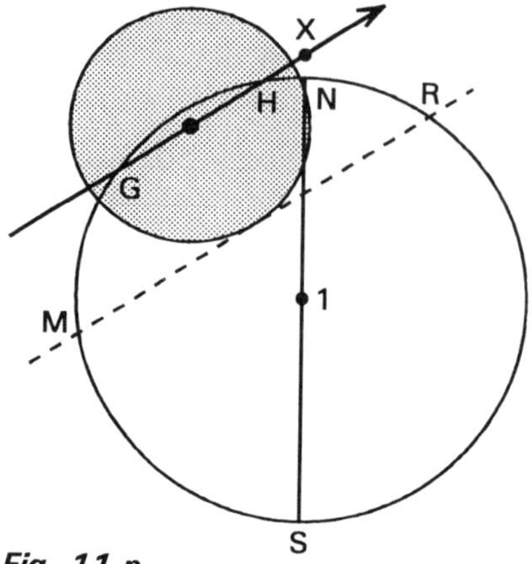

Fig. 11.p

In the case of an eclipse of type V, there is no central eclipse at true noon (or midnight): the segments *GH* and *NS* do not intersect. Along the whole central line central eclipse occurs either before, or after noon.

Type V can occur at total, annular or annular-total eclipses. The conditions for type V to take place are

$$0.997 \sin \psi \ < \ |\gamma| \ < \ 0.997$$

where the angle ψ is to be calculated in the same way as for type IV. Type V *can* take place only when the absolute value of γ is between 0.86 and 0.997.

The condition $0.997 \sin \psi < |\gamma|$ can be expressed in another way. At the instant when the center of the shadow is exactly on the segment *NS* or its extension (point *X* in Figure 11.*p*), its distance to the Earth's center must be larger than the Earth's radius. In other words, the distance *1-X* must be larger than the distance *1-N*. That is to say, at the time when the Besselian element *x* is zero, the value of the element *y* must be larger than the quantity ρ which is given by

$$\rho^2 \ = \ 1 \ - \ 0.006\,694\,385 \ \cos^2 d$$

where *d* is another of the Besselian elements (it is almost equal to the declination of the Sun). Here is the value of ρ for some values of *d* :

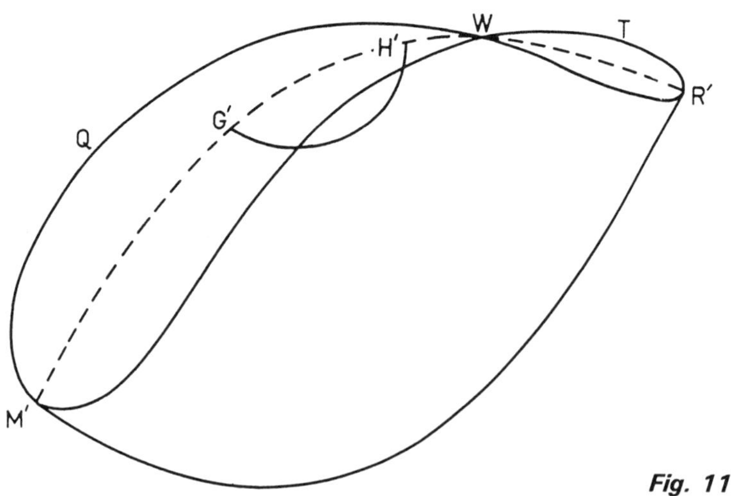

Fig. 11.q

$$|d| = 0° \qquad \rho = 0.99665$$
$$4° \qquad 0.99666$$
$$8° \qquad 0.99671$$
$$12° \qquad 0.99679$$
$$16° \qquad 0.99690$$
$$20° \qquad 0.99704$$
$$24° \qquad 0.99720$$

For any instant near the time when $x = 0$, calculate the values of the Besselian elements x, y, x', y', L and d. (For the meaning of x', y' and L, see type IV). Find ρ, and calculate y_0 from $y_0 = y - y'x/x'$. Then the conditions for the occurrence of type V are

$$|\gamma| < 0.997 \qquad \text{and} \qquad |y_0| > |\rho|$$

An example of an eclipse of type V is shown in the map below: the total eclipse of 1968 September 22 which was visible in Siberia. At this eclipse, we had $\gamma = +0.945$ and $y_0 = +1.081$. The central line began *and* ended at sunset.

At an eclipse of type V, the central line begins and ends at sunrise when y' and γ have the same sign, at sunset when they are of opposite signs.

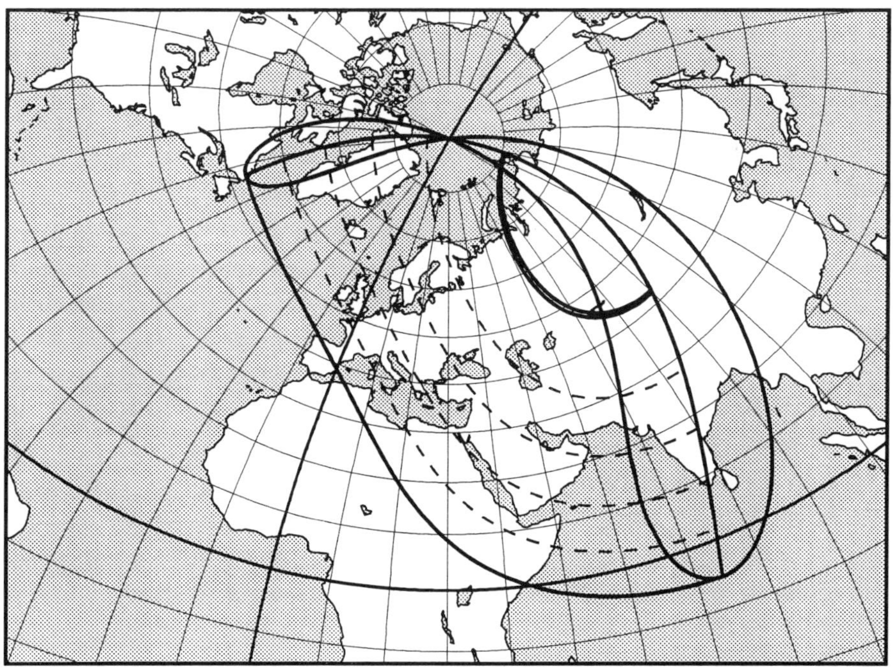

Fig. 11.r : *The region of visibility of the total solar eclipse of 1968 September 22*

During the period A.D. 1950 to 2030, the following solar eclipses are of type V. Note that the eclipses of $1950 - 1968 - 1986$ and those of $1990 - 2008 - 2026$ belong to a same Saros series. Although the annular eclipse of 2003 May 31 satisfies to the conditions we have described above, we prefer to consider it as being of type VII, because only a part of the (extension of the) umbral cone passes over the Earth.

1950	September 12	$\gamma = +0.8901$	total
1968	September 22	0.9449	total
1985	November 12	-0.9797	total
1986	October 3	$+0.9929$	annular-total
1990	January 26	-0.9459	annular
2003	November 23	-0.9640	total
2008	February 7	-0.9572	annular
2026	February 17	-0.9744	annular

Type VI

This type occurs when only a small part of the Moon's penumbral cone passes over the Earth, and in such a way that the southern (or northern) limit begins *and* ends at sunrise (or sunset). In Figure 11.*s* an event is represented for which the southern limit of partial eclipse begins and ends at sunset, in *M* and *R*, respectively.

The entire region of visibility is situated at one side of the central meridian *NS*. Nowhere is the eclipse visible at local true noon, since the penumbra does not touch the central meridian. In the case of Figure 11.*s*, the eclipse is visible only in the afternoon, in the small region *MR*.

On a geographical map, then, the region of visibility looks as in Figure 11.*t*. *M'XR'* is the southern limit of the region. (Of course, if the event occurs in the southern polar regions, it is the northern limit).

For all places in the oval region *M'QR'T* the Sun sets while the eclipse is still in progress. (For other eclipses, it can take place at sunrise). On the line *M'TR'* the eclipse begins at sunset. On *M'QR'* it ends at sunset. On the dashed line *M'Y'R'* maximum eclipse takes place at sunset. So, in the region *M'Y'R'Q* the begin-

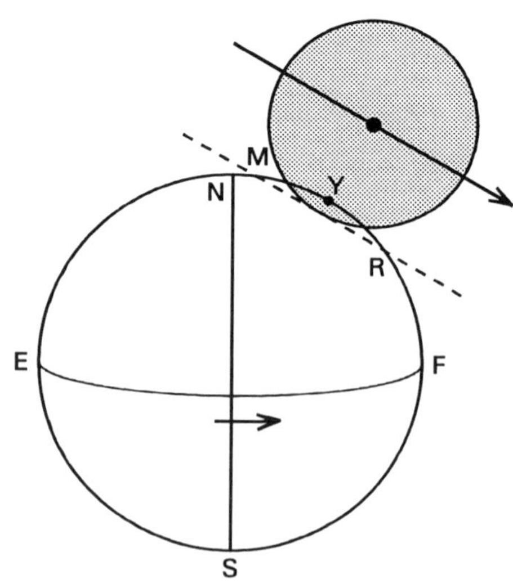

Fig. 11.s

Fig. 11.t

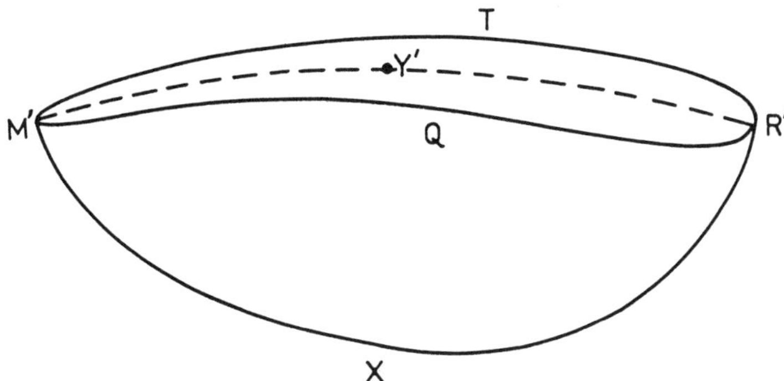

ning and the maximum can be seen, but last contact takes place after sunset. In the region $M'Y'R'T$ the beginning is visible, but the Sun already sets before maximum eclipse. The maximum magnitude occurs somewhere on the line $M'R'$, for instance at Y'. This is the point Y of the surface of the Earth that comes nearest to the axis of the shadow (Figure 11.*s*).

An example of an eclipse of type VI is shown in the map below: the small partial eclipse of 1931 September 12, which was the last eclipse of a Saros series.

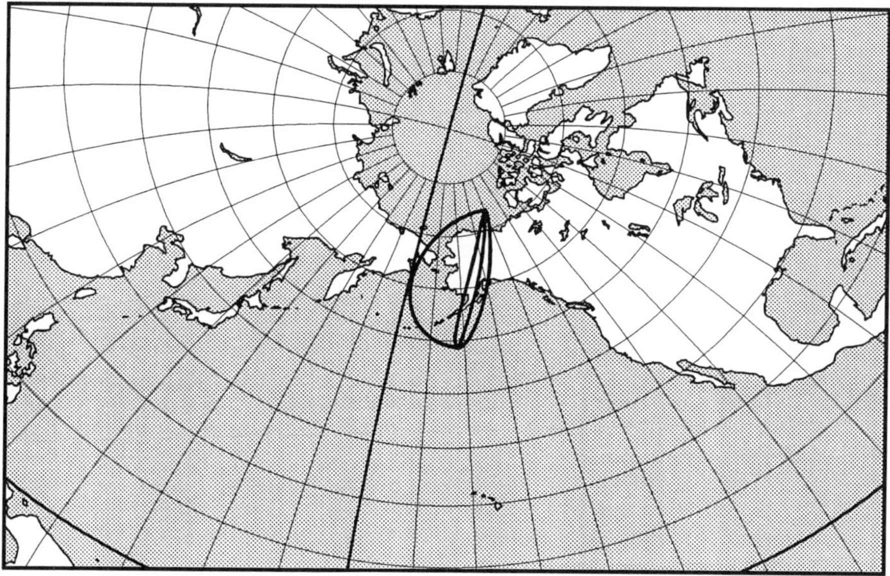

Fig. 11.u : The region of visibility of the partial solar eclipse of 1931 September 12

The conditions for occurrence of type VI are

$$0.997 \sin \psi \quad < \quad |\gamma| - L \quad < \quad 0.997$$

Type VI *can* occur only at partial eclipses with a maximum magnitude less than 0.25, or when the absolute value of γ is larger than 1.39.

The eclipse is visible from the northern polar regions if γ is positive, in the southern polar regions if γ is negative. The event is visible before noon when y' and γ have the same sign, in the afternoon when they have opposite signs.

Between A.D. 1900 and 2100 there are only eight eclipses of type VI:

Date	γ	*Max. magnitude*
1902 April 8	$+1.5022$	0.065
1913 August 31	$+1.4510$	0.152
1916 December 24	-1.5323	0.011
1931 September 12	$+1.5059$	0.047
1935 January 5	-1.5383	0.001
1971 July 22	$+1.5128$	0.069
2083 July 15	$+1.5463$	0.017
2098 October 24	-1.5409	0.005

Type VII

In this case, only a part of the Moon's *umbral* cone or of its extension passes over the Earth, as illustrated in Figure 11.*v*. So, the non-central annular and the non-central total eclipses belong to this type. But the eclipse *can* be central, as for instance the annular event of 2003 May 31. The region of annular phase of this eclipse will have a southern limit and a central line, but no northern limit.

In both cases, the area in which the eclipse is annular or total has the shape of a semi-circle, as depicted in Figure 11.*w*. Except for this semi-circular region, the region of visibility has the same appearance as that of an eclipse of type III.

In the semi-circular region, a total or annular eclipse can be seen at low altitude. At one side, the region is limited by the curve 'maximum on the horizon'. In the case depicted in Figure 11.*w*, an observer in the small semi-circular region can see the beginning of the partial phase, as well as the total

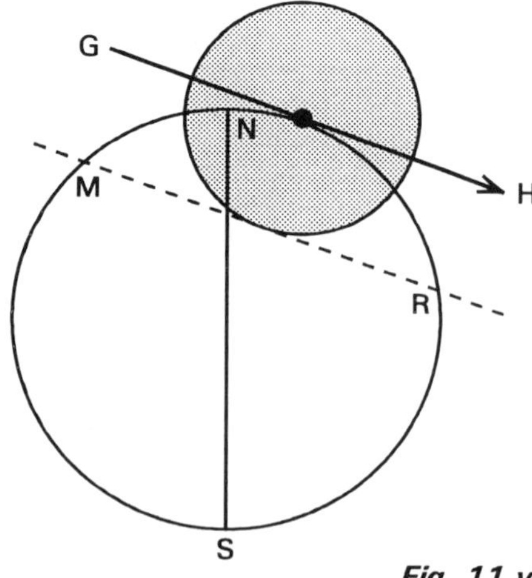

Fig. 11.v

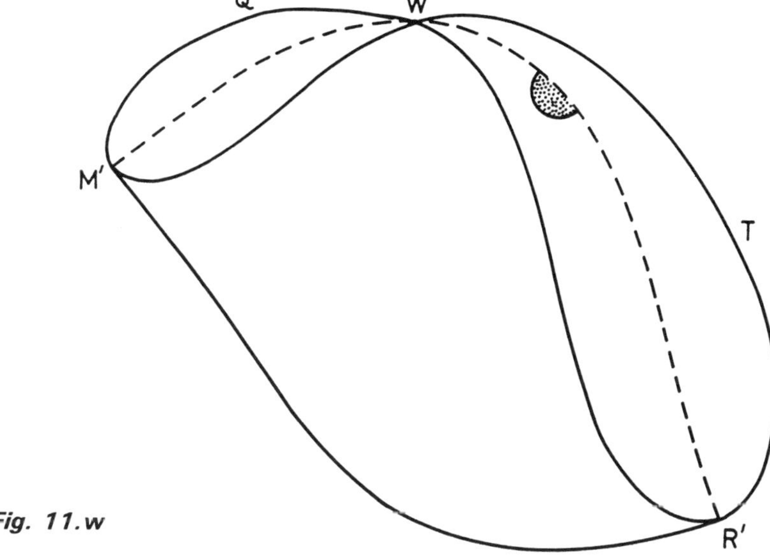

Fig. 11.w

(or annular) phase, while last contact occurs after sunset. Of course, the semi-circular region can also lie along the arc *M'W*, and in that case the event takes place near sunrise. Or the event can take place in the southern polar regions.

Another possibility is that the small semi-circular region of total or annular phase almost coincides with the node *W*; this happens when the direction of motion *GH* is approximately perpendicular to the central meridian *NS*. In such a case, uncommon situations happen near the node *W*, for example the beginning of the eclipse taking place before sunrise, the total (or annular phase) occurring above the horizon, and last contact after sunset! That will be the case at the (A) eclipse of 2104 December 17.

Near the node

The curve 'maximum on the horizon' (the dashed curve in Figures 11.*f*, *i*, *k*, *q*, and *w*) does *not* pass exactly through the node *W*, but at some distance from it, between the node and the nearer pole. The distance of the curve to the node is smaller when the magnitude of the eclipse at *W* is smaller, and is also smaller as the distance of the point *W* to the pole is smaller, that is, according as the declination of the Sun is closer to zero.

Let us now look at what happens near the node *W*. Here we must distinguish two cases, according as the (North or South) Pole *P*, which is in the vicinity of the node, is or is not illuminated by the Sun.

1. *Winter* (see Figure 11.*x*). — The curve 'maximum on the horizon' (*M'G'CH'R'*) does not pass through *W*, but at some distance from it, closer to the pole *P*. There exists a parallel of latitude *XY* which is tangent to the three curves :

— in *A*, to the curve *QAWV* ('eclipse ends on the horizon');
— in *C*, to the curve *G'ECFH'* ('maximum of the eclipse on the horizon');
— in *B*, to the curve *UWBT* ('eclipse begins on the horizon').

The geographical latitude of this parallel circle is equal to $90° - |\delta|$, where δ is the declination of the Sun. Here we may safely assume that δ is constant during the progress of the eclipse. The parallel of latitude *XACBY* and the three mentioned curves delineate three small 'triangular' regions, indicated by 1, 2 and 3 in Figure 11.*x*, where they have been drawn too large for clarity. For the observers situated inside of each of these small regions the daylight period is short and the Sun remains at low altitude; first contact takes place before sunrise, and last contact after sunset. So, for these observers the eclipse last 'the whole day'!

In the region 1 (*ECFW*) the maximum of the eclipse is visible. In region 2 (*AEC*) the Sun rises and then sets between maximum eclipse and last contact. In region 3, the Sun rises and sets between first contact and maximum.

On the arc *AEW* last contact occurs at sunset. On the arc *WFB* the eclipse begins at sunrise. On the arc *ECF* maximum eclipse takes place on the horizon.

On the arc *EW* the Sun rises between first contact and maximum eclipse, and it sets at last contact. On *EC* the Sun rises at maximum eclipse, and it sets before last contact, etc.

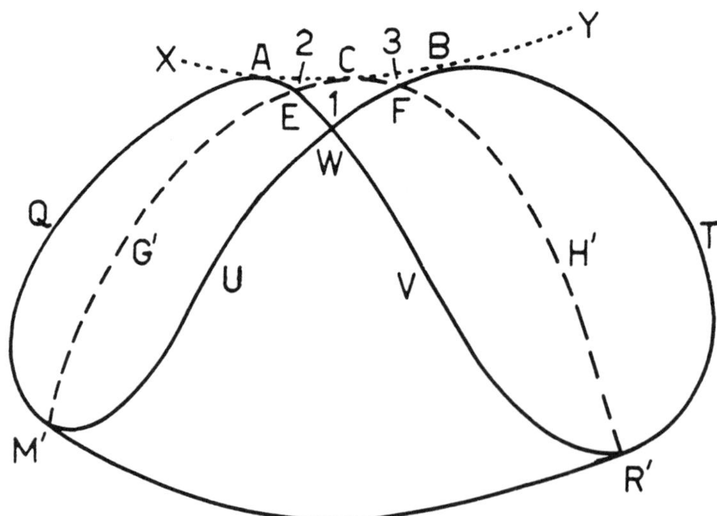

Fig. 11.x

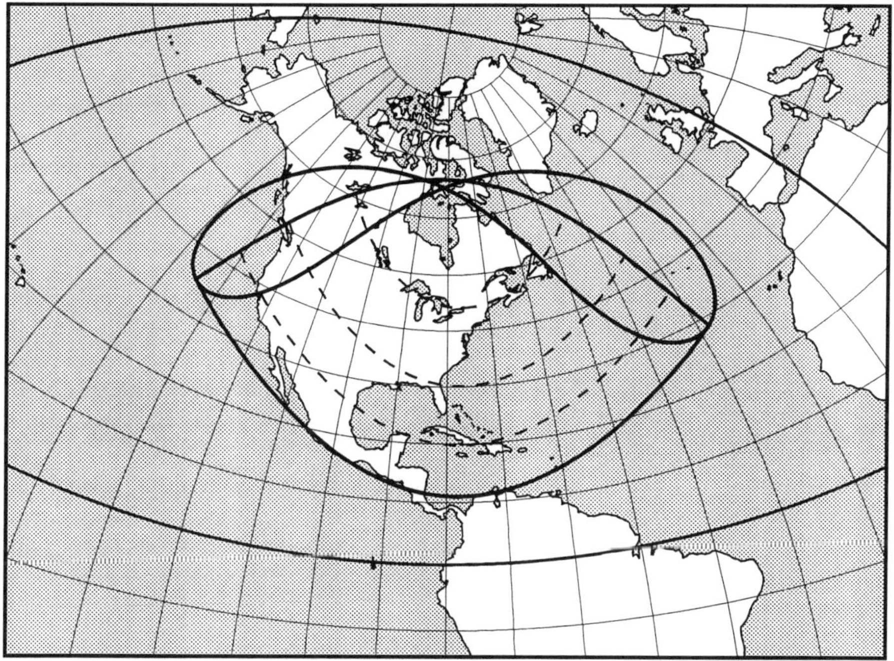

Fig. 11.y : *The region of visibility of the partial solar eclipse of 2000 December 25, which is visible in the United States. It is clearly seen that the curve 'maximum on the horizon' — the curve passing over Vancouver Island — doesn't pass over the node of the distorted curve '8', forming a small triangular region (called 1 in the text) north of the Hudson Bay.*

For all points between the pole P and the parallel circle XY the eclipse is not visible, as the Sun remains below the horizon there. Therefore, the arc ACB constitutes the northern (or southern) limit of the region of visibility. On the arc ACB itself the Sun is seen on the horizon at the instant of true noon, and this happens while the eclipse is in progress. In C, this occurs at the time of maximum eclipse; in B, it occurs at the instant of first contact; in A it takes place at last contact.

At the node W itself the eclipse begins at sunrise and ends at sunset, so the entire eclipse is just visible there. At point E, the Sun rises at maximum eclipse and sets at last contact.

2. *Summer*. — Here a similar description can be given, but the difference with the previous case is that the pole P now is illuminated by the Sun, and that it is situated inside of the region $M'UWVR'$. The curve 'maximum on the horizon' too passes a little through this area, between the node W and the pole P, while the parallel of latitude XY now is tangent *internally* to the three curves. So in this case this parallel of latitude is no longer the northern or southern limit of the region of visibility!

At *W* the eclipse begins at sunset and ends at sunrise, so nothing (just) of the event is seen there. In region 1 the Sun sets while it is partially eclipsed; when it rises 'the next day' (that is, perhaps one hour later) the eclipse is still in progress!

The readers are asked to draw themselves the regions around *W* for this case. What is seen by an observer at *P*, at *C*, in region 2, etc?

A final remark

Because the apparent diameter of the solar disk is not zero, still other complications do occur. If we call sunrise and sunset the times when the *center* of the solar disk is on the horizon, then we see some uncommon cases in Figure 11.*z*.

In case *a* the Sun is up because its center is up; nevertheless nothing is seen of the eclipse because the eclipsed part of the Sun is below the horizon! In case *b* just the opposite takes place.

Generally, however, sunrise and sunset are defined as the instants when the *upper limb* of the solar disk is on the horizon. Even then strange situations can occur, as in case *c* of the figure below: here the Sun is 'up' because its upper limb is above the horizon; but nothing is seen of the eclipse, although the magnitude is larger than 0.5.

Of course, all this is of little practical interest. But certainly it is interesting to look for such peculiarities. And often these exercises help us to better understand the events.

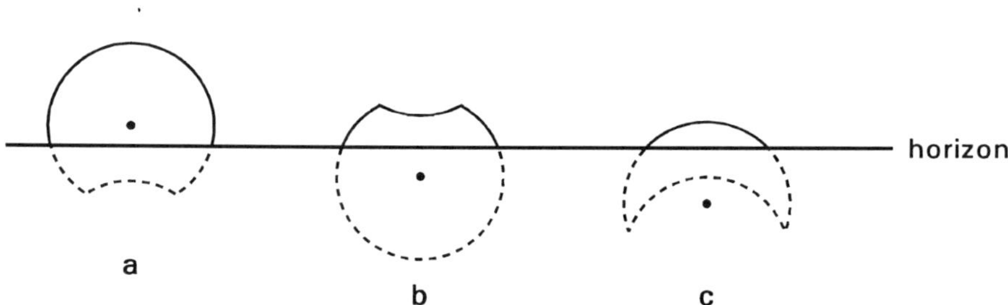

Fig. 11.z : *Some strange situations. In* a *and in* c *the eclipsed Sun is up though nothing is seen of the eclipse. In* b *the eclipse is seen, but the center of the solar disk is under the horizon.*

The original version of this chapter was published under the title 'Eclipsgebieden' in the April 1971 issue of the Dutch journal *Hemel en Dampkring*.

12. *When is the northern limit the southern one ?*

On 1985 November 12 a total solar eclipse took place in the southern hemisphere. The path of totality passed over the southern Pacific Ocean and a tiny part of Antarctica; it remained south of latitude 50° S. Along the central line the altitude of the Sun, at central eclipse, was less than 11°. An oddity of this eclipse was that, on the Earth's surface, the 'northern' limit of the path of totality was situated to the *south* of the 'southern' limit. Here is the explanation.

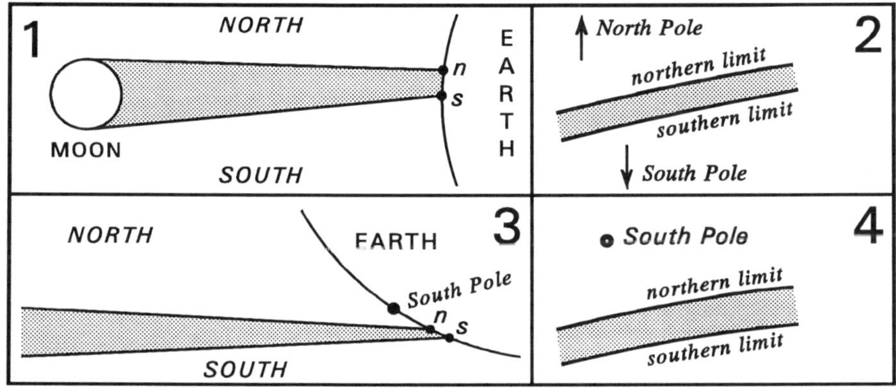

Fig. 12.a

At a total solar eclipse, the umbral cone of the Moon falls upon the surface of the Earth — see diagram 1 of Figure 12.*a*, where the circle at right is a north-south cross-section of the Earth. The lunar shadow runs over the Earth's surface, giving rise to a long path, the totality zone, within which a total solar eclipse is visible (diagram 2). This zone is limited by two curves: the northern limit and the southern limit of the path. The northern limit corresponds to the north direction in space (diagram 1); think, for instance, of the northern side of the plane of the ecliptic. Of course, the path of total eclipse lies in the hemisphere of the Earth that is illuminated by the Sun, and in most cases this path is situated somewhere between the North Pole and the South Pole, as illustrated in diagram 2. In diagram 1, *n* and *s* are points of the northern limit and of the southern limit, respectively.

However, in some cases the Moon's shadow happens to fall 'under' the South Pole; that is, south of it as seen from space, as in diagram 3. Here, *n* and *s* again are points of the northern and southern limits, respectively, but now *n* is nearer to the South Pole than *s*, and thus lies *geographically* south of it! Consequently, on the Earth's surface the northern limit of the path lies south of the southern limit, as illustrated in the fourth diagram of Figure 12.*a*. This was the case at the eclipse of 1985 November 12.

Of course, it also can happen that the path runs 'above' the North Pole. In such a case, on the Earth's surface the northern limit of the path will be south of the southern limit, too.

During the period A.D. 1950 to 2050, the following solar eclipses have that North ⟷ South oddity. The type of the eclipse is given (T = total, A = annular), and it is noted whether the event occurs in the northern (N) or in the southern (S) polar regions. The eclipses indicated by an asterisk belong to the same Saros series; the same holds for those marked † and ‡.

1950	Sep.	12		T	N	2021	Dec.	4 †	T	S
1954	Jan.	5 *	A	S	2026	Feb.	17 *	A	S
1972	Jan.	16 *	A	S	2026	Aug.	12 ‡	T	N
1985	Nov.	12 †	T	S	2033	Mar.	30	T	N
1990	Jan.	26 *	A	S	2039	Dec.	15 †	T	S
2003	Nov.	23 †	T	S	2044	Aug.	23 ‡	T	N
2008	Feb.	7 *	A	S				

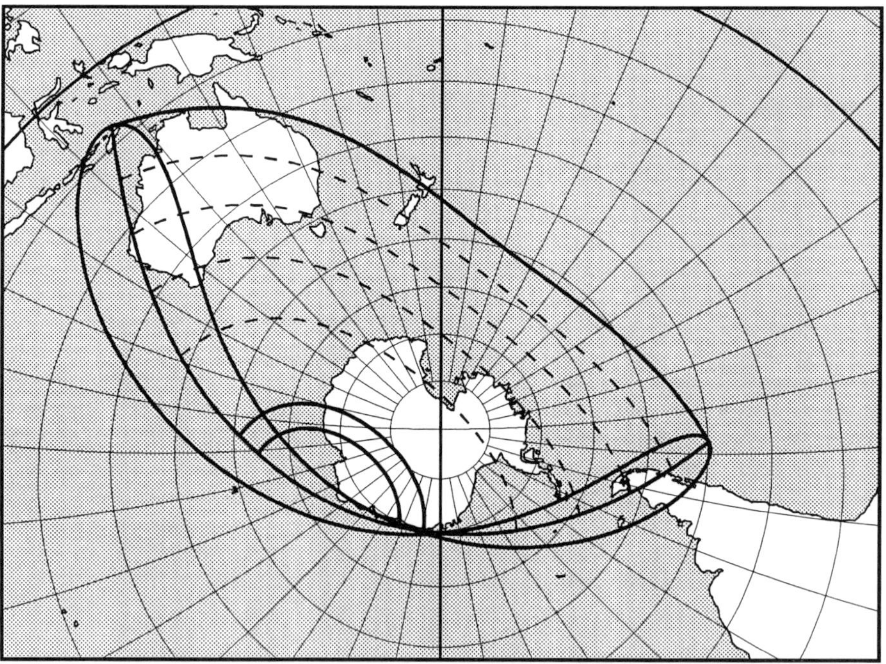

Fig. 12.b : *Region of visibility of the solar eclipse of 2003 November 23, which will take place one Saros after that of 1985 November 12. The lunar shadow moves from left to right in the drawing. Total eclipse is seen between the two parallel solid curves passing over Antarctica. The northern limit of this path is actually closer to the South Pole than the southern limit. The dashed lines are the curves of constant eclipse magnitude (at maximum eclipse) : from north to south 0.2, 0.4, 0.6 and 0.8, respectively.*

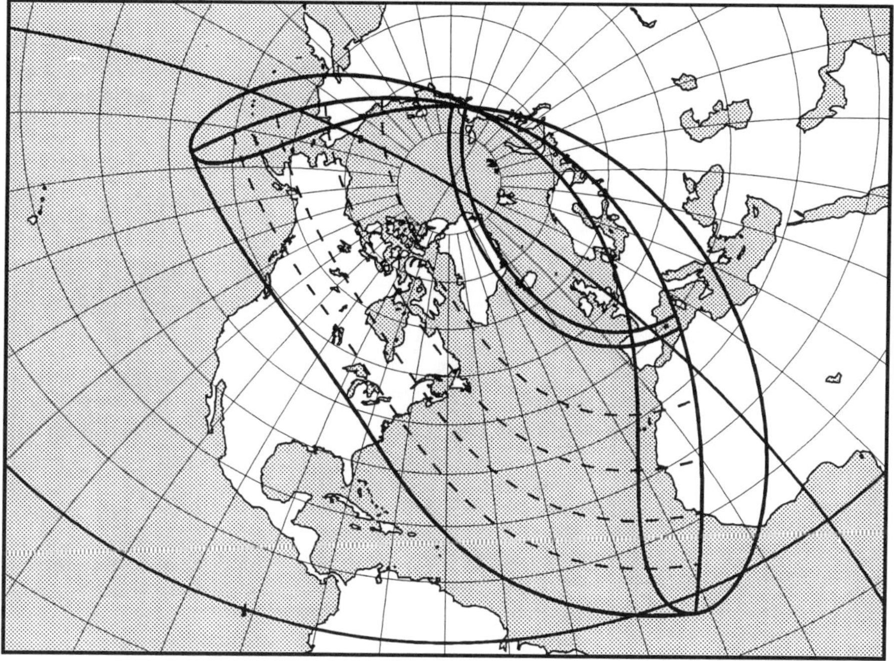

***Fig. 12.c** : Region of visibility of the solar eclipse of 2026 August 12.*

But this is not yet the end of the story. In the figure above the region of visibility of the solar eclipse of 2026 August 12 is shown. This event is seen as a total eclipse between the two parallel solid lines which cross Greenland and Spain. This path passes close to the North Pole, where the maximum magnitude is 0.986.

The central line of this eclipse begins at sunrise in northern Siberia, at latitude 75° north. After reaching the maximum northern latitude of 87°53′, the central line passes *southward* over Greenland. After reaching the westernmost longitude of 28° west, the central line runs southeastward, passes over Spain, and ends at sunset in the western part of the Mediterranean Sea, at latitude 39° north.

The first oddity is what we already described above : near the North Pole, what is *astronomically* the southern limit of the path of totality, is *geographically* the northern one (because there it lies closer to the North Pole than the northern limit). But now there is a second oddity : in Spain the (astronomically) southern limit has correctly become the southern limit geographically !

Still another oddness occurs in the polar regions when an eclipse is total or annular at the pole itself. Between A.D. 1900 and 2100, this happens on the following dates : 1917 December 14 (annular at the South Pole), 1939 April 19 (annular at the North Pole), 2021 June 10 (annular at the North Pole), 2094 January 16 (total at the South Pole), and 2097 November 4 (annular at the South Pole).

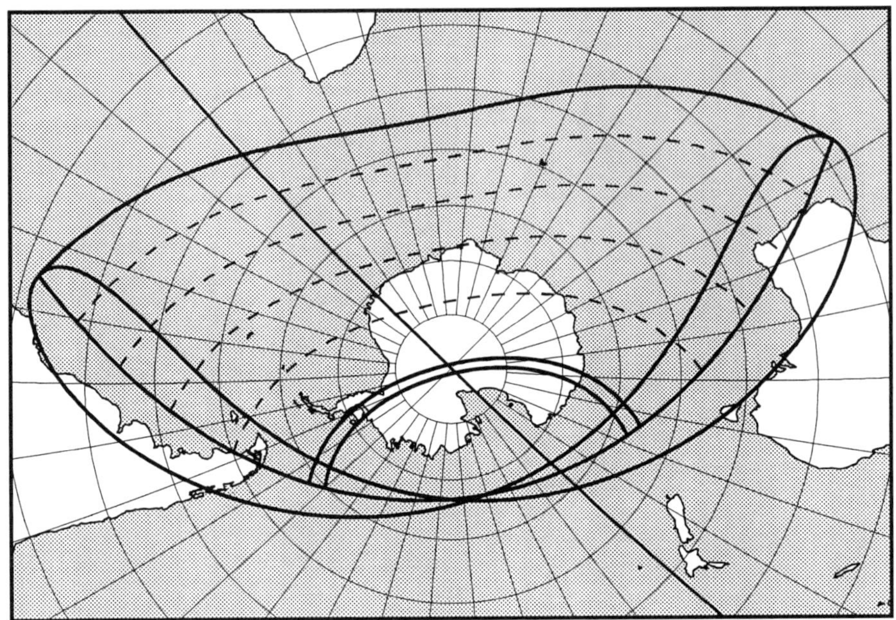

Fig. 12.d : *Area of visibility of the solar eclipse of 1917 December 14. The lunar umbra moves from the left to the right in the drawing. The path of annular eclipse crossed Antarctica and passed exactly over the South Pole.*

Now look at Figure 12.*e*. The shaded area represents the path of total (or annular) eclipse passing over the South Pole S. Start from point A, and go southward along the meridian. Then limit a is reached before arriving at S, so a is geographically the northern limit of the path. But now start at B, and go *southward* along the meridian. Then limit b is reached before you arrive at S, so b is geographically a *northern* limit of the path too. Consequently, in the vicinity of the pole S both a and b are northern limits of the path of totality! (But, of course, *astronomically* speaking, one of the limits is to be considered as being the northern one, and the other as the southern).

Concerning the annular eclipse of 1917 December 14, one reads the following on page 402 of the excellent *Text-Book on Spherical Astronomy* by W. M. Smart (Cambridge University Press, England; edition of 1956) :

"In 1917, within a few days of Dec. 21, an annular eclipse of the sun took place, visible near the South Pole. According to the *Nautical Almanac* the eclipse was exactly central at *midnight* in Latitude 89°57′ S, Longitude 142° W. According to the *Connaissance des Temps* the eclipse was central at *noon* also in Latitude 89°57′ S, but in Longitude 38 E. Show that the difference between the two statements would be accounted for if one almanac

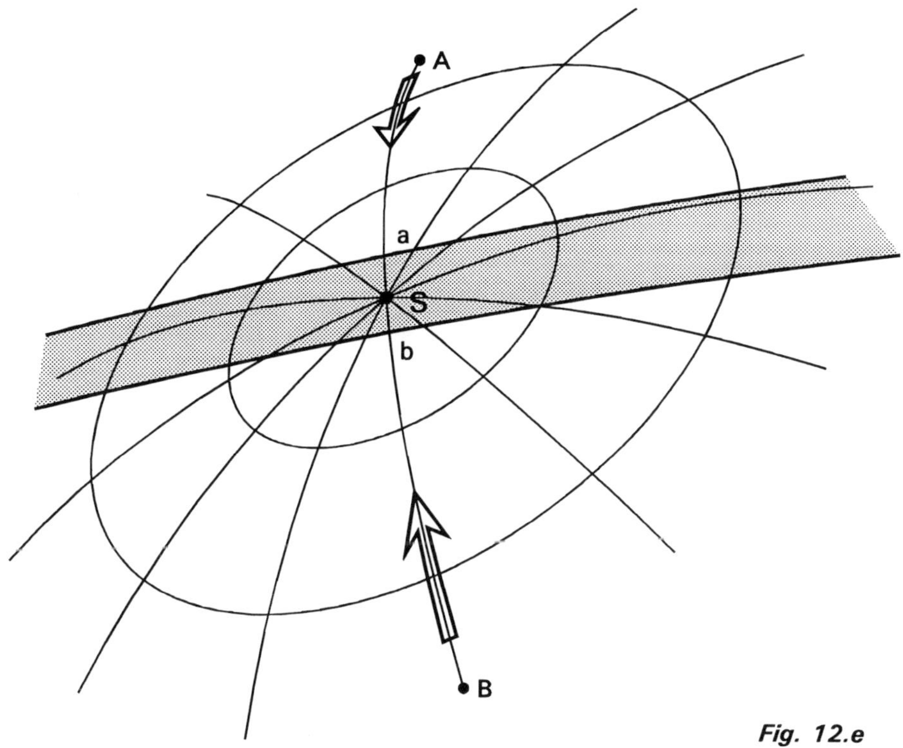

Fig. 12.e

had made its calculations for the sea-level, and the other for a plateau about 15,000 feet above sea-level. Show also that the difference would be accounted for if the positions of the moon adopted by the two almanacs differed by about 2½″ in declination, the moon's parallax being 57′.″

According to my calculation, using modern data, central eclipse took place at local apparent *midnight*, at 9^{h}23^{m}14^s UT, in longitude 142° West and latitude 89°57′ South. This is for sea-level, and in the calculation of the Besselian elements of the eclipse a correction of −0″.6 has been applied to the Moon's latitude to make allowance for the fact that the center of figure of the Moon does not exactly coincide with its center of mass.

13. The frequency of total and annular solar eclipses for a given place

The following text was first published in the Journal *of the British Astronomical Association, Vol. 92, No. 3, pages 124-126 (April 1982).*

A classical question concerning solar eclipses is the following one: How often can a total or an annular eclipse of the Sun be expected at a given point on the Earth's surface? In their classical textbook *Astronomy* (Boston, 1926), H. N. Russell, R. S. Dugan and J. Q. Stewart write (vol. I, page 227):

> "Solar eclipses that are *total* somewhere or other on the earth's surface are not very rare, averaging one for about every year and a half. But *at any given place* the case is very different. Since the track of a solar eclipse is a very narrow path over the earth's surface, averaging only 60 or 70 miles in width, we find that in the long run a total eclipse happens at any given station only once in about 360 years."

These authors, however, give no details about how this mean frequency of one total eclipse every 360 years has been found. The theoretical calculation is not easy, by reason of the large variations of the length and width of the path from one eclipse to another. Even for a given eclipse, the width can vary widely along the path. For the total eclipse of 1981 July 31, for instance, the width was 57 kilometers at the beginning of the path, reached a maximum value of 108 km near the middle, and then decreased to 51 km towards the end.

In order to recompute the answer given by Russell, Dugan and Stewart, and to find the mean frequency for annular eclipses too, we attacked the problem statistically. For every solar eclipse in the period A.D. 1700 to 2299, the local circumstances at 408 'standard points' on the Earth's surface were calculated. The following standard points have been chosen: the points at latitudes $+80°$, $+70°$, $+60°$, etc., to $-80°$ on the 24 meridians of longitudes $+180°$, $+165°$, $+150°$, etc., to $-165°$.

The calculation, made on an HP-85 microcomputer, proceeded automatically. After calculating the Besselian elements of an eclipse, the machine examined for each of the 408 standard points whether or not there was a total or annular eclipse there, after which the next eclipse was calculated, etc. A total or annular eclipse at a standard point was considered to be visible if, and only if, at the time of maximum eclipse, the geometric altitude of the center of the Sun's disk was positive.

A part of the computer outprint looks as shown in the box on the next page. The second column gives the longitude (measured positively westward from the meridian of Greenwich), and the third column the latitude.

1973	JUN	30	+45	+10	TOTAL
1973	JUN	30	+15	+20	TOTAL
1973	JUN	30	+0	+20	TOTAL
1973	JUN	30	−60	−10	TOTAL
1973	DEC	24	+90	+10	ANNULAR
1973	DEC	24	+60	+0	ANNULAR
1973	DEC	24	+30	+10	ANNULAR
1973	DEC	24	+15	+20	ANNULAR
1976	APR	29	+30	+10	ANNULAR
1976	APR	29	−45	+40	ANNULAR
1976	OCT	23	−75	−20	TOTAL
1977	APR	18	−15	−20	ANNULAR
1979	FEB	26	+60	+70	TOTAL
1979	AUG	22	+135	−60	ANNULAR
1979	AUG	22	+120	−60	ANNULAR
1979	AUG	22	+105	−60	ANNULAR
1979	AUG	22	+105	−70	ANNULAR
1979	AUG	22	+90	−70	ANNULAR
1980	AUG	10	+60	−20	ANNULAR
1981	FEB	4	−135	−40	ANNULAR
1983	JUN	11	−90	−20	TOTAL
1983	JUN	11	−105	−10	TOTAL
1983	JUN	11	−150	−10	TOTAL

As mentioned above, the investigation was made over a time-period of six centuries, this value having been chosen in order to avoid any possible effect with a period of six centuries. It is well known, for instance, that the mean frequency of total *lunar* eclipses varies with time, the period being 586 years. [See Chapter 16 and, for solar eclipses, see Chapter 10].

For the period 1700-2299, we found 665 total and 1208 annular eclipses at standard points, distributed over the various latitudes as indicated in Table 13.A, columns (2) and (3). These values are plotted in Figure 13.*a* — see the two solid lines.

There is an evident *latitude effect* in the distribution of these total and annular eclipses. For instance, we see at once that total eclipses are less frequent in the zone of southern latitudes 30° to 80° than in the northern hemisphere. Annular eclipses are more frequent at latitudes 50°S to 80°S than near the equator.

These distributions are explained by the combination of the following effects:

(i) in the equatorial regions, total eclipses are more frequent, and annular eclipses are less frequent, than at higher latitudes because the equatorial regions are generally closer to the Moon, whose disk thus appears larger;

(ii) on the other hand, the lunar shadow moves in approximately the same direction as the rotating surface of the Earth. Therefore, the mean duration of solar eclipses at a given place will be longer, and their mean frequency less, than it would be for a non-rotating Earth. This is connected with the fact that the probability for the Sun to be eclipsed (expressed, for instance, in minutes per century) remains independent of the speed of rotation of the Earth. As a consequence, near the equator a solar eclipse will be somewhat rarer;

(iii) in the summer months, the Sun is for a longer time above the horizon, increasing the frequency of visible eclipses there. In the northern hemisphere, this occurs around the time when the Earth is near the aphelion of its orbit, resulting in a smaller-than-average solar disk, thus favouring the occurrence of a total eclipse, and disfavouring that of an annular one; the opposite holds for the southern hemisphere.

TABLE 13.A

Mean frequencies of eclipses

Latitude	Number of eclipses at standard points		Mean time interval (years)	
	Total	Annular	Total	Annular
(1)	(2)	(3)	(4)	(5)
+80°	57	92	254	166
+70	62	76	275	185
+60	41	67	295	205
+50	34	55	315	226
+40	48	58	333	246
+30	42	64	350	262
+20	47	52	364	274
+10	32	63	377	279
0	32	42	388	275
−10	44	56	398	264
−20	44	67	407	247
−30	33	54	417	226
−40	34	59	427	203
−50	28	94	441	180
−60	26	86	458	159
−70	25	107	481	140
−80	36	116	513	122

As a consequence, the mean frequency of a total eclipse at a given point is higher in the northern hemisphere than in the southern one. That of an annular eclipse is higher in the northern hemisphere than in the equatorial regions, and still higher in the southern hemisphere.

Using the method of least squares, we fitted a polynomial of the third degree to our data — see the dashed curves in the figure. This led us to the mean frequencies given in Table 13.A, columns (4) and (5). For instance, at latitude 50° north, one can expect one total solar eclipse every 315 years, and one annular eclipse every 226 years.

In order to obtain the mean frequencies at a given point on the Earth's surface *generally*, we must take into account the variation of the circumference of a parallel of latitude. Having chosen the same number of standard points, namely 24, on each parallel of latitude, these points are closer to each other at high latitudes than near the equator. This fact cannot be neglected due to the latitude effect discussed above. As a consequence, the number of points and the number of events at each latitude φ should be weighted by $\cos \varphi$ before the general mean can be deduced — see the Appendix.

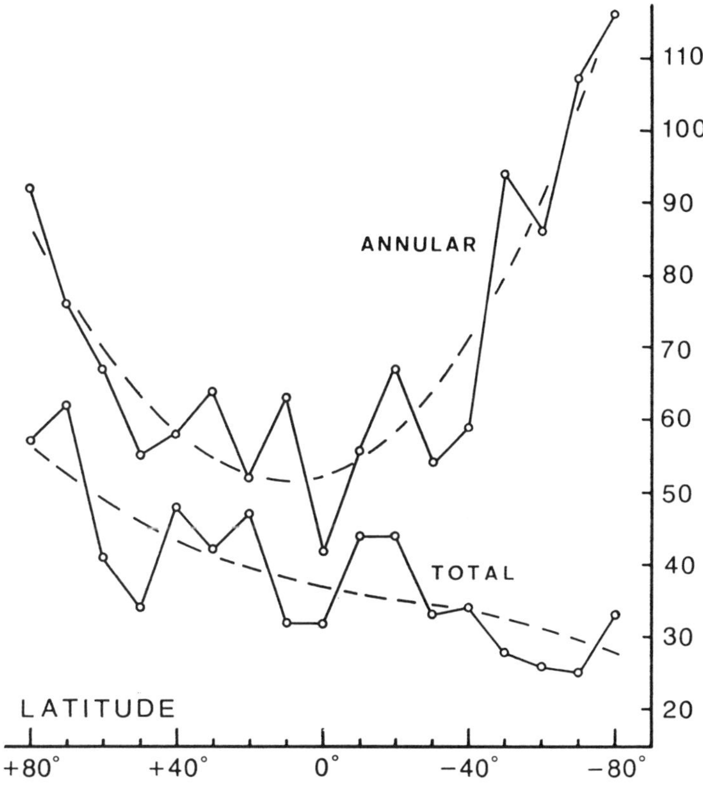

Figure 13.a : *Number of total and annular eclipses at the standard points for the period A.D. 1700 to 2299.*

We then, finally, obtain the following mean frequencies for any given point at the Earth's surface :

a total eclipse once in 375 years,
an annular eclipse once in 224 years,

which, by combination, gives an annular *or* a total eclipse every 140 years for a place chosen at random. Because our results are based on a sample of observable eclipses, they are subject to uncertainty of about 16 years' standard error for total eclipses, 7 years for annular, and 4 years for both. Further error may arise because of the limited number of points on the Earth's surface used in the analysis. However, if the 600-year interval examined and the points considered are representative, total eclipses of the Sun visible at an average point on the Earth's surface may be a little less frequent than stated by Russell, Dugan and Stewart.

It should be noted that the obtained values are the *mean* frequencies of total and annular eclipses for a given place. Actually, these events take place at very irregular intervals for a given place. For instance, the eclipse of 2142 May 25 will be the first total solar eclipse at Antwerp, Belgium, for at least seven centuries. But only nine years later, on 2151 June 14, there will be another total one for the same city. — See also the next Chapter.

Appendix

The calculation of the general mean frequency of eclipses

Let $F(\Omega)$ be the mean frequency of an eclipse of a certain kind for the point $\Omega(\lambda, \varphi)$ on the Earth. Then, the mean frequency for any given station on the Earth — thus averaged over all possible positions of such stations — is given by

$$\overline{F} = \frac{\int F(\Omega)\, d\Omega}{\int d\Omega}$$

where $d\Omega = \cos \varphi \, d\varphi \, d\lambda$. Since the frequency is independent of the longitude λ, we have $F(\lambda, \varphi) = F(\varphi)$, and the formula above reduces to

$$\overline{F} = \frac{\int_{-90}^{+90} F(\varphi)\, \cos \varphi \, d\varphi}{\int_{-90}^{+90} \cos \varphi \, d\varphi}$$

In our case, $F(\varphi)$ is given for a limited number of latitudes $\varphi_1 \ldots \varphi_n$. Thus, the last formula can be approximated by

$$\overline{F} \approx \frac{\sum F(\varphi_i)\, \cos \varphi_i}{\sum \cos \varphi_i}$$

since the chosen latitudes are uniformly distributed over the Earth's surface ($+80°$, $+70°$, etc. to $-80°$). The last formula leads to the procedure of averaging as indicated in the text.

14. Total and annular solar eclipses in close succession at a given place

For a given place on the surface of the Earth, a total or an annular solar eclipse is a rare event. In the preceding Chapter their *mean* frequencies were discussed.

However, it is not rare at all that the path of one event crosses that of another one after a short time interval. When this happens, people living in the common region of the two paths can experience two total or annular eclipses in an unusually short interval.

The table which follows lists all such cases, between the years 1900 and 2100, when for a given area two total or annular eclipses are visible *within a period less than 18 months*. The first column gives the type and the date of the first eclipse (T = total, A = annular). Similar data are given in the second column for the second eclipse. The next column is the number N of lunations between the two eclipses; a lunation is the synodic revolution period of the Moon, or the time interval between two successive New Moons. Finally, the last column states in which region the paths of the two eclipses cross each other.

The list totals 56 events, twelve of which involve two total eclipses. It appears that some events have *two* crossings! We also note that many cases occur in series: the well-known Saros. See, for instance, the events of $1937 - 1955 - 1973 - 1991/2 - 2009/10 - 2027/8 - 2045/6 - 2063/4 - 2081/2 - 2099/2100$. However, other series are much shorter, for instance $2008/10 - 2026/28$, and we even see an 'isolated' case, that of $2052/2053$.

First eclipse	Second eclipse	N	Common area
A 1900 Nov. 22	T 1901 May 18	6	South of Madagascar
T 1903 Sep. 21	A 1905 Mar. 6	18	Southern Indian Ocean
A 1907 July 10	A 1908 Dec. 23	18	Pacific Ocean off coast of Chile; the first path passes over the beginning of the second path.
A 1915 Feb. 14	A 1916 July 30	18	Australia
A 1925 July 20	A 1927 Jan. 3	18	Near New Zealand
A 1933 Feb. 24	T 1934 Aug. 10	18	Southern Atlantic Ocean
A 1933 Aug. 21	T 1934 Feb. 14	6	In the sea between the Malaya peninsula and Borneo. The first path passes over a tiny part at the beginning of the second.
T 1937 June 8	A 1937 Dec. 2	6	Pacific Ocean

First eclipse	Second eclipse	N	Common area
A 1943 Aug. 1	A 1945 Jan. 14	18	South of Australia
T 1947 May 20	T 1948 Nov. 1	18	A narrow strip in Kenya, from Lake Victoria to just south of Nairobi. The central lines of the two eclipses do *not* cross each other.
A 1951 Sep. 1	T 1952 Feb. 25	6	Gulf of Guinea (Africa)
T 1955 June 20	A 1955 Dec. 14	6	*Two* crossings : Indian Ocean, and Cambodia / Thailand / Laos / Vietnam / Burma
T 1958 Oct. 12	A 1959 Apr. 8	6	Pacific Ocean. The first path passes over the end of the second path.
A 1961 Aug. 11	A 1963 Jan. 25	18	South Atlantic Ocean
T 1965 May 30	T 1966 Nov. 12	18	Pacific Ocean off coast of Peru
A 1969 Sep. 11	T 1970 Mar. 7	6	Pacific Ocean
T 1973 June 30	A 1973 Dec. 24	6	*Two* crossings : Venezuela and W. Africa
T 1976 Oct. 23	A 1977 Apr. 18	6	Indian Ocean off coast of Tanganyika
T 1983 June 11	T 1984 Nov. 22	18	New Guinea
A 1987 Sep. 23	T 1988 Mar. 18	6	Pacific Ocean
T 1991 July 11	A 1992 Jan. 4	6	Pacific Ocean
T 1994 Nov. 3	A 1995 Apr. 29	6	Pacific Ocean
T 2001 June 21	T 2002 Dec. 4	18	In a small part of Angola (Africa), just north and northeast of Lobito, and in the near Atlantic Ocean
A 2005 Oct. 3	T 2006 Mar. 29	6	Libya
T 2006 Mar. 29	A 2006 Sep. 22	6	Atlantic Ocean
T 2008 Aug. 1	A 2010 Jan. 15	18	China. The second path passes over the end of the first.
T 2009 July 22	A 2010 Jan. 15	6	China
T 2012 Nov. 13	A 2013 May 10	6	Queensland (Australia)
T 2019 July 2	T 2020 Dec. 14	18	Pacific (113° W, 18° S)

First eclipse	Second eclipse	N	Common area
A 2019 Dec. 26	A 2020 June 21	6	*Two* crossings : Arabia and Pacific Ocean
A 2023 Oct. 14	T 2024 Apr. 8	6	Texas
T 2024 Apr. 8	A 2024 Oct. 2	6	Pacific Ocean
T 2026 Aug. 12	A 2028 Jan. 26	18	Spain (near sunset at both eclipses)
T 2027 Aug. 2	A 2028 Jan. 26	6	Atlantic Ocean, Spain
T 2027 Aug. 2	T 2028 July 22	12	Indian Ocean (90°E, 13°S). Only a very small part at the end of the path of the first eclipse lies inside the path of the second.
T 2037 July 13	T 2038 Dec. 26	18	*Two* crossings: western Australia (near 124° E, 26° S), and in Pacific Ocean at 40° S just east of New Zealand.
A 2038 Jan. 5	A 2038 July 2	6	*Two* crossings: Atlantic Ocean near Barbados, and Niger (Africa)
A 2041 Oct. 25	T 2042 Apr. 20	6	Pacific Ocean southeast of Japan
T 2042 Apr. 20	A 2042 Oct. 14	6	Borneo
T 2045 Aug. 12	A 2046 Feb. 5	6	California
T 2045 Aug. 12	T 2046 Aug. 2	12	Atlantic Ocean off coast of Brazil
T 2052 Mar. 30	T 2053 Sep. 12	18	Atlantic Ocean south of Azores (31° to 28° W, 36° N)
T 2055 July 24	T 2057 Jan. 5	18	*Two* crossings: southeastern Atlantic Ocean, and south-western Indian Ocean
A 2056 Jan. 16	A 2056 July 12	6	*Two* crossings in Pacific Ocean
A 2059 Nov. 5	T 2060 Apr. 30	6	Libya, Egypt
T 2060 Apr. 30	A 2060 Oct. 24	6	Ivory Coast
T 2063 Aug. 24	A 2064 Feb. 17	6	China

First eclipse	Second eclipse	N	Common area
T 2066 Dec. 17	T 2068 May 31	18	Indian Ocean near coast of Australia (33°S, 112° to 114°E)
A 2077 Nov. 15	T 2078 May 11	6	Gulf of Mexico near Texas
T 2078 May 11	A 2078 Nov. 4	6	Pacific Ocean
T 2081 Sep. 3	A 2082 Feb. 27	6	Switzerland, Italy, Austria
T 2084 Dec. 27	T 2086 June 11	18	*Two* crossings : southern Atlantic Ocean and southern Indian Ocean
A 2089 Apr. 10	T 2089 Oct. 4	6	Pacific Ocean
A 2095 Nov. 27	T 2096 May 22	6	Pacific Ocean
T 2096 May 22	A 2096 Nov. 15	6	Celebes Sea
T 2099 Sep. 14	A 2100 Mar. 10	6	USA : North Dakota, Minnesota

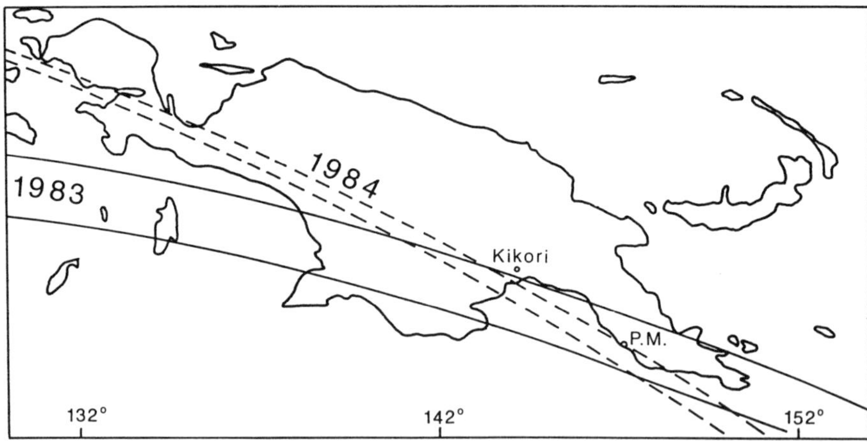

Fig. 14.a : *In New Guinea two total solar eclipses were visible within a period of 18 months: that of 1983 June 11 (solid lines) and that of 1984 November 22 (dashed). For both eclipses, the northern and southern limits of the path of totality are drawn. The two paths crossed each other on and near New Guinea. In both cases, the lunar shadow moves from West (left) to East (right). Port Moresby ('P.M.') was inside of the totality path of the 1983 eclipse, and just on the northern limit of the path of the 1984 eclipse.*

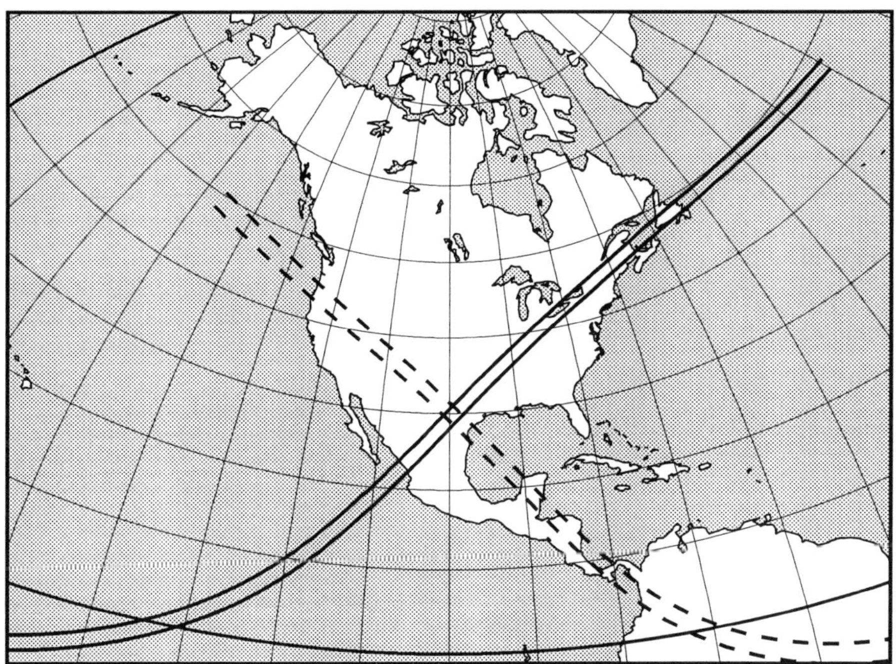

Fig. 14.b : *The path of the annular solar eclipse of 2023 October 14 (dashed) and that of the total eclipse of 2024 April 8 (solid lines) will cross in Texas. In the common area of these paths, an annular and a total eclipse will be experienced at an interval of just under six months. The central lines of the eclipses will cross each other at longitude 99°30' west, latitude 29°46' north. This point lies about 10 miles north of Utopia, which is west of San Antonio. At that point, the duration of the annular phase of the eclipse of 2023 will be 300 seconds, that of the totality of the eclipse of 2024 will be 266 seconds; the widths of the two paths will be almost equal there, 192 and 194 kilometers, respectively.*

It is even possible that *three* total or annular solar eclipses are visible from the same place in a rather short time interval. We found the following cases accidentally, and a systematic search certainly would reveal many more :

— in China, at longitude 104° E, latitude 29° N, a total solar eclipse will be seen on 2009 July 22, an annular eclipse six lunations later on 2010 January 15, and another annular on 2020 June 21 ;

— at longitude 25° E, latitude 30° N, near the Libya – Egypt border, the following three eclipses will be visible within a time span shorter than seven years : 2053 September 12 (total), 2059 November 5 (annular), and 2060 April 30 (total);

— at St. Paul, Minnesota, the following three eclipses will be visible within a time span of twelve years : 2099 September 14 (total), 2106 May 3 (total), and 2111 August 4 (annular).

Even more unexpected are the following cases involving *four* total or annular eclipses in a rather short time interval :

— in Laos, at latitude 16° N, longitude 106° to 107° E, an annular solar eclipse was seen on 1944 July 20, a total eclipse on 1955 June 20, an annular on 1955 December 14, and again an annular eclipse on 1958 April 19 ;

— the following four eclipses will be visible at 178°00' east longitude, 40°20' south latitude, which is a point in the Pacific Ocean just off the coast of New Zealand : 2035 March 9 [March 10, local time] (annular), 2037 July 13 (total), 2038 December 26 (total), and 2045 February 16 [February 17, local time] (annular).

15. *Nearly-zenithal central solar eclipses*

The following text was first published in the Journal *of the British Astronomical Association, Vol. 100, No. 5, pages 227-228 (October 1990).*

At the famous total solar eclipse of 1991 July 11, the axis of the Moon's shadow passed so nearly centrally over the Earth, that for some places central eclipse occurred with the Sun almost exactly in the zenith. The maximum altitude of the center of the solar disk, at central eclipse, was 89°53'. (The zone of totality was so large, however, that at some places total eclipse took place with the Sun *exactly* in the zenith.)

Because it is interesting to know the frequency of such nearly-zenithal central eclipses, I searched all solar eclipses in the period A.D. 1700 to 2399 where the maximum altitude of the Sun at central eclipse is larger than 89°. The calculations were based on the modern theories of the Sun and the Moon constructed by Bretagnon and Chapront, respectively, at the Bureau des Longitudes of Paris. [The VSOP87 and the ELP-2000 theories]. My results are given in Table 15.A, where the third column mentions the type of the eclipse (T = total, A = annular).

It is evident from the table that the frequency of these events varies widely in the course of the centuries. There are only two cases during the 20th century, and none between A.D. 2000 and 2099. On the other hand, there were ten cases during the eighteenth century, and there will be six ones between the years 2300 – 2399. (Sometimes there are two cases during the same calendar year, such as in 1803, 2255 and 2389.)

To investigate the subject further, I counted the number of the nearly-zenithal central solar eclipses in the longer period A.D. 1000 to 2999. The results are shown in Table 15.B.

TABLE 15.A

Nearly-zenithal central solar eclipses,
1700 — 2399

Date	Max. altitude of Sun at central eclipse	Type of eclipse	Place
1716 Apr 22	89° 32′	T	Pacific Ocean
1723 Nov 27	89 09	A	Pacific Ocean
1741 Dec 8	89 43	A	Indian Ocean
1756 Mar 1	89 59	A	Near New Guinea
1759 Dec 19	89 52	A	Atlantic Ocean
1763 Apr 13	89 59	A	Sudan
1767 Jan 30	89 01	T	NW Australia
1777 Dec 29	89 31	A	Pacific Ocean
1785 Feb 9	89 38	T	Atlantic Ocean
1796 Jan 10	89 07	A	Pacific Ocean
1803 Feb 21	89 30	T	Pacific Ocean
1803 Aug 17	89 48	A	Near Socotra
1915 Aug 10	89 12	A	Pacific Ocean
1991 Jul 11	89 53	T	Mexico
2132 Jun 13	89 03	T	N. of Haiti
2208 May 15	89 38	A	Yucatan
2255 May 7	89 40	T	Bay of Bengal
2255 Oct 31	89 24	A	Gulf of Carpentaria
2313 Mar 27	89 57	A	Pacific Ocean
2331 Oct 2	89 24	T	Pacific Ocean
2342 Mar 8	89 38	T	Pacific Ocean
2371 Feb 16	89 29	A	Atlantic Ocean
2389 Feb 26	89 37	A	Pacific Ocean
2389 Aug 22	89 10	T	Pacific Ocean

TABLE 15.B

The frequency of nearly-zenithal central solar eclipses (max. alt. > 89°), A.D. 1000 to 2999

Years	Number
1000 — 1099	2
1100 — 1199	12
1200 — 1299	6
1300 — 1399	0
1400 — 1499	1
1500 — 1599	1
1600 — 1699	3
1700 — 1799	10
1800 — 1899	2
1900 — 1999	2
2000 — 2099	0
2100 — 2199	1
2200 — 2299	3
2300 — 2399	6
2400 — 2499	1
2500 — 2599	1
2600 — 2699	1
2700 — 2799	2
2800 — 2899	6
2900 — 2999	3

And indeed, we find other centuries with a large number of cases, namely A.D. 1100–1299 and 2800–2899. Evidently, these 'rich' periods repeat at intervals of 600 years, or a little less. They surely are related to the period of 586 years which governs the frequency of the tetrads, of the total lunar eclipses, and of the total penumbral eclipses — see Chapters 16 and 17.

The correspondence between the frequencies of the four events is remarkable. The number of nearly-zenithal central solar eclipses, that of the tetrads, that of total umbral eclipses of the Moon, and that of total penumbral lunar eclipses, vary with a period of a little less than six centuries. In centuries when there are *more* tetrads, there are more total umbral eclipses of the Moon, and more total penumbral lunar eclipses, but *less* nearly-zenithal central solar eclipses.

16. Curious and interesting facts about lunar eclipses

Tetrads

Four successive lunar eclipses can all be total ones. In such a case, they occur at intervals of six lunations (177 days). Examples are the group of the four total eclipses of 1985 and 1986 and, one Saros later, those of 2003 and 2004.

> *See, page 44, the Note about lunar eclipses*

However, the frequency of these *tetrads* (as these groups are called) is very variable with time. G. Schiaparelli (1835 – 1910) discovered that during a period of about three centuries tetrads occur rather frequently, about 17 times; then in the next three hundred years tetrads never occur at all. Then again they reappear during another period of nearly 300 years, and so on.

For example, no tetrad at all took place from A.D. 1582 to 1908, while 16 ones occur from 1909 to 2156. Then there will be again an 'empty' period until the year 2448. The cycle of this variation has a mean period of 586 years

How can this remarkable phenomenon be explained? Schiaparelli did not give an answer to this question. The solution of the problem, which is by no means simple, was found by Pannekoek [1], while G. van den Bergh [2] has given a half-theoretical, half-empirical solution.

The distribution of the tetrads during twenty centuries is given in the sixth column of Table 16.A. Another distribution of these 52 tetrads is as follows:

Date of the first eclipse of the tetrad	Number of tetrads
January 16 to February 15	5
February 16 to March 15	9
March 16 to April 15	13
April 16 to May 15	12
May 16 to June 15	10
June 16 to July 15	3
July 16 to August 15	0
August 16 to September 15	0
September 16 to October 15	0
October 16 to November 15	0
November 16 to December 15	0
December 16 to January 15	0

It thus appears that, in order that a tetrad be possible, the first of the four total eclipses should take place between mid-January and mid-July, and preferably in March, in April, or in May.

So, presently we are living in a period where tetrads take place, while no tetrads at all occurred at the time when Louis XIV was king of France. One might presume that the number of total lunar eclipses does not vary much from one century to another, and that only their grouping in tetrads is variable. This is not the case, however : during a period of approximately three centuries there really are *more* total lunar eclipses than during the preceding or the next three centuries.

This brings us to the subject of the distribution of the lunar eclipses over the centuries.

Distribution of lunar eclipses

From the year 0 to A.D. 2999 inclusively, there are 7245 lunar eclipses, of which 2089 are total in the umbra, 2536 are partial in the umbra, and 2620 are penumbral eclipses. From this we derive that in one century there are 241 or 242 lunar eclipses, of which 70 are total, 84 or 85 are partial, and 87 are penumbral eclipses.

These are *mean* values, however. The actual numbers of lunar eclipses per century are given in Table 16.A for the period A.D. 1000 to 2999. The 'deep' total eclipses are the umbral eclipses with a magnitude larger than 1.50. The variations of these quantities, and that of the tetrads, are illustrated in Figure 16.*a*.

All the curves in the figure exhibit the famous period of 586 years! It is true that the number of *all* umbral eclipses does not vary much from one century to another, but that of total eclipses is subject to an important cyclic variation : it varies from about 60 during the 17th, the 18th and the 19th centuries to 86 for the period 2000 – 2099.

The number of deep total eclipses is even more variable, but the surprising fact is that these eclipses exhibit the opposite trend : their number per century is *least* at the times when the number of total eclipses is greatest! During the period 1700 – 1799, when there were 62 total eclipses, 45 of them had a magnitude larger than 1.50. On the contrary, during the period 2000 – 2099, when there will be 86 total eclipses, only 11 of them will be deep eclipses! This remarkable fact was mentioned in 1982 by H. Feijth [3].

This can be illustrated in another way. Figure 16.*b* shows a 'panorama' of all umbral eclipses from A.D. 1500 to 2500. Every dot represents an eclipse. The years are read vertically ; the horizontal scale indicates the magnitude of the eclipses, from zero at the left border to 1.88 (the greatest possible magnitude of a lunar eclipse) at right. All eclipses with a magnitude larger than 1 are total. So, along the left border are plotted the very small partial eclipses, while the deep totals are situated at extreme right. The dot *A* represents the lunar eclipse of 1953 July 26, which had a magnitude of 1.864 and was the deepest total lunar eclipse of the 20th century : the center of the Moon's disk passed almost exactly over the center of the Earth's shadow.

This panorama was first published by the author of this book in 1982 [4].

TABLE 16.A

Number of lunar eclipses per century, from A.D. 1000 to 2999

Years	Pen-umbral eclipses	umbral eclipses	total umbral eclipses	deep total umbral eclipses	tetrads	(a)	(b)
1000 – 1099	91	157	62	35	0	4	10
1100 – 1199	98	160	58	45	0	0	13
1200 – 1299	91	159	62	41	0	1	9
1300 – 1399	85	148	75	22	5	14	1
1400 – 1499	80	148	85	11	5	20	0
1500 – 1599	80	153	76	23	6	15	6
1600 – 1699	93	157	59	39	0	1	11
1700 – 1799	98	158	62	45	0	0	11
1800 – 1899	88	161	62	37	0	2	9
1900 – 1999	85	144	79	17	5	16	3
2000 – 2099	84	144	86	11	7	20	0
2100 – 2199	84	154	69	30	4	13	3
2200 – 2299	94	158	59	42	0	0	12
2300 – 2399	91	159	62	44	0	0	11
2400 – 2499	87	153	70	28	4	11	8
2500 – 2599	82	145	85	13	7	19	2
2600 – 2699	78	151	81	15	8	17	1
2700 – 2799	91	153	62	36	0	3	6
2800 – 2899	100	159	59	45	0	0	12
2900 – 2999	86	162	63	39	1	2	11

(a) Number of cases a penumbral eclipse is followed, after six lunations, by a total eclipse in the umbra, or vice versa

(b) Number of cases there is no total umbral eclipse in three successive years

It is at once evident that the dots are not evenly distributed in the panorama. About the year 1700, for instance, there were many partial eclipses with a magnitude near 0.6 *and* many deep total eclipses with a magnitude of 1.6 or larger; but there were relatively few total eclipses with a magnitude between 1.0 and 1.5, and also few very small partial eclipses. The same holds around A.D. 2300.

On the contrary, around the year 2000 there are many total eclipses with a magnitude between 1.0 and 1.4, and many small partial eclipses, but few eclipses having a magnitude near 0.5 or larger than 1.5.

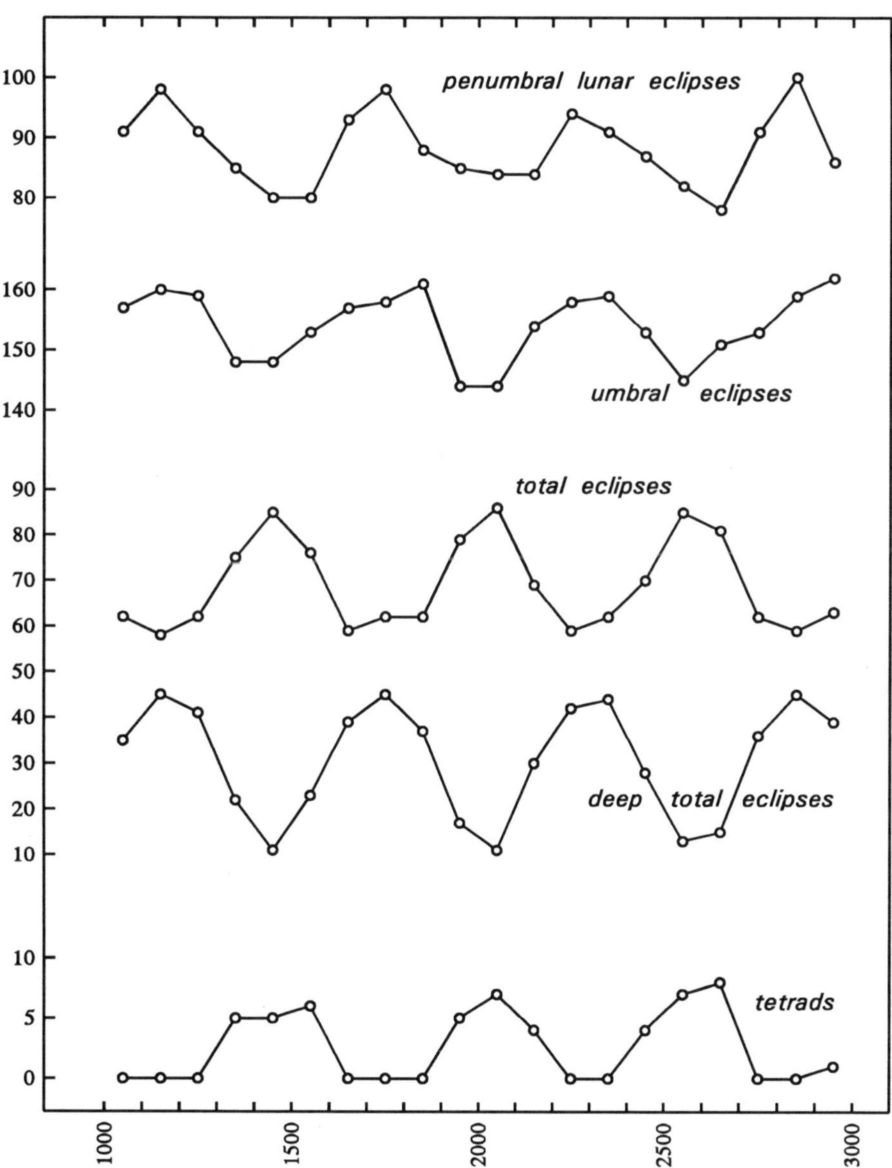

Fig. 16.a : *The number of lunar eclipses per periods of 100 years*

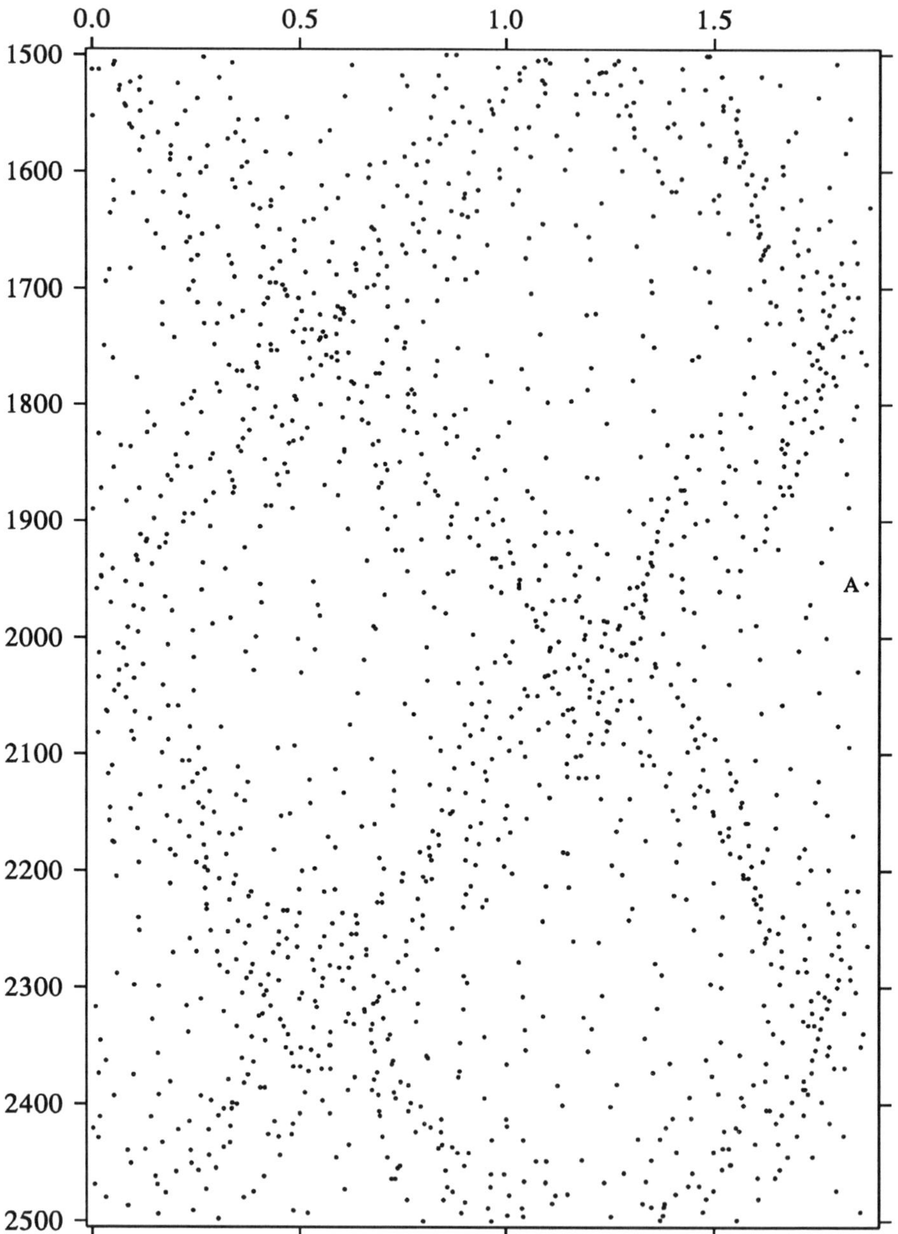

Fig. 16.b : *Panorama of the umbral lunar eclipses, from A.D. 1500 to 2500*

From what has been said, it appears that it is dangerous to perform statistics on a too short base. From the fact that during the period 1900–2099 there are 165 total lunar eclipses and only 123 partial ones, one might generalize that when the Moon passes through the Earth's umbra it will, on average, be fully eclipsed more often than partially. Actually, as we have seen, in the long run there are more partial eclipses than totals. This is also evident from Figure 16.c.

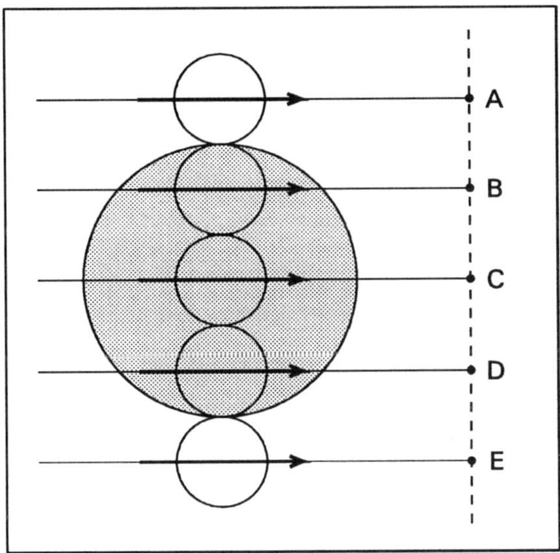

Fig. 16.c : The grey circle represents a section through the Earth's umbral cone by a plane perpendicular to its axis at the distance of the Moon. The small circles represent five positions of the Moon, and the arrows indicate its direction of motion. If the diameter of the umbra was 3 times that of the Moon, as shown, an eclipse would be total if the center of the Moon passes between B and D ; the eclipse would be partial if the Moon's center passes between A and B, or between D and E. But the distance BD is equal to the sum of the distances AB and DE, and therefore the mean number of total eclipses would be equal to that of partial eclipses. In fact, however, the diameter of the Earth's umbra at the distance of the Moon is not 3, but merely 2.7 times as large as the Moon, and therefore in the long run there are less *total eclipses than partials.*

Some other interesting facts

1. Sometimes two successive Full Moons both give rise to a lunar eclipse. In such a case, both eclipses are nearly always penumbral ones ; at one of the eclipses the Moon passes to the north of the Earth's umbra, and at the other eclipse to its south. Examples are the penumbral eclipses of 1991 June 27 and July 26, and these of 1998 August 8 and September 6.

On very rare occasions one of the two eclipses is a very small eclipse in the *umbra*, and the other eclipse a very small (not observable !) eclipse in the penumbra. Between the years 1600 and 2200, this occurs only seven times, and it is remarkable that in all these cases the first eclipse occurs during the period April to August :

Date	Magnitude
1608 July 27	0.05 in the umbra
1608 August 25	0.01 in the penumbra
1694 June 7	0.11 in the penumbra
1694 July 7	0.03 in the umbra
1749 June 30	0.03 in the umbra
1749 July 29	0.07 in the penumbra
1835 May 12	0.05 in the penumbra
1835 June 10	0.07 in the umbra
1958 April 4	0.01 in the penumbra
1958 May 3	0.01 in the umbra
2013 April 25	0.02 in the umbra
2013 May 25	0.02 in the penumbra
2147 August 11	0.09 in the umbra
2147 Sept. 9	0.01 in the penumbra

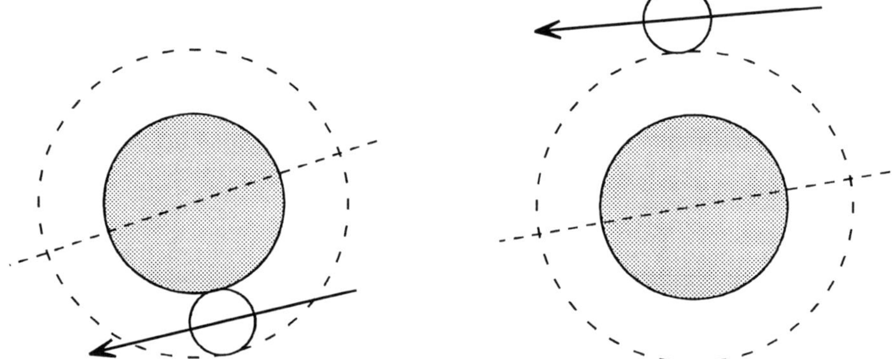

Fig. 16.d : The very small umbral lunar eclipse of 2013 April 25 (left) will be followed, after one lunation, by the very small (and inobservable) penumbral eclipse of 2013 May 25. In each drawing, the grey circle represents a section through the umbral cone of the Earth by a plane perpendicular to its axis at the distance of the Moon. The outer, dashed circumference is the edge of the penumbra. The small circle indicates the position of the Moon at maximum eclipse. The dashed line through the center of the shadow is a part of the ecliptic. North is up.

2. A penumbral lunar eclipse can be followed, six lunations later, by a total eclipse in the umbra, or vice versa. This is not at all rare, for instance:

1995 October 8	penumbral
1996 April 3	total in the umbra
1997 September 16	total in the umbra
1998 March 13	penumbral
2002 November 19	penumbral
2003 May 16	total in the umbra

But here again the famous period of 586 years plays a role, as can seen from column *(a)* in Table 16.A!

3. Sometimes there is *no total* lunar eclipse (in the umbra) in the course of three successive years, so for instance in 1903 – 1904 – 1905, and in 1914 – 1915 – 1916. Between the years 1700 and 1800 this occurred not less than eleven times. To my surprise I found that the last time it took place was in 1932 – 1933 – 1934, and that it will not occur again before 2160 – 2161 – 2162! Therefore I strongly suspected that here too we have to deal with the famous period of 586 years. And indeed this appears to be the case from the data in column *(b)* of Table 16.A.

4. It is *not* possible that during two successive years there are only penumbral lunar eclipses. At least, this holds for the period from A.D. 0 to 3000. However, it nearly happened in 1947 – 1948, when the only lunar eclipses were two penumbral ones and two very small eclipses in the umbra:

1947 June 3	magnitude 0.020 in the umbra
1947 November 28	penumbral eclipse
1948 April 23	magnitude 0.023 in the umbra
1948 October 18	penumbral eclipse

And in A.D. 1495 – 1496 the only lunar eclipses were three penumbrals and one very small umbral eclipse of magnitude 0.010.

REFERENCES

1. A. Pannekoek, Periodicities in Lunar Eclipses, *Koninklijke Nederlandse Akademie van Wetenschappen*, 1951.
2. G. van den Bergh, *Periodicity and Variation of Solar (and Lunar) Eclipses*, Vol. 1, page 95 (Haarlem, Netherlands; 1955).
3. *Heelal* (Belgium), March 1982, pages 59 – 60.
4. *Heelal*, March 1982, page 61.

17. Total penumbral lunar eclipses

The original version of the following text was published in the *Journal of the Royal Astronomical Society of Canada*, Vol. 74, No. 5, pages 291-295 (1980).

As seen from the Earth's center, the extreme values of the apparent diameter of the Moon are 1763″ (for an extreme apogee) and 2012″ (for an extreme perigee). On the other hand, the extreme values of the width of the penumbral corona at the Moon's distance — the difference between the radii of penumbra and umbra — are 1888″ and 1952″. Consequently, it is possible that the Moon be totally immersed in the Earth's penumbral cone without touching the umbra at all. Such total penumbral eclipses are not possible, however, when the Moon is near to its perigee.

It should be pointed out that total penumbral eclipses are only of theoretical interest. For the observer, the important point for a penumbral eclipse is the least separation between the Moon's limb and the edge of the umbra. Very small values for this separation are possible even when the Moon is *not* totally immersed in the penumbra. Therefore, unlike umbral eclipses, the 'total' type of a penumbral eclipse should not be considered to be an important and beautiful phenomenon! In any case, the edge of the penumbra just is not observable.

The maximum phase of the total penumbral lunar eclipse of 2006 March 14 (illustrated in Figure 17.*a*) will take place at 23^h47^m UT, and will be visible in the eastern part of the United States, in Europe, Africa, and the western half of Asia. The magnitude of the eclipse in the penumbra will be 1.030. In the umbra the magnitude is -0.060, *negative* since the Moon does not touch the umbra. At the time of maximum eclipse, the angular separation between the Moon's limb and the edge of the umbra will be 167″.

These values have been calculated using the 'French' rule for the radii of the penumbra and the umbra (see the Note on page 44). The traditional rule consisting to increase by 1/50 the radius of the umbra certainly is not valid in the case of the penumbra. To complicate the matter further, the Earth's umbra is not a perfect circle; it is somewhat flattened by reason of the flattening of the Earth's globe, and the calculation even shows that the flattening of the umbra is *larger* than that of the Earth. Of course, the flattening of the Earth's umbra should be taken into account in the reduction of observations, for example of the eclipses of lunar craters in the umbra. However, in the general calculation of the phases of lunar eclipses the umbra and the penumbra are supposed to be circular, and we have followed this traditional rule here.

Table 17.A lists all total penumbral eclipses from 1600 to 2500. The magnitude in the penumbra is given. Eclipses belonging to the same Saros series are indicated by *A*, *B*, *C*, or *D*. It is remarkable that there are so few cases between the years 1600 and 1900, and between 2160 and 2420. This gave us the idea to look somewhat closer at the distribution of the total penumbral lunar eclipses with time.

TABLE 17.A : *Total penumbral lunar eclipses, 1600 to 2500*

Date	Magnitude	Saros series	Date	Magnitude	Saros series
1607 Mar. 13	1.050		2070 Apr. 25	1.052	
1637 July 7	1.008		2082 Aug. 8	1.001	
1665 July 27	1.006		2099 Sept. 29	1.034	
1806 June 30	1.031		2103 Jan. 23	1.009	C
1900 June 13	1.001 *		2121 Feb. 2	1.030	C
1901 May 3	1.043		2128 Mar. 16	1.009	
1908 Dec. 7	1.034	A	2139 Feb. 13	1.058	C
1926 Dec. 19	1.027	A	2157 Feb. 24	1.094 *	C
1944 Dec. 29	1.022	A	2222 Aug. 23	1.052	
1948 Oct. 18	1.014		2429 Dec. 11	1.084	D
1963 Jan. 9	1.018	A	2447 Dec. 22	1.066	D
1981 Jan. 20	1.014	A	2458 May 28	1.002	
1988 Mar. 3	1.092 *	B	2466 Jan. 1	1.054	D
1999 Jan. 31	1.003	A	2484 Jan. 13	1.043	D
2006 Mar. 14	1.030	B	2498 Sep. 30	1.006	
2053 Aug. 29	1.020				

** This would be a small eclipse in the* umbra *if the calculation is made using the traditional rule of 1/50 for the increase of the radius of the umbra (see p. 44).*

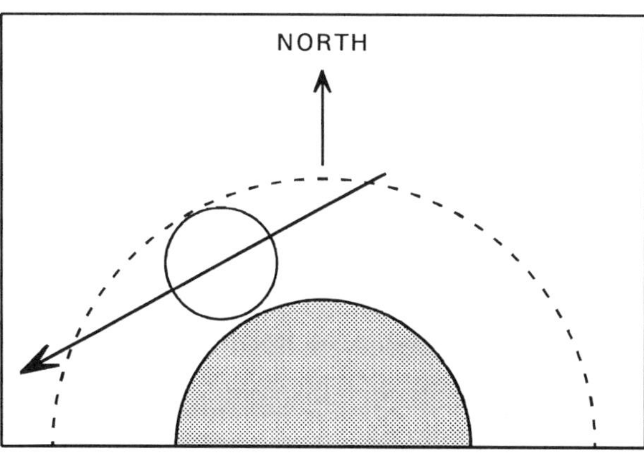

Fig.17.a : *The total penumbral lunar eclipse of 2006 March 14. The dashed circle is the edge of the penumbra; the grey circle is the umbra. Near the time of maximum eclipse, 23 h 47 m UT, the Moon (the smaller circle) is fully immersed in the penumbral cone of the Earth but does not touch the umbra.*

Table 17.B gives the number of total penumbral eclipses per 100 years. We note a cycle of a little less than six centuries, giving alternatively 'rich' and 'poor' periods. Here again we find the famous periodicity of 586 years cited in the previous chapter. It appears that the number of total penumbral eclipses, that of total umbral eclipses, and that of the tetrads, vary in phase with a period of 586 years. In centuries when there are more tetrads, there are not only more total umbral eclipses, but also more total penumbral eclipses.

TABLE 17.B

Number of total penumbral eclipses of the Moon, from A.D. 1000 to 2999

Years	Number	Years	Number
1000 – 1099	3	2000 – 2099	5
1100 – 1199	0	2100 – 2199	5
1200 – 1299	1	2200 – 2299	1
1300 – 1399	8	2300 – 2399	0
1400 – 1499	4	2400 – 2499	6
1500 – 1599	2	2500 – 2599	4
1600 – 1699	3	2600 – 2699	1
1700 – 1799	0	2700 – 2799	3
1800 – 1899	1	2800 – 2899	1
1900 – 1999	10	2900 – 2999	0

18. The half-saros

The original version of the following text was published in the journal Hemel en Dampkring *(Netherlands), Vol. 63, pages 141-143 (1965).*

On pages 181-182 of his *Kalender für Sternfreunde 1965*, Dr. Paul Ahnert (1897-1989) draws the attention on a remarkable fact related to the Saros period, which he had never seen mentioned in the literature.

As is well known, the Saros is equal to 223 synodic months; this is 6585.32 days, or 18 years + approximately 10 days. This period is almost equal to 242 draconic months, and to 239 anomalistic months (*). If we divide the length of this period by 2, we obtain 111.5 synodic, 121 draconic, and 119.5 anomalistic months.

(*) A synodic month is the time interval between two successive New Moons; its mean length is equal to 29.530589 mean solar days.

The draconic month is the time interval between two successive passages of the Moon at the same node of its orbit around the Earth (27.212221 days).

The anomalistic month is the time interval between two successive passages of the Moon at the perigee of its orbit (27.554550 days).

So we have: 223 synodic months   = 6585.321 days
                242 draconic months   = 6585.357 days
                239 anomalistic months = 6585.537 days

After one Saros period the solar and lunar eclipses repeat in approximately the same order and under rather similar circumstances.

Consequently, after a time interval of half a Saros, that is after 9 years and 5 days, a *solar* eclipse is followed by a *lunar* eclipse, and vice versa. This occurs at the *same* node of the Moon's orbit, but with a difference of approximately 180° with respect to the perigee of the lunar orbit. For instance, a solar eclipse with the Moon near perigee is followed, half a Saros later, by a lunar eclipse near the apogee, and vice versa.

A solar eclipse with $\gamma = 0$, hence visible in the equatorial regions (*), is followed, half a Saros later, by a total lunar eclipse. A solar eclipse visible in the northern (or southern) hemisphere is followed by a lunar eclipse at which the Moon passes through the northern (or southern) part of the Earth's umbral cone. Half a Saros after a *partial* eclipse of the Sun, there occurs a *penumbral* lunar eclipse, or at most a partial eclipse in the umbra. Let us give some examples:

1. Total solar eclipse of 1955 June 20 ($\gamma = -0.15$); this is the solar eclipse with the longest duration of totality during the 20th century, namely 7 minutes 08 seconds. The Moon was near the *perigee* of its orbit. — Half a Saros later, on 1964 June 25, there was a deep total lunar eclipse, with a magnitude of 1.56, the Moon being near the *apogee*.

2. Solar eclipse of 1972 July 10, total in the *northern* hemisphere (Alaska, Canada), $\gamma = +0.69$. It was followed, half a Saros later, by a partial lunar eclipse, magnitude 0.55, in the *northern* part of the Earth's umbra, on 1981 July 17.

3. Partial solar eclipse of 2000 December 25, maximum magnitude 0.723; $\gamma = +1.14$, visible in *North*-America. — Half a Saros later, partial lunar eclipse on 2009 December 31, magnitude 0.08, in the *northern* part of the Earth's umbra.

4. On 1953 August 9, partial solar eclipse, maximum magnitude 0.373. On 1962 August 15, penumbral lunar eclipse; magnitude in the penumbra = 0.60.

However, sometimes a *small* partial solar eclipse is followed by 'no eclipse', for instance:

1946 June 29, partial solar eclipse, max. magnitude 0.180, $\gamma = +1.44$;
1955 July 5, Full Moon, *no* eclipse (†);
1964 July 9, partial solar eclipse, maximum magnitude 0.322, $\gamma = +1.36$.

The opposite too can happen:

1951 August 17, penumbral lunar eclipse, magnitude 0.12;
1960 August 22, New Moon, *no* solar eclipse;
1969 August 27, penumbral lunar eclipse, magnitude 0.01.

(*) In the theory of solar eclipses, γ is the least distance of the axis of the Moon's shadow cone to the center of the Earth's globe, expressed in units of the equatorial radius of the Earth. The quantity γ is positive (negative), if the axis of the shadow passes to the north (south) of the center of the Earth. At a central solar eclipse, the absolute value of γ is less than 0.997. If the absolute value of γ is larger than 1.58, no solar eclipse takes place.

(†) But one Saros later, on 1973 July 15, there was a small penumbral lunar eclipse, the magnitude in the penumbra being 0.10.

Hence, from a list of solar eclipses it is possible to predict lunar eclipses approximately. We will not look deeper into this question, and leave it as an amusement for the reader. It remains to choose a suitable name for the half Saros. The name *Sar* seems to us the most obvious one! So we have:

$$1 \text{ Sar} = 3292.661 \text{ days}$$
$$= 9 \text{ calendar years} + 5 \text{ or } 6 \text{ or } 7 \text{ days } (*) - 8 \text{ hours}$$

This is a mean value, however. Let us look at some Sar periods. In the list below, the times of maximum eclipse are expressed in decimals of a day (Universal Time):

Solar eclipse	Lunar eclipse	Difference (days)
1995 Apr. 29.73	2004 May 4.85	3293.12
* 1995 Oct. 24.19	2004 Oct. 28.13	3291.94
1996 Apr. 17.94	2005 Apr. 24.41	3293.47
* 1996 Oct. 12.58	2005 Oct. 17.50	3291.92
1997 Mar. 9.06	2006 Mar. 14.99	3292.93
1997 Sep. 2.00	2006 Sep. 7.79	3292.79
* 1998 Feb. 26.73	2007 Mar. 3.97	3292.24
1998 Aug. 22.09	2007 Aug. 28.44	3293.35
* 1999 Feb. 16.27	2008 Feb. 21.14	3291.87
1999 Aug. 11.46	2008 Aug. 16.88	3293.42

As you see, the length of a Sar can differ from its mean value by as much as 0.81 day, or 19 hours. The reason for this is easily found: 1 Sar is equal to 119 *and a half* anomalistic revolutions of the Moon. Hence, from a given eclipse till its 'successor' one Sar later, the Moon describes 119 times its orbit + another *half* of this orbit. According as this half contains the perigee or the apogee, the Moon will need less or more time to reach the next conjunction or opposition with the Sun.

In the list above, an asterisk indicates the solar eclipses at which the Moon is going from apogee to perigee. As you see, for all these cases the length of the Sar (given in the last column) is shorter than its mean value.

(*) According as there are 3, 2, or 1 leap (bissextile) days in the Sar.

19. Series of occultations

> After a description of series of occultations of stars by the Moon,
> a process for an approached calculation of these series is given.
> This method is also significative for long-term predictions. — This
> text was originally published in *Ciel et Terre*, Vol. 87, No. 2, pages
> 240-252 (March-April 1971).

Successive occultations of a star by the Moon (for the Earth generally, not for a given place) occur in series, which are separated by periods during which the Moon does not occult the star. Within a series, there is an occultation at each conjunction of the Moon with the star, hence at intervals of 27.3 days. For example, a series of occultations of Regulus (α Leonis) began on 1969 December 1 :

Conjunctions of the Moon with Regulus

1969 Oct. 7	no occultation
1969 Nov. 3	no occultation
1969 Dec. 1	occultation (NE Europe)
1969 Dec. 28	occultation (NE Asia, North America)
1970 Jan. 24	occultation (Scandinavia, Asia, Alaska)
etc . . .	

Empirical description

It is well-known that the plane of the lunar orbit is inclined to that of the Earth's orbit (the ecliptic), the mean value of the inclination being $5°08'.7$. The intersection of the two planes is the *line of nodes*, which intersects the celestial sphere in two opposite points: the ascending node and the descending node. So, as seen from the center of the Earth, the Moon describes on the celestial sphere a great circle which is inclined on the ecliptic by $5°08'.7$.

The line of nodes is not fixed with respect to the stars, but regresses slowly as is seen from the following values of the longitude of the ascending node (given for 0^h UT) :

337° 47'	on	1989 January 1
318° 27'	on	1990 January 1
299° 08'	on	1991 January 1
279° 48'	on	1992 January 1
260° 25'	on	1993 January 1, and so on.

The nodes regress by $19°21'$ per year, and one complete revolution takes 18.6 years.

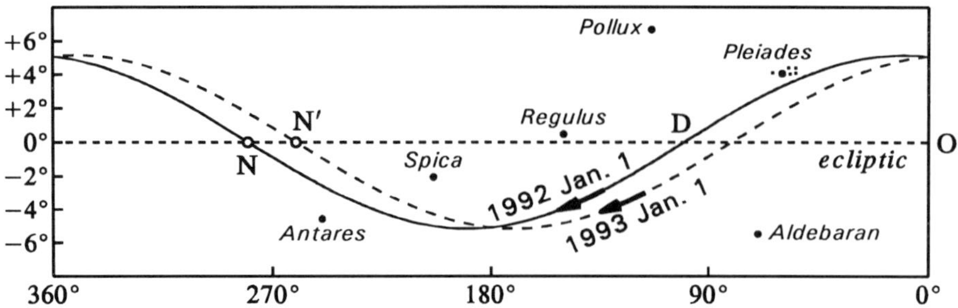

Fig. 19.a : *The path of the Moon on the sky at the beginning of two successive years. Horizontal scale: celestial (ecliptical) longitude. Vertical scale : ecliptical latitude. For clarity, the vertical scale has been exaggerated. Some bright stars are shown.*

Consequently, the path described by the Moon on the celestial sphere is slightly displaced from one revolution to the next. While the inclination remains unchanged, the nodes are shifted 19 degrees per year towards the west.

Figure 19.*a* represents the ecliptical zone of the sky. The central line *ODN* is the ecliptic; *O* is the vernal equinoctial point (longitude 0°). The undulating line is the Moon's path; it is a great circle inclined at 5°08′.7 to the ecliptic. *N* is the ascending node, and *D* the descending node. *ON* is the longitude Ω of the ascending node.

Let us consider the position of the lunar orbit on 1992 January 1 (the solid undulating curve in the drawing), and let us suppose that a star coincides exactly with the ascending node *N*. In other words, we suppose that the celestial longitude and latitude of the star are $\lambda = 279°48′$ and $\beta = 0°$, respectively. So, when the Moon passes at its ascending node, it is occulting the star as seen by an observer on the Earth. One year later, on 1993 January 1, the node has regressed by 19°21′ and is then in position *N'*; the Moon then is describing the path shown as the dashed undulating line, and therefore no longer can occult the star.

However, in order that an occultation of the star be possible, it is not needed that the Moon passes exactly over the point *N*. Indeed, the Earth and the Moon are not dots, but bodies which are several thousands of kilometers across. For this reason, occultations of the star *N* already take place several months before 1992 January 1, and they continue during several months after this date : in total, about twenty occultations of the star will occur during a time period of 17 months. The first of these occultations are visible from the southern polar regions, those near the middle of the series concern the equatorial regions, and the last occultations are visible from the northern regions of the Earth.

Then the series has come to an end, and during several years the Moon will pass, at each conjunction with the star, to the north of it. After nine years, it is the descending node *D* which will pass over the star *N*, and a new series of occultations will take place. But now the first occultations will be visible from the northern hemisphere, and the last ones from the southern hemisphere.

So, the occultations of star N occur in series. Each series has a duration of 17 months and contains about 20 events. The time interval between the middle of a series and that of the next is 9.3 years. But the series are alternated: in one of them, the zones of visibility of the successive occultations are displaced southward, and in the next series they gradually shift northward. So, the period of 18.6 years (that of one complete revolution of the lunar nodes) contains one series of each type.

The occultation series and the periodicity of 18.6 years

One can easily understand that the same rule holds for stars such as Regulus and Spica, which are not situated exactly on the ecliptic, but are not far off. For stars lying less than $3° 56'$ from the ecliptic, there are *two* series of occultations by the Moon per period of 18.6 years. However, the larger the latitude β of the star, the longer is the length of each series. For a star situated exactly on the ecliptic ($\beta = 0°$), each series has a length of 1.4 years. But the duration is

$$1.5 \text{ years for a star at latitude } 2° 00' \text{ (north or south)}$$
$$1.8 \quad - \quad - \quad - \quad - \quad 3° 00'$$
$$2.2 \quad - \quad - \quad - \quad - \quad 3° 40'$$

Moreover, for a star situated exactly on the ecliptic, the middles of the series are separated by 9.3 years. This is no longer the case for stars whose latitude is not zero. The larger the star's latitude, the more different are the intervals. To illustrate this point, here are some data about the occultations of Regulus and Spica:

Regulus ($\beta = +0° 28'$)		Spica ($\beta = -2° 03'$)	
Instant of the middle of the series	*Difference (years)*	*Instant of the middle of the series*	*Difference (years)*
1980.4		1976.1	
	8.7		11.8
1989.1		1987.9	
	9.9		6.8
1999.0		1994.7	
	8.7		11.8
2007.7		2006.5	
	9.9		6.8
2017.6		2013.3	

The middles of the occultation series of Regulus are separated by alternatively 8.7 and 9.9 years. For Spica, the intervals are 11.8 and 6.8 years. Each time we find again the periodicity of 18.6 years: $8.7 + 9.9 = 18.6$; $11.8 + 6.8 = 18.6$.

For the stars whose latitude is situated between $3°56'$ and $6°21'$ (north or south), there is only *one* series of occultations per period of 18.6 years. This is the case, for instance, for the Pleiades, Aldebaran and Antares. This can easily be deduced from Figure 19.*a*. Let us imagine that the sinusoidal path of the Moon glides to the right, so that point N moves from the left to the right border; that is, we consider the regression of the ascending node of the lunar orbit along the ecliptic. We then see that Antares and Aldebaran will be reached only by the lower (southern) part of the Moon's track, and the Pleiades only by its upper (northern) part.

For the stars that lie at ecliptical latitude $+4°$ or $-4°$, the occultation series is of very long duration, almost six years. Stars at higher northern or southern latitudes undergo shorter series :

5.9 years	for stars at latitude	$4°00'$	(north or south)
4.9 years	—	—	$4°40'$
3.8 years	—	—	$5°20'$
2.2 years	—	—	$6°00'$

The series then starts with occultations visible near one of the geographic poles. From one occultation to the next, the region of visibility moves toward the other pole, but during the second half of the series it returns to the first pole. If the star's latitude is positive (north), the first and the last occultations of each series are visible from the southern hemisphere. For stars at southern latitude, the series begins and ends in the northern hemisphere.

As an example, let us consider the series of occultations of the Pleiades which took place from 1986 to 1992. This series started with occultations visible in the southern hemisphere. From January 1987 on, the occultations were visible from the northern hemisphere; some of them were observable from North America, others from Europe or Japan. The last occultations of the series, in 1992, again concerned the southern hemisphere of the Earth.

The brightest star of the Pleiades, *Alcyone* (η Tauri), is situated at latitude $\beta = +4°03'$. For stars with higher latitudes, the zone of occultation cannot longer reach the other pole. This is the case for Aldebaran ($\beta = -5°28'$). For this star the series of occultations begin in the northern hemisphere (for instance in 1996); then for the successive occultations, the region of visibility is displaced towards the south, but their southern limit reaches at most latitude $13°S$; thereafter, the zones are again displaced toward the north (1999 – 2000). Consequently, a resident of Montevideo, Uruguay, never can see the Moon occult Aldebaran.

Finally, the stars with a latitude larger than $6°21'$ cannot be occulted by the Moon. Indeed, for a geocentric observer the maximum latitude of the center of the lunar disk is $5°09'$. But, as we shall see, for some places on the Earth's surface the Moon can still occult stars which are situated $72'$ from its center. And $5°09' + 72' = 6°21'$. (At the end of this chapter we will see that the extreme latitude is, in fact, $6°36'$ instead of $6°21'$).

Mathematical approach of the problem

Let α and δ be the right ascension and declination of a star referred to the mean equinox of 2000.0. We intend to search the periods during which there are occultations of this star by the Moon.

We first calculate the ecliptical coordinates λ and β by means of the classical formulae

$$\tan \lambda = (\sin \varepsilon \, \tan \delta + \cos \varepsilon \, \sin \alpha) / \cos \alpha$$
$$\sin \beta = \cos \varepsilon \, \sin \delta - \sin \varepsilon \, \cos \delta \, \sin \alpha$$

where ε is the obliquity of the ecliptic, and $\cos \lambda$ has the same sign as $\cos \alpha$. As all the coordinates refer to the mean equinox of 2000.0, we have $\sin \varepsilon = 0.397777$ and $\cos \varepsilon = 0.917482$.

If L is the ecliptical longitude of the Moon's center, and Ω is the longitude of the ascending node of the lunar orbit, the geocentric latitude B of the Moon, expressed in degrees, can be obtained with sufficient accuracy from

$$B = 5.145 \sin (L - \Omega)$$

where $5°.145$ is the inclination of the Moon's orbit. (Actually, the exact formula is $\tan B = \tan i \sin (L - \Omega)$, but the error made by using the first expression never exceeds $19''$, which is negligible here). When the Moon is in conjunction with the star, we have $L = \lambda$. Consequently, at the time of conjunction

$$B = 5°.145 \sin (\lambda - \Omega).$$

In order that the conjunction gives rise to an occultation, the difference between the latitudes of the two celestial bodies must be smaller than the sum of the

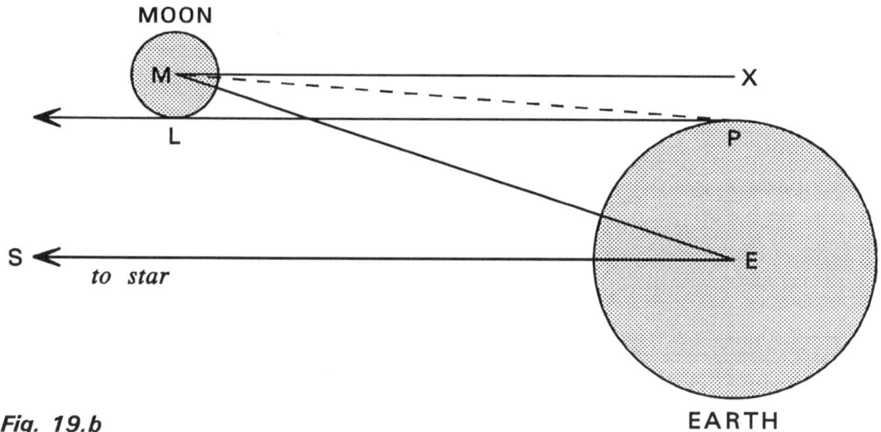

Fig. 19.b

Moon's horizontal parallax π and apparent semidiameter s. In other words, we should have $|B - \beta| < \pi + s$. This is shown in Figure 19.b, where there is just occultation for point P of the Earth. In this case, the difference $|B - \beta|$ is equal to the angle SEM, from which we deduce:

$$SEM = EMX = EMP + PMX = EMP + MPL = \pi + s.$$

For the Moon, the mean values are $\pi = 57'02".6$ and $s = 15'32".6$, and hence the mean value of the sum $\pi + s$ is 72.6 arcminutes, or 1.21 degree.

Consequently, there is occultation of the star by the Moon somewhere on the Earth's surface when, at the time of conjunction, the difference $|B - \beta|$ is less than $1°.21$. So, in order to find the times of beginning and end of the occultation series, we have to solve the equation $|B - \beta| = 1°.21$. We have successively, all values being expressed in degrees (and decimals):

$$B - \beta = \pm 1.21$$

$$B = \beta \pm 1.21$$

$$5.145 \sin (\lambda - \Omega) = \beta \pm 1.21$$

$$\sin (\lambda - \Omega) = \frac{\beta \pm 1.21}{5.145} \tag{1}$$

If T is the time measured from the beginning of the year 2000 in Julian years of 365.25 days, we have

$$\Omega = 125.0446 - 19.3553\,T$$

where Ω is expressed in degrees and refers to the fixed equinox 2000.0. This last expression can be written as follows, k being an integer:

$$T = \frac{125.04 - \Omega + 360\,k}{19.3553} \tag{2}$$

Consequently, in order to find the beginning and the end of occultation series of a given star, one has to calculate Ω by means of formula (1). Then, with the values of Ω so obtained, T is found by formula (2).

It should be remembered that each value for $\sin (\lambda - \Omega)$ gives *two* values for $\lambda - \Omega$ in the interval 0 – 360 degrees, and hence two values for Ω. Consequently, according as the absolute value of β is smaller than $3°56'$, is between $3°56'$ and $6°21'$, or is larger than $6°21'$, formula (1) yields 4, 2, or 0 values for Ω, resulting in 2, 1, or 0 occultation periods per 18.6 years, respectively.

The value $k = 0$ in formula (2) will give an instant near A.D. 2000. Earlier occultation series are found by making k equal to -1, -2, etc. Future series are found with the values $k = +1, +2$, etc.

As an exemple, let us calculate the times of the occultation series of Spica (α Virginis). The ecliptical coordinates of this star, referred to the equinox 2000.0, are (in degrees) $\lambda = 203.841$ and $\beta = -2.054$. Whence, by formula (1),

$$\sin (\lambda - \Omega) = -0.63440 \quad \text{and} \quad -0.16404$$

Each sine gives two values for Ω, and so we obtain the following four values, again in degrees and decimales, the first one having been increased by 360° for the sake of the case :

$$\Omega = 374.40, \quad 344.47, \quad 243.22, \quad 213.28.$$

For these values of Ω, and with $k = 0$, formula (2) gives

$$T = -12.88, \quad -11.34, \quad -6.11, \quad \text{and} \quad -4.56,$$

corresponding to the years 1987.12, 1988.66, 1993.89, and 1995.44. Consequently, there were occultations of Spica from 1987.12 to 1988.66, and then again from 1993.89 to 1995.44. Remember that an expression such as 1987.12 means 0.12 year after the beginning of 1987, that is, mid-February 1987.

The next series are obtained by putting $k = 1, 2$, etc, but it may be easier to add 18.60 years (more exactly, 18.5995) to the instants just obtained. This gives the series 2005.72 – 2007.26, 2012.49 – 2014.04, 2024.32 – 2025.86, and so on. However, it is important to note that it is not allowed to make extrapolations too far into the past or into the future, for the following two reasons :
— the stars are subject to proper motions, so their longitudes and latitudes, even referred to a fixed equinox such as 2000.0, are slowly changing in the course of the centuries ;
— the plane of the ecliptic is not fixed in space, but has a slow rotation along a 'line of nodes' and which presently amounts to 47″ per century. From this it results that the latitude of a star is slowly variable, even if its proper motion is zero. We will discuss in the next chapter the case of Pollux (β Geminorum).

Another method for finding the series of occultations

Instead of performing the calculation by means of formulae (1) and (2), it is possible to proceed as follows.

First, find τ by means of the formula

$$\tau = 1992.51 - 0.051665\,\lambda \tag{3}$$

where λ is the star's ecliptical longitude referred to the equinox of 2000.0, and is expressed in degrees.

Add or subtract from the time τ a convenient multiple of 9.30 years. It *must* be an *even* multiple if the star's latitude is positive (north), or an *odd* multiple if this latitude is negative (south). Consequently:

Number of years to subtract from or to add to τ

β positive	β negative
0.00	9.30
18.60	27.90
37.20	46.50
55.80	65.10
74.40	83.70
etc.	etc.

As a function of the star's latitude β, take from Table 19.A the corrections to obtain the times of beginning and end of the occultation series. In the table, the (positive or negative) latitudes β are expressed in degrees and *decimals*. The corrections a and b are given in years and decimals; they are positive or negative.

If the latitude of the star is between 6.355 and 3.935 degrees, there is only one series of occultations per 18.6 years. The correction $-a$ gives the beginning of that series, while correction $+a$ corresponds to its end.

If the star's latitude is less than $3°.935$ (north or south), there are two series of occultations per 18.6 years, and the times are obtained as follows:

correction	$-a$	beginning of the first series
correction	$-b$	end of the first series
correction	$+b$	beginning of the second series
correction	$+a$	end of the second series

Let us mention, for the interested reader, that the values a and b have been calculated as follows.

$$\cos x = 0.1943635 \ (\beta \pm 1.21)$$

$$a \text{ or } b = 0.051665 \, x$$

In the first formula, the star's latitude β should be expressed in degrees. Between the parentheses, the minus sign is used for the calculation of a, and the $+$ sign for the calculation of b. In the second formula, the angle x should be expressed in degrees and decimals.

Let us take again the example of the occultation series of Spica. The ecliptical coordinates of this star, referred to the equinox of 2000.0, are $\lambda = 203°.841$ and $\beta = -2°.054$. Formula (3) gives

$$\tau = 1992.51 - (0.051665 \times 203.841) = 1981.98.$$

TABLE 19.A

$\mid \beta \mid$	a	$\mid \beta \mid$	a	b
6.355	0.00	3.935	3.00	0.00
6.35	0.13	3.93	3.00	0.13
6.34	0.23	3.92	3.01	0.23
6.33	0.29	3.91	3.01	0.29
6.32	0.35	3.90	3.02	0.35
6.30	0.43	3.88	3.03	0.43
6.25	0.60	3.85	3.05	0.54
6.20	0.73	3.80	3.09	0.68
6.15	0.84	3.75	3.12	0.80
6.10	0.94	3.70	3.15	0.90
6.05	1.02	3.65	3.19	0.99
6.00	1.11	3.60	3.22	1.07
5.90	1.25	3.50	3.28	1.23
5.80	1.39	3.40	3.35	1.36
5.70	1.51	3.20	3.47	1.60
5.60	1.62	3.00	3.60	1.81
5.50	1.73	2.75	3.75	2.05
5.40	1.83	2.50	3.90	2.27
5.30	1.93	2.25	4.05	2.47
5.20	2.02	2.00	4.19	2.66
5.10	2.11	1.75	4.34	2.84
5.00	2.20	1.50	4.48	3.01
4.80	2.36	1.25	4.63	3.17
4.60	2.52	1.00	4.77	3.34
4.40	2.67	0.75	4.91	3.49
4.20	2.81	0.50	5.06	3.65
4.00	2.95	0.25	5.21	3.80
3.935	3.00	0.00	5.35	3.95

Since β is negative, let us add 9.30 years, which yields 1991.28. (Of course, we also might have added or subtracted 27.90 or 46.50 years, etc.). Then, from Table 19.A we find for $\beta = 2.054$, by linear interpolation, $a = 4.16$ and $b = 2.62$. From which

beginning of the series of occultations :	1991.28 − 4.16 =	1987.12
end of this series :	1991.28 − 2.62 =	1988.66
beginning of the next series :	1991.28 + 2.62 =	1993.90
end of this series :	1991.28 + 4.16 =	1995.44

These values are practically identical to those found before (page 119).

In the same way, we made the calculations for three other zodiacal first-magnitude stars (Aldebaran, Regulus, Antares), and for Alcyone (η Tauri), the brightest star of the Pleiades. Here are the occultation series of these stars found for the period 1920 – 2030 :

<table>
<tr><td colspan="4">Regulus</td><td colspan="4">Spica</td></tr>
<tr><td>b</td><td>1923.89</td><td>–</td><td>1925.30</td><td>b</td><td>1919.50</td><td>–</td><td>1921.04</td></tr>
<tr><td>a</td><td>1932.64</td><td>–</td><td>1934.05</td><td>a</td><td>1931.32</td><td>–</td><td>1932.85</td></tr>
<tr><td>b</td><td>1942.49</td><td>–</td><td>1943.90</td><td>b</td><td>1938.10</td><td>–</td><td>1939.64</td></tr>
<tr><td>a</td><td>1951.24</td><td>–</td><td>1952.65</td><td>a</td><td>1949.92</td><td>–</td><td>1951.46</td></tr>
<tr><td>b</td><td>1961.09</td><td>–</td><td>1962.50</td><td>b</td><td>1956.70</td><td>–</td><td>1958.24</td></tr>
<tr><td>a</td><td>1969.84</td><td>–</td><td>1971.25</td><td>a</td><td>1968.52</td><td>–</td><td>1970.06</td></tr>
<tr><td>b</td><td>1979.69</td><td>–</td><td>1981.10</td><td>b</td><td>1975.29</td><td>–</td><td>1976.84</td></tr>
<tr><td>a</td><td>1988.44</td><td>–</td><td>1989.85</td><td>a</td><td>1987.12</td><td>–</td><td>1988.66</td></tr>
<tr><td>b</td><td>1998.29</td><td>–</td><td>1999.70</td><td>b</td><td>1993.89</td><td>–</td><td>1995.44</td></tr>
<tr><td>a</td><td>2007.04</td><td>–</td><td>2008.45</td><td>a</td><td>2005.72</td><td>–</td><td>2007.26</td></tr>
<tr><td>b</td><td>2016.89</td><td>–</td><td>2018.30</td><td>b</td><td>2012.49</td><td>–</td><td>2014.04</td></tr>
<tr><td>a</td><td>2025.64</td><td>–</td><td>2027.05</td><td>a</td><td>2024.32</td><td>–</td><td>2025.86</td></tr>
</table>

<table>
<tr><td colspan="3">Aldebaran</td><td colspan="3">Antares</td><td colspan="3">Alcyone</td></tr>
<tr><td>1922.04</td><td>–</td><td>1925.57</td><td>1930.56</td><td>–</td><td>1935.65</td><td>1930.69</td><td>–</td><td>1936.53</td></tr>
<tr><td>1940.64</td><td>–</td><td>1944.17</td><td>1949.16</td><td>–</td><td>1954.25</td><td>1949.29</td><td>–</td><td>1955.13</td></tr>
<tr><td>1959.24</td><td>–</td><td>1962.77</td><td>1967.76</td><td>–</td><td>1972.85</td><td>1967.89</td><td>–</td><td>1973.73</td></tr>
<tr><td>1977.84</td><td>–</td><td>1981.37</td><td>1986.36</td><td>–</td><td>1991.45</td><td>1986.49</td><td>–</td><td>1992.33</td></tr>
<tr><td>1996.44</td><td>–</td><td>1999.97</td><td>2004.96</td><td>–</td><td>2010.05</td><td>2005.09</td><td>–</td><td>2010.93</td></tr>
<tr><td>2015.04</td><td>–</td><td>2018.57</td><td>2023.56</td><td>–</td><td>2028.65</td><td>2023.69</td><td>–</td><td>2029.53</td></tr>
</table>

For instance, occultations of Aldebaran are expected to occur from November 1977 to May 1981, then from June 1996 to December 1999, etc.

For Regulus and Spica, the series indicated by *a* start in the northern hemisphere and end in the southern one. The series marked by *b* begin in the southern hemisphere and end in the northern regions. For Aldebaran and Antares, each series begins *and* ends in the northern hemisphere. Those of the Pleiades begin and end in the southern hemisphere.

The reader can easily check our results, or make the calculations for other stars. For this purpose, we give in Table 19.B the ecliptical longitude and latitude of the 20 brightest stars which can be occulted by the Moon as seen from the Earth's surface. The first column gives the number of the star in the *Zodiacal Catalog* [*Catalog of 3539 Zodiacal Stars*, by James Robertson, *Astronomical Papers*, Vol. X, Part II (Washington, 1940)]. The visual magnitudes, given in the third column, are those of the Revised Harvard Photometry (RHP).

TABLE 19.B

The twenty brightest zodiacal stars

The longitudes λ and latitudes β are for epoch and equinox 2000.0

Z.C.	Star	Magnitude	λ	β	Name
552	η Tau	3.0	59°.992	+4°.051	Alcyone
692	α Tau	1.1	69.789	−5.467	Aldebaran
810	β Tau	1.8	82.575	+5.385	
847	ζ Tau	3.0	84.785	−2.196	
1172	β Gem	1.2	113.216	+6.684	Pollux
1487	α Leo	1.3	149.829	+0.465	Regulus
1821	γ Vir	2.9	190.141	+2.790	
1925	α Vir	1.2	203.841	−2.054	Spica
2118	α Lib	2.9	225.083	+0.333	
2287	π Sco	3.0	242.940	−5.475	
2290	δ Sco	2.5	242.571	−1.986	
2302	β Sco	2.9	243.190	+1.008	
2349	σ Sco	3.1	247.800	−4.037	
2366	α Sco	1.2	249.762	−4.570	Antares
2383	τ Sco	2.9	251.457	−6.120	
2655	δ Sgr	2.8	274.581	−6.472	
2672	λ Sgr	2.9	276.317	−2.136	
2750	σ Sgr	2.1	282.385	−3.450	
2797	π Sgr	3.0	286.252	+1.437	
3190	δ Cap	3.0	323.543	−2.602	

How accurate is the method of calculation described here? The exact dates of the beginning and end of several series are given in the next chapter, so the reader can compare. In some cases, the error is as large as 0.2 year. For instance, while the method predicts that a series of occultations of Aldebaran started at the epoch 1977.84, the first event of that series actually took place on 1978 January 19, that is 1978.05. However, the mean error of the results is about 0.06 year. This is a rather excellent accuracy, if we consider that the length of one sidereal revolution of the Moon is equal to 27.32 days, or 0.075 year, and that the Moon can be in conjuncton with a given star shortly before the beginning of a favorable period for occultations, or shortly after its end. So, even in the case of a 'rigorous theory', the mean deviation of the results is necessarily equal to 0.04 year.

In fact, the mean deviation of our results is still somewhat larger. The reason for this is that in the calculation we made some simplifications, namely :

(a) We considered the *mean* lunar node, while this node actually is affected by inequalities of which the most important has a period of 173.31 days and a semi-

amplitude of 1°30' (see Chapter 1). And an error of 1°30' in the position of the node yields an error of 28 days (0.08 year) in the predicted times of beginning or end of a series of occultations;

(b) The inclination of the lunar orbit on the ecliptic is affected by an inequality of the same period as that of the node, 173.31 days. This inclination actually varies between 5°00' and 5°18', while we adopted the mean value 5°08'.7;

(c) For the sum $\pi + s$ we used the mean value 72'.6. Actually, by reason of the variable Earth – Moon distance, this quantity can vary between the extremes 68'.6 and 78'.3.

From this it results that the Moon can occult stars situated up to latitude 5°18' + 78' = 6°36', not 6°21' as stated earlier in this chapter.

20. *Occultations of bright stars by the Moon*

The table on the next pages mentions the first and last occultations of the series for stars of magnitude 3.0 and brighter, during the years 1950 to 2050. These dates have been obtained by an accurate calculation, not by the approximate method described in the previous chapter.

Column N gives the number of occultations in the series. The last column mentions the direction of the displacement of the region of visibility on the Earth's surface in the course of the series. For instance, N – S means that the series begins with occultations visible in the northern hemisphere and ends in the southern one. See more about this in the previous chapter.

Of course, not all occultations of a series are visible from a given place on the surface of the Earth, far from that. It even can happen that an occultation is visible nowhere, by reason of the vicinity of the Sun. This was the case for the Spica occultation of 1993 October 15, which took place near New Moon; the angular separation between the star and the Sun was only 3 degrees. However, such inobservable cases are taken into account in the list for reason of completeness.

It may happen that some occultations are missing near the beginning or the end of a series. This is due to the periodic oscillation in the inclination of the lunar orbit on the ecliptic. For example, the 1959–1962 series of Aldebaran began on 1959 March 16 with an occultation in northwestern Siberia. Then there was an occultation on April 12 visible from Greenland, and on May 9 a very short event visible from a part of the Arctic Ocean. But at the next conjunction, on 1959 June 6, no occultation took place, the 'shadow' of the Moon not touching the Earth. (It happened that it was also New Moon that day, but this is irrelevant here). Then there was again an occultation on July 3, and the series further went on without any other interruption.

These 'missing occultations' are *not* included in the totals N given in the list.

TABLE 20.A

Occultation series of bright stars, 1950 to 2050

Star	First occultation of the series	Last occultation of the series	N	
η Tauri (*Alcyone*)	1949 Apr. 2	1955 Feb. 28	77	S – S
	1968 Jan. 11	1973 Sep. 17	76	S – S
	1986 July 30	1992 Mar. 10	76	S – S
	2005 Feb. 16	2010 Dec. 19	78	S – S
	2023 Sep. 5	2029 July 7	79	S – S
	2042 Mar. 25	2048 Jan. 25	78	S – S
α Tauri (*Aldebaran*)	1959 Mar. 16	1962 Sep. 19	47	N – N
	1978 Jan. 19	1981 Apr. 8	44	N – N
	1996 Aug. 8	2000 Feb. 14	48	N – N
	2015 Jan. 29	2018 Sep. 3	49	N – N
	2033 Aug. 18	2037 Feb. 23	48	N – N
β Tauri	1949 Mar. 8	1952 Oct. 8	49	S S
	1967 Sep. 26	1971 Aug. 15	50	S – S
	1986 Aug. 1	1990 Mar. 4	49	S – S
	2005 Feb. 18	2008 Sep. 21	49	S – S
	2023 Sep. 7	2027 Apr. 11	49	S – S
	2042 Feb. 28	2045 Oct. 29	50	S – S
ζ Tauri	1956 Mar. 19	1957 Aug. 20	20	N – S
	1962 Oct. 17	1964 Mar. 20	20	S – N
	1974 Oct. 7	1976 Mar. 9	20	N – S
	1981 May 6	1982 Nov. 4	21	S – N
	1993 Apr. 26	1994 Sep. 27	20	N – S
	1999 Nov. 24	2001 Aug. 14	24	S – N
	2012 Feb. 3	2013 July 6	20	N – S
	2018 June 13	2020 Feb. 5	22	S – N
	2030 Aug. 23	2032 Jan. 24	20	N – S
	2037 Mar. 24	2038 Aug. 25	20	S – N
	2049 Mar. 11	2050 Aug. 12	20	N – S
α Leonis (*Regulus*)	1951 May 14	1952 June 27	16	N – S
	1961 Feb. 2	1962 May 12	18	S – N
	1969 Dec. 1	1971 Apr. 6	19	N – S
	1979 Nov. 12	1980 Dec. 26	16	S – N
	1988 June 19	1989 Oct. 24	19	N – S
	1998 June 1	1999 Oct. 5	19	S – N
	2007 Jan. 7	2008 May 12	19	N – S
	2016 Dec. 18	2018 Apr. 24	19	S – N
	2025 July 26	2026 Dec. 27	20	N – S
	2035 June 11	2036 Nov. 11	20	S – N
	2044 May 5	2045 Oct. 7	20	N – S

TABLE 20.A (Cont.)

Star	First occultation of the series	Last occultation of the series	N	
γ Virginis	1960 June 4	1961 Dec. 2	21	S – N
	1966 Feb. 9	1967 Nov. 27	25	N – S
	1978 Dec. 23	1980 June 21	21	S – N
	1984 Aug. 29	1986 June 16	25	N – S
	1997 Aug. 8	1999 Apr. 28	24	S – N
	2003 Mar. 19	2005 Jan. 3	25	N – S
	2016 Feb. 25	2017 Nov. 14	24	S – N
	2022 Jan. 23	2023 June 26	20	N – S
	2034 Dec. 5	2036 June 3	21	S – N
	2040 Aug. 12	2042 Jan. 13	20	N – S
α Virginis (Spica)	1949 Dec. 15	1951 June 15	21	N – S
	1956 Sep. 7	1958 Mar. 8	21	S – N
	1968 July 31	1970 Jan. 1	20	N – S
	1975 Mar. 28	1976 Nov. 19	23	S – N
	1987 Feb. 18	1988 June 24	19	N – S
	1993 Oct. 15	1995 June 9	21	S – N
	2005 Sep. 7	2007 Jan. 11	19	N – S
	2012 July 25	2013 Dec. 27	20	S – N
	2024 June 16	2025 Nov. 17	20	N – S
	2031 Feb. 12	2032 July 15	20	S – N
	2043 Jan. 4	2044 June 6	20	N – S
	2049 Sep. 1	2051 Feb. 2	20	S – N
α Librae	1957 Feb. 20	1958 July 24	20	S – N
	1966 Feb. 11	1967 June 19	19	N – S
	1975 Sep. 9	1977 Feb. 10	20	S – N
	1984 Aug. 31	1986 Jan. 6	19	N – S
	1994 Mar. 30	1995 Aug. 31	20	S – N
	2003 Mar. 21	2004 July 26	19	N – S
	2013 Jan. 7	2014 May 14	19	S – N
	2021 Oct. 8	2023 Feb. 12	19	N – S
	2031 July 28	2032 Nov. 30	19	S – N
	2040 July 18	2041 Sep. 1	16	N – S
	2050 Feb. 14	2051 June 20	19	S – N
π Scorpii	1950 Mar. 9	1953 Sep. 14	48	N – N
	1968 Sep. 26	1972 Apr. 3	47	N – N
	1987 Aug. 4	1991 Jan. 11	45	N – N
	2006 Feb. 21	2009 Aug. 27	48	N – N
	2024 Sep. 9	2028 Mar. 16	48	N – N
	2043 Mar. 30	2046 Aug. 10	45	N – N

TABLE 20.A (Cont.)

Star	First occultation of the series	Last occultation of the series	N	
δ Scorpii	1954 Oct. 1	1956 Mar. 3	20	S – N
	1966 July 27	1968 Jan. 24	21	N – S
	1973 Apr. 20	1974 Oct. 18	21	S – N
	1985 Feb. 12	1986 Aug. 13	21	N – S
	1992 Feb. 25	1993 July 1	19	S – N
	2003 Sep. 2	2005 Feb. 3	20	N – S
	2010 Sep. 13	2012 Feb. 15	20	S – N
	2022 Mar. 23	2023 Aug. 24	20	N – S
	2029 Mar. 6	2030 Sep. 4	21	S – N
	2040 Nov. 6	2042 June 3	22	N – S
	2047 Sep. 24	2049 Feb. 25	20	S – N
β Scorpii	1956 Sep. 10	1958 Jan. 15	19	S – N
	1964 Oct. 9	1966 Feb. 13	19	N – S
	1975 Mar. 4	1976 Aug. 4	20	S – N
	1983 Apr. 1	1984 Sep. 1	20	N – S
	1993 Sep. 20	1995 Feb. 22	20	S – N
	2001 Oct. 19	2003 Mar. 22	20	N – S
	2012 Apr. 9	2013 Dec. 2	21	S – N
	2020 July 29	2021 Dec. 31	20	N – S
	2031 Jan. 19	2032 June 21	20	S – N
	2039 Feb. 16	2040 July 20	20	N – S
	2049 Sep. 4	2051 Jan. 9	19	S – N
α Scorpii (*Antares*)	1949 Feb. 20	1954 Feb. 26	68	N – N
	1967 Sep. 10	1972 Sep. 14	68	N – N
	1986 Mar. 30	1991 Apr. 4	68	N – N
	2005 Jan. 7	2010 Feb. 7	69	N – N
	2023 Aug. 25	2028 Aug. 27	68	N – N
	2042 Mar. 13	2047 Mar. 17	68	N – N
τ Scorpii	1950 Aug. 21	1952 Aug. 28	26	N – N
	1969 Feb. 11	1971 Mar. 18	27	N – N
	1987 Sep. 1	1989 Sep. 8	23	N – N
	2006 Aug. 4	2008 Mar. 27	21	N – N
	2025 Feb. 21	2026 Sep. 17	19	N – N
	2043 Sep. 10	2045 Aug. 20	21	N – N
δ Sagittarii	1950 Mar. 12	1950 Sep. 19	3	N – N
	1968 Sep. 29	1969 Apr. 8	4	N – N
	1987 Mar. 22	1987 Sep. 30	4	N – N
	2006 Mar. 22	2006 Mar. 22	1	N – N
	2024 Sep. 12	2024 Oct. 9	2	N – N

TABL E 20.A (Cont.)

Star	First occultation of the series	Last occultation of the series	N	
λ Sagittarii	1952 Nov. 20	1954 Apr. 23	20	S – N
	1964 Nov. 8	1966 Mar. 15	19	N – S
	1971 June 10	1973 Jan. 31	23	S – N
	1983 May 29	1985 Jan. 19	23	N – S
	1990 Mar. 20	1991 Aug. 20	20	S – N
	2002 Mar. 7	2003 Aug. 9	20	N – S
	2008 Nov. 3	2010 Mar. 9	19	S – N
	2020 Sep. 24	2022 Feb. 26	20	N – S
	2027 Apr. 26	2028 Sep. 25	20	S – N
	2039 Apr. 14	2040 Sep. 15	20	N – S
	2045 Nov. 12	2047 May 13	21	S – N
σ Sagittarii	1951 Jan. 7	1953 Apr. 6	28	S – N
	1965 Apr. 22	1967 June 23	27	N – S
	1969 Nov. 13	1971 Oct. 25	27	S – N
	1983 Nov. 9	1986 Feb. 6	30	N – S
	1988 June 2	1990 May 14	27	S – N
	2002 Aug. 19	2004 Aug. 26	28	N – S
	2006 Dec. 21	2009 Feb. 20	28	S – N
	2021 Mar. 8	2023 Feb. 17	27	N – S
	2025 Nov. 24	2027 Sep. 10	25	S – N
	2039 Sep. 25	2041 Sep. 6	27	N – S
	2044 May 16	2046 Mar. 29	26	S – N
π Sagittarii	1954 Sep. 8	1956 Feb. 8	20	S – N
	1962 Mar. 29	1963 Aug. 30	20	N – S
	1973 Apr. 24	1974 Aug. 28	19	S – N
	1980 Nov. 12	1982 Mar. 19	19	N – S
	1991 Nov. 11	1993 Mar. 16	19	S – N
	1999 June 2	2000 Oct. 5	19	N – S
	2010 May 31	2011 Oct. 4	19	S – N
	2017 Dec. 19	2019 Apr. 25	19	N – S
	2028 Dec. 17	2030 July 14	22	S – N
	2036 Sep. 28	2038 Feb. 1	19	N – S
	2047 Sep. 27	2049 Feb. 28	20	S – N
δ Capricorni	1950 Jan. 20	1951 Oct. 10	24	S – N
	1962 Sep. 12	1964 Apr. 8	22	N – S
	1968 Aug. 8	1970 Apr. 29	23	S – N
	1981 Apr. 28	1982 Oct. 26	21	N – S
	1987 May 19	1988 Nov. 16	21	S – N
	1999 Nov. 16	2001 May 15	21	N – S
	2005 Dec. 6	2007 June 6	21	S – N
	2018 June 4	2020 Feb. 22	23	N – S
	2024 June 25	2025 Dec. 24	21	S – N
	2037 Jan. 18	2038 Sep. 11	22	N – S
	2043 Jan. 13	2044 Oct. 3	24	S – N

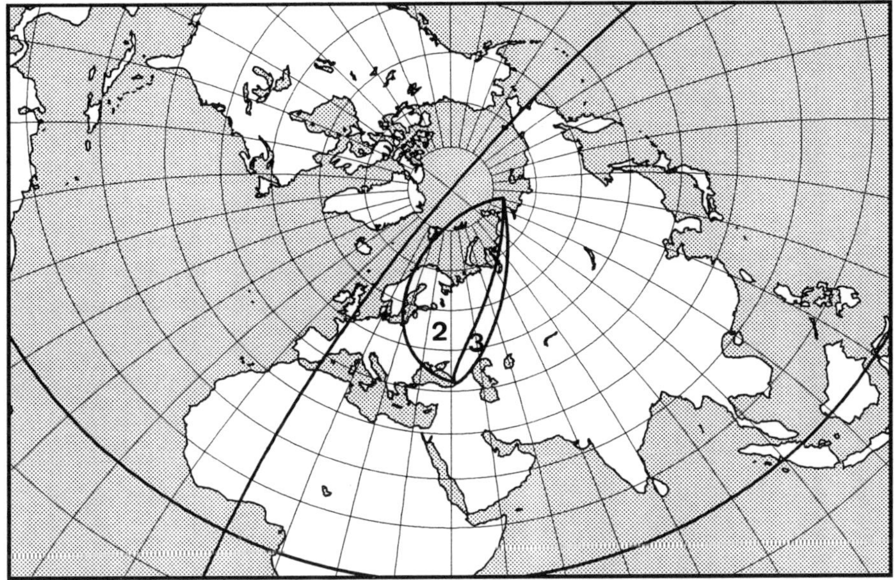

Fig. 20.a : *Area of visibility of the occultation of Regulus, 2007 January 7*

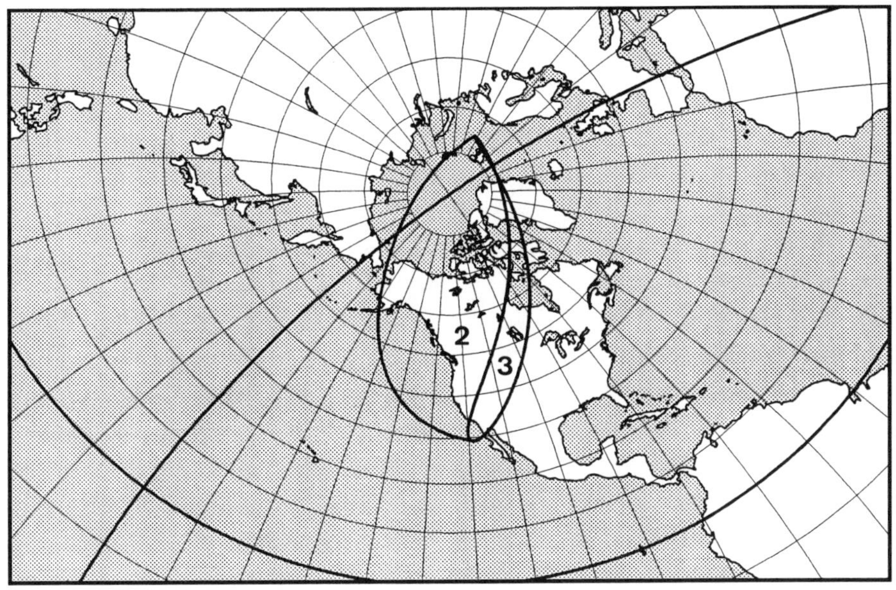

Fig. 20.b : *Area of visibility of the occultation of Regulus, 2007 February 3*

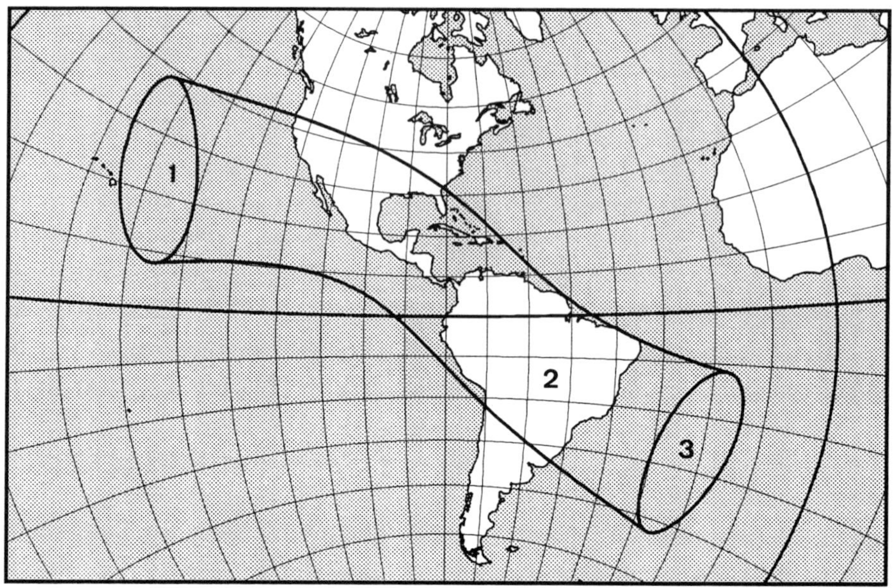

Fig. 20.c : *Area of visibility of the occultation of Regulus, 2007 November 3*

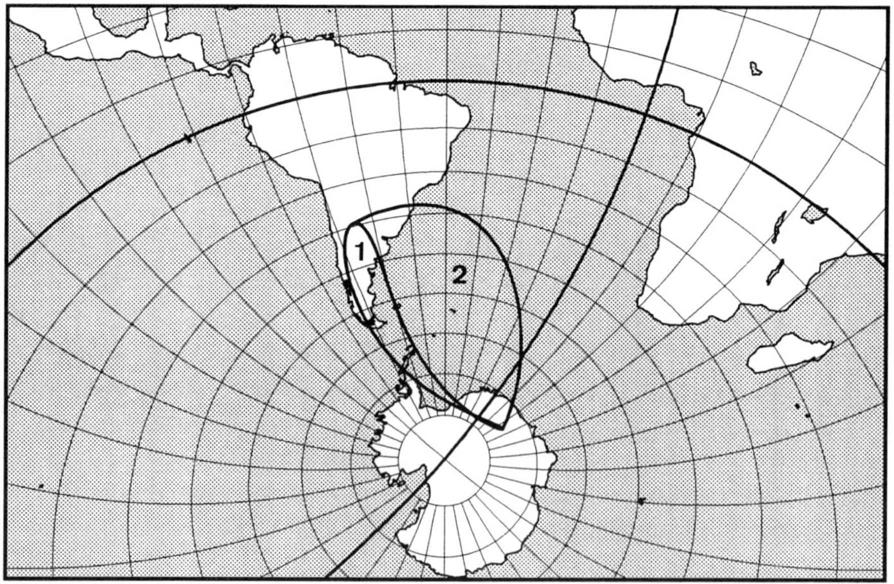

Fig. 20.d : *Area of visibility of the occultation of Regulus, 2008 May 12*

The occultation series of Regulus, 2007–2008

As an example, we give on the preceding pages maps showing the regions of visibility of four occultations of Regulus of the series 2007–2008.

The occultation of 2007 January 7 (Fig. 20.*a*) will be the first of the series. The area of visibility concerns mainly eastern and northeastern Europe. The southern limit of the zone — the line crossing Scandinavia and the Black Sea — reaches latitude 42° N.

One sidereal revolution of the Moon later, the event of 2007 February 3 (Fig. 20.*b*) will be observable from the arctic regions and the northwestern half of North America. Because the axis of the Moon's 'shadow' comes closer to the center of the Earth than at the preceding occultation, the area of visibility is larger; the southern limit now reaches latitude 29° N.

The 12th event of the series is the occultation of 2007 November 3 (Fig. 20.*c*). Now the Moon's shadow passes almost centrally over the Earth. There is a northern limit (which passes over the United States) *and* a southern limit. The southern part of North America, Central America, and the northern half of South America enjoy the occultation.

The 19th and last occultation of the series, that of 2008 May 12, will be visible from the southern part of South America and a small part of Antarctica (Fig. 20.*d*).

On the maps, in region 1 only the reappearance of the star is visible: the star is behind the Moon at moonrise. In region 2 the entire occultation is visible, although the event occurs in daylight in a part of that area. In region 3 only the star's disappearance may be seen, as the star is still behind the Moon at moonset. Compare this with the regions of visibility we described in Chapter 11 for solar eclipses. In the case of the occultation of 2007 January 7, there is no region 1; this is comparable to what we called solar eclipses of type VI in Chapter 11. At the occultation of 2007 February 3, region 1 is very small (near Spitsbergen), and on 2008 May 12 it's region 3 which is very small (over Antarctica).

Pollux, Delta Sagittarii, and the period of 183 years

As was mentioned at the end of the preceding chapter, stars with a celestial latitude larger than 6°36′ cannot be occulted by the Moon. The latitude of Pollux (β Geminorum) is +6°41′. So this star is just outside this limit and hence cannot be occulted for observers on the Earth's surface. However, due to the star's proper motion and the rotation of the ecliptic, the latitude of Pollux is presently increasing at the rate of +26″ per century. Near the year −400, the latitude of the star was +6°30′, so in B.C. years occultations of Pollux were indeed possible.

We calculated all occultations of Pollux taking place between the years −600 and +2100. Fifteen cases were found, which are listed in Table 20.B. So, the very last occultation of β Gem took place on −116 September 30. It was a very short, almost grazing occultation, visible from a very small part of the southeastern Pacific Ocean, near longitude 90° W, latitude 55° S. See Figure 20.*e*.

TABLE 20.B

The occultations of Pollux since the year −600

−599 March 24	−469 September 6	−320 March 28
−506 March 16	−469 October 4	−302 September 18
−488 October 2	−450 February 27	−283 September 18
−487 March 16	−450 March 26	−264 March 11
−469 March 26	−432 September 16	−116 September 30

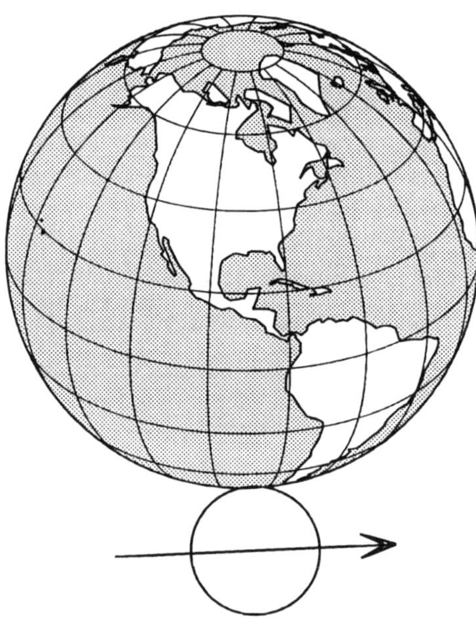

Fig. 20.e : *The last occultation of Pollux by the Moon. It took place on September 30 of the year −116. The drawing shows the Moon (the smaller circle) passing in front of the Earth as seen from Pollux.*

The 18.6-year period of revolution of the nodes of the lunar orbit is clearly visible from the dates mentioned in Table 20.B. For instance, there was an occultation of Pollux in the year −506, then two others in −488 and −487, followed by three in −469, two in −450, and one in −432.

Moreover, we see that the occultations of Pollux took place in 'groups'. For instance, one group of nine occultations lasted from −506 to −432. After a gap of more than one century, we have a smaller group of four occultations, from −320 to −264; and, finally, there is a 'group' of one isolated occultation in the year −116.

These groups occur at intervals of 183 years. The reason for this is as follows. One sidereal revolution of the lunar nodes has a duration of 6793.477 days, or 18.59953 Julian years. During this time interval, the motion of the Moon's perigee with respect to the stars is 756.55745 degrees, or 720° + 36.55745 degrees. For a star lying near the above-mentioned latitude limit of 6°36′, occultations are possible only when, at the time of its conjunction with such a star, the Moon is near its greatest celestial latitude *and* near its perigee, because in this last case the sum $\pi + s$, mentioned in the previous chapter, has its maximum value. If a conjunction of the Moon with, say, Pollux takes place with the Moon at its maximum northern latitude, and simultaneously with the Moon in perigee, this situation will repeat after

$$\frac{360 \times 18.59953}{36.55745} = 183 \text{ years}$$

Another interesting case is that of δ Sagittarii. This star is presently at celestial latitude $\beta = -6°28'$, and hence is just inside of the limit $6°36'$. It is, however, receding from the ecliptic, and hence its occultations by the Moon will be less and less frequent in future years, and finally come to an end.

Table 20.C lists all occultations of δ Sgr from A.D. 1900 to 3000. For this star, too, we find 'groups' at intervals of 183 years. One group begins on 1913 March 29 (this was the first occultation of δ Sgr since that of 1857 March 19); it will end with the occultation of 2024 October 9. This group contains 18 events. After a gap of almost one century, there will be a group of 10 events, from 2117 to 2192. Finally, there will be a group of two grazing occultations, in 2322 and 2359. That of 2359 October 1 will be the last occultation of δ Sgr before many centuries!

TABLE 20.C

The occultations of δ Sagittarii from A.D. 1900 to 3000

1913 March 29	1969 April 8	2136 April 21
1931 September 19	1987 March 22	2136 October 2
1931 October 16	1987 April 19	2154 October 13
1932 March 28	1987 September 3	2155 March 26
1950 March 12	1987 September 30	2173 September 16
1950 April 8	2006 March 22	2173 October 13
1950 September 19	2024 September 12	2192 March 8
1968 September 29	2024 October 9	2192 April 4
1968 October 26	2117 October 2	2322 April 9
1969 March 12	2136 March 25	2359 October 1

We leave it as an exercise to the reader to explain why the occultations of Pollux listed in Table 20.B took place only during the months February – March and September – October, and why those of δ Sagittarii (Table 20.C) take place only in March – April and September – October.

21. Series of occultations of Saturn

The original article on this subject was published in the journal Heelal *of October 1980, pages 232–234.*

As we have seen in chapters 19 and 20, the occultations of a given star by the Moon occur in series. Planets too can be occulted by the Moon, and when it concerns a planet which moves *slowly* on the starry background, its occultations too take place in series. So, then, these are the planets Jupiter, Saturn, Uranus, and Neptune. For instance, a series of occultations of Saturn began on 1979 September 20; it ended on 1980 August 13 and included a total of 13 events.

On the contrary, there are no occultation *series* for the fast moving planets Mercury, Venus, and Mars. So, for instance, there was an isolated occultation of Mercury on 1995 June 26, and an isolated occultation of Mars on 1995 August 30. Indeed, these planets move so rapidly with respect to the nodes of the lunar orbit, that in most cases an occultation cannot be repeated at the next conjunction with the Moon, about 28 days later.

And now we propose the following problem: find the series of occultations of Saturn during the next twenty or thirty years. No screams of despair, please! Even when you have never calculated an occultation, it's rather easy to obtain *approximate* results. We shall do the following simplifications:

— we will neglect the eccentricity of the orbit of Saturn. In other words, we will consider a 'fictive' Saturn describing its orbit around the Sun at a constant speed;

— because Saturn is much more distant from the Sun than the Earth is, we shall work with positions of that planet as seen from the Sun, not from the Earth. In other words, we will neglect the difference between the heliocentric and the geocentric positions of Saturn. (This difference amounts to 6 degrees at most).

These simplifications are allowed, because what we want are *approximate* results, not the exact dates of beginning and end of the several occultation series.

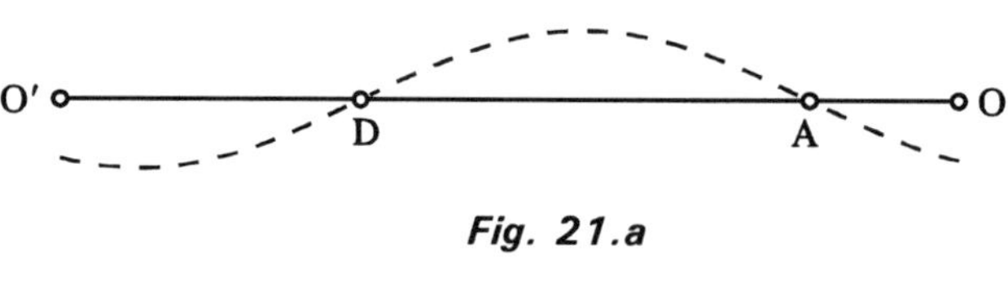

Fig. 21.a

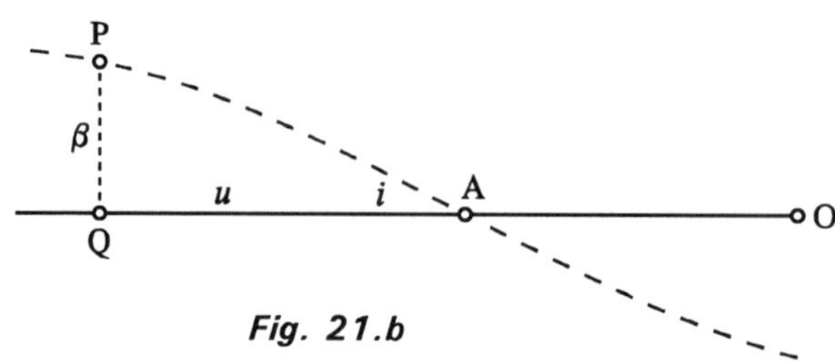

Fig. 21.b

Now, suppose we cut the ecliptic at the vernal equinox, and then spread it out on a plane surface. We so obtain Figure 21.*a*, which is to be compared with Figure 19.*a*. The straight line *OO'* is the ecliptic, with at extreme right the vernal point O, which we find back at left in O'. So the distance from O to O' corresponds to 360 degrees. (Think of a Mercator projection of the Earth's surface, in which the equator runs centrally from the left to the right). In Figure 21.a, the dashed curve represents the orbit of the planet projected on the sky, A being the ascending node, and D the descending node. Of course, the distance AD amounts to exactly 180 degrees, and the distance from O to A is the longitude Ω of the ascending node of the planet's orbit. In the drawing, the dashed line is almost a sine curve, because the orbital inclination of the planet is small.

The vicinity of the node A is shown on a larger scale in Figure 21.b. P is the planet's position at a given instant, while Q is its projection on the ecliptic. Angle $i = PAQ$ is the orbital inclination of the planet. The planet's (mean) longitude is equal to $OA + AP$. Because the inclination i is small, the arc AP is almost equal to the arc $AQ = u$. So, with a sufficient accuracy for our purpose we may write that the longitude of the planet is equal to

$$\lambda = OA + AQ = \Omega + u$$

whence $u = \lambda - \Omega$. The latitude $\beta = PQ$ of the planet is given by the approximate formula $\beta = i \sin u$. (The correct formula is $\tan \beta = \tan i \sin u$).

At the beginning of the year 2000, the mean longitude of Saturn is equal to 50.08 degrees, and in the course of one Julian year (365.25 days) this longitude increases by 12.2351 degrees. So, for a given instant Y, expressed as years and decimals, for instance 1996.3 (which means 0.3 year after the beginning of A.D. 1996), the mean longitude λ of Saturn is equal to

$$\lambda = 50.08 + 12.2351 \, (Y - 2000) \text{ degrees} \tag{1}$$

The (heliocentric) latitude of Saturn is then given by

$$\beta = 2°489 \sin (\lambda - 113°67)$$

where $i = 2°489$ and $\Omega = 113°67$ are the values near the year 2000. We may consider i and Ω as constants (again a simplification!), because it's not our purpose to do calculations for epochs far in the past or into the future. [Otherwise, the values for i and Ω should be changed accordingly]. Combining the two previous expressions, we obtain

$$\beta = 2°489 \sin (296°41 + 12°2351 \, (Y - 2000)) \tag{2}$$

Until now, we have talked only about Saturn. Let us now consider the Moon. For this body the reasoning given above is also valid, although it is just a little more complicated :

— the inclination of the lunar orbit on the ecliptic oscillates between the extremes $5°00'$ and $5°18'$. Here we shall adopt the mean value $i' = 5°09' = 5°15$;

— the nodes of the lunar orbit are not fixed in space. They are regressing, making a complete revolution in 18.6 years. In other words, the longitude Ω' of the ascending node of the lunar orbit decreases by 19.3414 degrees per Julian year. This should be taken into account in our calculation, otherwise meaningless results would be obtained

At the beginning of the year 2000, the longitude Ω' of the ascending node of the Moon's orbit is 125.04 degrees. So, for the year Y we have, in degrees

$$\Omega' \;=\; 125.04 \;-\; 19.3414 \,(Y - 2000) \tag{3}$$

Actually, this formula yields the longitude of the *mean* ascending node of the lunar orbit. The true node oscillates around the mean node (see Chapter 1), but this effect will be ignored here.

When the Moon is in conjunction with Saturn, its longitude λ' is equal to the longitude λ of the planet. So we have for the Moon, at the times of its conjunctions with Saturn, $u' \;=\; \lambda' - \Omega' \;=\; \lambda - \Omega'$. This gives, using (1) and (3),

$$u' \;=\; 285°\!.04 \;+\; 31°\!.5765 \,(Y - 2000)$$

Then the latitude of the Moon, at the instant of its conjunction with Saturn, is given by the formula $\beta' \;=\; i' \sin u'$, or

$$\beta' \;=\; 5°\!.15 \, \sin\,(285°\!.04 + 31°\!.5765\,(Y - 2000)) \tag{4}$$

So the difference of the celestial latitudes of the two bodies, at the time of conjunction in longitude, is $\Delta\beta \;=\; \beta' - \beta$, and it's *this difference* of the latitudes we need. If $\Delta\beta$ is precisely zero, then a geocentric observer (a fictive observer at the center of the Earth) sees the center of the Moon passing exactly over Saturn. But Earth and Moon are not dots, and an occultation somewhere on the Earth's surface is possible as long as the absolute value of $\Delta\beta$ is less than the sum of the horizontal parallax and the semidiameter of the Moon (see Chapter 19), the mean value of this sum being $1°\!.21$. This yields the following condition for the occurrence of an occultation of Saturn somewhere on the surface of the Earth :

$$\left|\, \beta' - \beta \,\right| \;<\; 1°\!.21$$

where β' and β are obtained by expressions (4) and (2), respectively.

As an example, let us make the calculation for the middle of the year 1981. Then $Y = 1981.5$, and $Y - 2000 \;=\; -18.5$. Then formula (4) gives

$$285.04 \;-\; (31.5765 \times 18.5) \;=\; -299.125$$

Of course, -299.125 degrees is the same as $-299.125 + 360 \;=\; 60.875$ degrees. We further find

Year	$\Delta\beta$
1979.0	-3.18
1979.5	-1.98
1980.0	-0.75
1980.5	$+0.41$
1981.0	$+1.41$
1981.5	$+2.16$
1982.0	$+2.59$
1982.5	$+2.68$
1983.0	$+2.40$
1983.5	$+1.79$
1984.0	$+0.88$
1984.5	-0.26
1985.0	-1.52
1985.5	-2.82
1986.0	-4.05
1986.5	-5.10
1987.0	-5.88
1987.5	-6.33
1988.0	-6.40
1988.5	-6.06
1989.0	-5.34
1989.5	-4.26
1990.0	-2.89
1990.5	-1.32
1991.0	$+0.35$
1991.5	$+2.01$
1992.0	$+3.55$
1992.5	$+4.88$
1993.0	$+5.90$
1993.5	$+6.56$
1994.0	$+6.82$
1994.5	$+6.68$
1995.0	$+6.15$
1995.5	$+5.29$
1996.0	$+4.17$
1996.5	$+2.88$
1997.0	$+1.53$
1997.5	$+0.22$
1998.0	-0.95
1998.5	-1.89
1999.0	-2.52
1999.5	-2.82
2000.0	-2.74
2000.5	-2.32
2001.0	-1.59
2001.5	-0.62
2002.0	$+0.52$

$$\beta' = 5.15 \times \sin(60°75) = +4.50 \text{ degrees}$$

$$\beta = 2.489 \times \sin(70°06) = +2.34 \text{ degrees}$$

from which we find that the difference of the latitudes is

$$\Delta\beta = \beta' - \beta = +2.16 \text{ degrees}.$$

Because $|\Delta\beta|$ is larger than $1°21$, no occultation of Saturn by the Moon could take place near the middle of A.D. 1981. Because $\Delta\beta$ is *positive*, the Moon passed to the *north* of Saturn at the times of its conjunctions with the planet.

We performed similar calculations for the years 1979 to 2002, at intervals of half a year. Our results are given in the table at left. Remembering that $\Delta\beta$ should lie between -1.21 and $+1.21$ in order that an occultation of Saturn take place somewhere on the Earth's surface, we find by interpolation that a series lasted from about 1979.81 to 1980.89, that is, from October 1979 to November 1980. Actually, that series started on 1979 September 20 and ended on 1980 August 13. So not a bad result, considering the several simplifications we made.

We also see from the table that in 1987 and 1988 the Moon passed 6 degrees south of Saturn at each of its conjunctions with this planet. In 1993 and 1994 the Moon passed 6° or 7° north of Saturn, etc. As you see, all this can be found by means of a rather easy calculation!

One should bear in mind that the method described here does *not* give the dates themselves of the conjunctions, but merely the epochs during which occultations of the planet are possible since, just as for the method described in Chapter 19, we did not consider the position of the Moon itself on its orbit, but we used only the position of the Moon's *path* on the sky.

Table 21.A notes the first and last occultations of Saturn of each series from 1970 to 2030. These dates have been obtained by an accurate calculation, not by the approximate method described above. As for Table 20.A, column N gives the number of occultations in the series, and the last column gives the direction of displacement of the region of visibility on the Earth's surface in the course of the series; for instance, the series of 1979–1980 began in the southern hemisphere and ended in the northern. And, here again, some events are visible nowhere because they are hidden by the Sun' glare. This was the case for the Saturn occultation of 1983 November 4, which took place near New Moon. Such unobservable cases are included in the list only for completeness.

TABLE 21.A

The occultation series of Saturn, 1970 to 2030

First occultation of the series	Last occultation of the series	N	
1973 September 20	1974 June 20	11	N – S
1979 September 20	1980 August 13	13	S – N
1983 November 4	1984 October 25	14	N – S
1990 October 25	1991 March 12	6	S – N
1997 April 7	1998 March 29	14	N – S
2001 May 23	2002 May 14	14	S – N
2006 December 10	2007 October 7	12	N – S
2013 December 1	2014 October 25	13	S – N
2018 December 9	2019 November 29	14	N – S
2024 April 6	2025 February 1	12	S – N

Why were there only six occultations during the 1990–1991 series, while for the other series the number of events are 11 to 14? The reason is that Saturn was in conjunction with the Sun on 1991 January 18, near the middle of the series. When Saturn is in conjunction with the Sun, the planet has a rapid eastward motion relative to the stars, and it is still more rapidly moving eastward with respect to the regressing lunar nodes, thus shortening the period favorable for occultations.

On the contrary, for the other cases Saturn reaches *opposition* in the course of the series. The planet then has during several months a retrograde motion on the sky, lengthening its favorable position relative to the lunar nodes. So, for instance, Saturn will be in opposition with the Sun on 1997 October 10, 2001 December 3, 2007 February 10, 2014 May 10, etc.

As we have seen in Chapter 19, for stars with a celestial latitude $|\beta| <$ 3°56′ there are two occultation series during a period of 18.60 years, the sidereal revolution period of the lunar nodes. In one of the series, the gradual displacement of the region of visibility is S – N, and during the next series it is N – S.

Because the latitudes of the planets Jupiter to Neptune are smaller than 3°, these planets too undergo two series of occultation per 'cycle', one having a S – N displacement, the other N – S. However, because the planets have an eastward (direct) motion with respect to the stars while the lunar nodes are regressing, the mean length of the 'cycle' is smaller than for the stars, namely

$$7.24 \text{ years for Jupiter}$$
$$11.40 \text{ years for Saturn}$$
$$15.23 \text{ years for Uranus}$$
$$16.71 \text{ years for Neptune}$$

22. *Occultations of bright stars by the eclipsed Moon*

This is a revised version of a paper by G. P. Können and Jean Meeus, which was published in the Journal *of the British Astronomical Association, Vol. 85, No. 1, pages 17-24 (December 1974).*

If we neglect the Earth's flattening and the enlargement of the Earth's umbra by the atmosphere, the radius of this umbra at the distance of the Moon is given by $\sigma = \pi' - s + \pi$, where π', s and π are the horizontal parallax of the Moon, the semidiameter of the Sun, and the horizontal parallax of the Sun, respectively.

To find the greatest possible value for σ, we take the maximum value of π' (61'32"), the smallest value of s (15'44"), while π is ever 9" to the nearest second. This yields $\sigma_{max} = 2757" = 0°766$.

Consequently, as seen from the center of the Earth only stars whose ecliptical latitude β is between -0.766 and $+0.766$ degrees may be reached by the Earth's shadow, and thus may be occulted by the *eclipsed* limb of the Moon — either by the totally eclipsed Moon, or at the eclipsed limb of the partially eclipsed Moon. As seen from the polar regions, however, the limiting latitudes are

$$\pm \, (\pm \, \sigma_{max} - \pi'_{max}) \, ;$$

this is $+0°766 - 1°026 = -0°260$ and $-0°766 - 1°026 = -1°792$ for the northern polar regions, and $+0°260$ and $+1°792$ for the southern polar regions. Thus, as seen from the northern polar regions of the Earth, the umbra may reach stars up to celestial latitude $-1°792$, but stars north of the ecliptic, such as Regulus ($\beta = +0°46$) may then no longer be occulted at the eclipsed lunar limb.

In Table 22.A, some data for the eleven brightest stars lying less than $1°792$ from the ecliptic are given. The dates in the last column are those of the conjunction of the anti-sun (the point opposite to the Sun on the celestial sphere) with the star. These dates are actually varying with time, by reason of the precession of the equinoxes. Moreover, there are jumps of $^1/_4$ or $^1/_2$ day by reason of the bissextile (leap) years.

In Table 22.B, the celestial longitude, latitude and conjunction date of the three brightest stars of the list as well as for ε Cancri, the brightest star of the Praesepe cluster, are presented as a function of time. The longitudes and latitudes are referred to the equinox and ecliptic of the date.

In this chapter, we will discuss first the occultations of the first-magnitude star Regulus in detail for the period A.D. 0–2450. Secondly, the occultations of the Praesepe cluster, β Sco and α Lib will be given for 1600–2150, and the results for the remaining stars of Table 22.A will be mentioned briefly for 1900–2050. Finally, we give results for stars which can be occulted only on the bright limb of the partially eclipsed Moon.

TABLE 22.A

Star	Magnitude	Long. and Latit., 1950		Date, 1950
η Gem	var.	92.°74	−0.°89	Dec. 25.1
μ Gem	3.2	94.60	−0.83	Dec. 27.0
δ Gem	3.5	107.82	−0.18	Jan. 8.7
α Leo	1.3	149.13	+0.46	Feb. 18.4
ρ Leo	3.8	155.69	+0.15	Feb. 24.9
β Vir	3.8	176.45	+0.69	Mar. 17.6
α Lib	2.9	224.39	+0.34	May 5.5
β Sco	2.9	242.49	+1.01	May 24.2
ξ Sgr	3.6	282.75	+1.67	July 5.4
π Sgr	3.0	285.55	+1.44	July 8.3
λ Aqr	3.8	340.88	−0.39	Sept. 4.0

TABLE 22.B

Star	Year	Long.	Latit.	Date of conjunction with anti-sun
α Leonis	1650	144.°97	+0.°45	Feb. 13.4
(*Regulus*)	1950	149.13	+0.46	Feb. 18.4
	2250	153.31	+0.47	Feb. 22.3
α Librae	1650	220.21	+0.38	Apr. 30.6
	1950	224.39	+0.34	May 5.5
	2250	228.57	+0.30	May 9.4
β Scorpii	1650	238.30	+1.05	May 19.3
	1950	242.49	+1.01	May 24.2
	2250	246.68	+0.98	May 28.1
ε Cancri	1650	122.51	+1.13	Jan. 22.3
	1950	126.70	+1.16	Jan. 27.2
	2250	130.89	+1.18	Jan. 31.1

Regulus

As this is the only first-magnitude star lying close to the ecliptic, its occultations by the eclipsed Moon have been investigated in detail for the period A.D. 0 to 2450.

Several types of occultations of a star by the eclipsed Moon can be considered, and we have chosen the following symbols — see also Figure 22.*a*.

T	For some places on the Earth's surface, immersion *and* emersion of the star by the totally eclipsed Moon;
t	For some places, immersion *or* emersion by the totally eclipsed Moon (and for *no* places immersion *and* emersion during the total phase);
t′	For some places, immersion *and* emersion at the *dark* limb during the partial phase preceding or following the totality (and nowhere *T* nor *t*, etc.);
(t)	The same, but at the *bright* limb;
(t′)	For some places, immersion *or* emersion at the *bright* limb during the partial phase preceding or following the totality;
P	For some places, immersion *and* emersion at the *dark* limb by the partially eclipsed Moon; the eclipse is a partial one;
(p)	The same, but at the *bright* limb;
(p′)	For some places, immersion *or* emersion at the *bright* limb by the partially eclipsed Moon; again, the eclipse is a partial one.

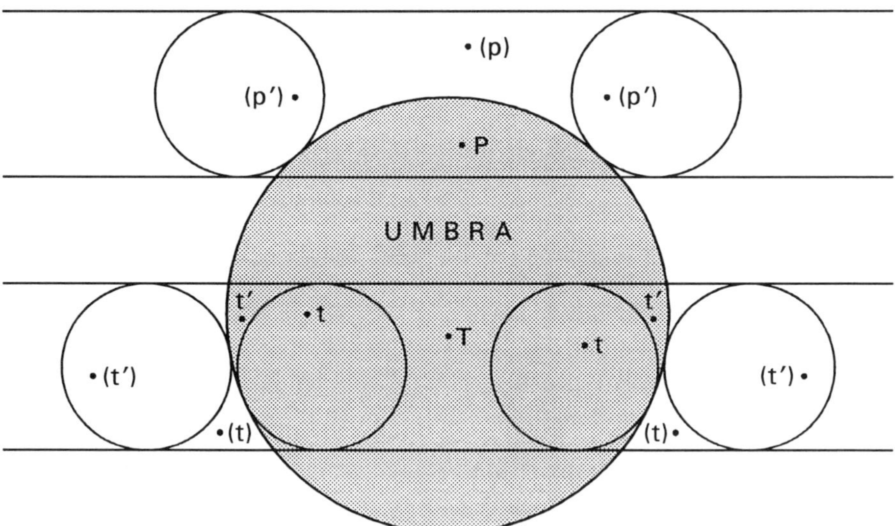

Fig. 22.a : *Path of the eclipsed Moon with respect to the occulted star*
for various types of occultations

TABLE 22.C

Occultations of Regulus by the eclipsed Moon, A.D. 0–2300

(p′)	64	Jan. 22	(t′)	864	Jan. 27	P	1710	Feb. 13
t	83	Jan. 22	t	929	Jan. 27	T	1729	Feb. 13
(p′)	102	Jan. 22	t′	948	Jan. 28	P	1775	Feb. 15
P	129	Jan. 23	T	994	Jan. 30	T	1794	Feb. 14
T	148	Jan. 23	T	1013	Jan. 29	P	1813	Feb. 15
P	194	Jan. 24	P	1059	Jan. 31	P	1840	Feb. 17
T	213	Jan. 24	T	1078	Jan. 30	T	1859	Feb. 17
T	232	Jan. 25	(p)	1124	Feb. 1	P	1878	Feb. 17
P	259	Jan. 26	T	1143	Feb. 1	(t′)	1924	Feb. 20
T	278	Jan. 26	P	1162	Feb. 1	(p)	1943	Feb. 20
P	297	Jan. 25						
(t′)	343	Jan. 27						
T	362	Jan. 26						
P	381	Jan. 26						
(t′)	408	Jan. 29						

In Table 22.C our results are given for Regulus for the years 0 to 2300. The event of 1859 February 17 was very favorable; Regulus was occulted by the totally eclipsed Moon as seen from the central and south Pacific Ocean (Samoa, Tahiti). The occultations of 1878 and 1924 were visible from the southern hemisphere.

Although the phenomena in the Earth's *penumbra* have not generally been taken into consideration in our investigation, the following events are worth mentioning. Regulus has been occulted during the first half of the penumbral eclipse of 1962 February 19 (in the United States). It was occulted (for some places) during the initial *penumbral* phase preceding the total lunar eclipse of 1989 February 20, and that will happen again on 2008 February 21.

Considering the results in Table 22.C, we find some interesting facts. The most remarkable one is that the occultations of Regulus by the eclipsed Moon occur in groups. So we have 15 events in the period 64–408, then no other phenomenon before A.D. 864, and so on. The event of 1943 February 20 was the last one of a group which began in 1710. The next two occultations of α Leonis by the totally eclipsed Moon will occur on 2445 February 22 (type *t*, and the first event of a new group) and on 2510 February 25 (type *T*).

Secondly, there are periodicities of 19 and 65 years, for instance 1775–1794–1813 and 1710–1775–1840. The period of 19 years is known as the Metonic cycle (235 lunations), which should not be confused with the Saros (223 lunations). A Full Moon will recur on the same date ($\pm$ 1 day), and thus near the same star, after 19 years because the lengths of the two periods, 19 sidereal years and 235 lunations, differ by only 0.18 day, having durations of 6939.87 and 6939.69 days, respectively. However, after 235 lunations the displacement of the Moon with respect to the node of its orbit is 7°.57. Consequently, there is not a lunar eclipse every 19 years. For instance, there was a small partial eclipse on 1970 February 21, preceding the total ones of 1989 February 20 and 2008 February 21; the eclipse of 2027 February 20 will be only penumbral, and on 2046 February 20 there will be no eclipse.

On the other hand, the period of 65 years corresponds to 804 lunations. After such a period, the displacement of the Moon with respect to its node is only 0.92 degree. Thus the period of 65 years is much more exact for predicting eclipses than the cycle of Meton. For instance, there were or will be lunar eclipses on the following dates, and all are total eclipses:

1664 February 11	1924 February 20
1729 February 13	1989 February 20
1794 February 14	2054 February 22
1859 February 17	2119 February 25

Unfortunately, the displacement with respect to the stars is more important, because 65 sidereal years = 23741.66 days, while 804 lunations = 23742.59 days. The difference is 0.93 day, corresponding to a displacement of 0.92 degree relative to the stars. Thus, while the 19-year period is fairly exact for conjunctions of the Full Moon with a given star, but not good for eclipses, it's just the opposite for the 65-year period.

Now, how can the existence of the groups separated by long 'empty' periods, which appear in Table 22.C, be explained?

In fact, in the table between successive cases some intervals of 27 and 46 years are found, for example 1813–1840 and 1878–1924. And we note that $46 = 65 - 19$, and $27 = 65 - (2 \times 19)$. These intervals are equal to 334 and 569 lunations, respectively. After the several periods, the displacements of the Moon with respect to the nodes of its orbit and with respect to a given star are as follows:

Years	*Lunations*	*Displacement of Moon with respect to*	
		node ($\Delta\Omega$)	*star* ($\Delta\lambda$)
19	235	$+7.57°$	$-0.18°$
27	334	-16.05	$+1.28$
46	569	-8.48	$+1.10$
65	804	-0.92	$+0.92$

There is a change of node (ascending ↔ descending) after a period of 27, 46, or 65 years

Now, suppose that a star is occulted by the eclipsed Moon. When will this occur again for that same star? We have to choose an integer number of lunations which is almost equal to an integer number of sidereal revolutions of the Moon (and, hence, of sidereal years) *and* to an integer number ($\pm$ 0.5) of draconic revolutions. In such a case, only one of the above-mentioned periods (19, 27, 46, or 65 years) can be considered. For instance, a period of 12 lunations (354.37 days) would not suit; it's true that in many cases a lunar eclipse is followed by another eclipse twelve lunations later (so, there was an eclipse on 1860 February 7, twelve lunations after that of 1859 February 17), but the displacement $\Delta\lambda$ relative to the stars would then be -10.7 degrees, which would prevent the repetition of the event.

After two periods of 19 years, the displacement relative to the node would be $\Delta\Omega = +15°.14$. Adding a new period of 19 years again would yield a too large displacement relative to the node, so instead we have to choose either 27 or 46 years to compensate with a negative value of $\Delta\Omega$. This is why we find in Table 22.C the combinations of $19 + 19 + 27$ and of $19 + 19 + 46$ years.

However, each time when one of the periods of 27, 46 or 65 years is involved, there is a displacement $\Delta\lambda$ of approximately $+1°$ of the eclipsed Moon with respect to the stars. This displacement cannot be compensated by the small displacement $\Delta\lambda = -0°.18$ of the 19-year period.

The consequence of this is that finally, after several periods of 19, 27, 46 and 65 years, the group of events inexorably comes to an end. The effect of the periodicities of 19, 27, 46 and 65 years on the distribution in time of the lunar eclipses is best illustrated by the 'panorama' of all umbral lunar eclipses from A.D. 1500 to 2500 shown on the next page (Figure 22.*b*). Every dot represents an eclipse. The years are read vertically. But contrarily to the panorama given earlier (Figure 16.*b*), the horizontal scale now indicates the day of the year, from day 1 (January 1) near the left border to day 365 or 366 (December 31) at right.

Clearly the dots are distributed in a number of slightly inclined 'columns'. For example, such a column starts near the label '300' at the upper border, and runs downward to the 'A'. In total, there are 35 columns in the panorama.

Now, if near a given epoch the date of the conjunction of a given star with the anti-sun happens to fall *between* two columns of the panorama, that star cannot then be occulted during a lunar eclipse. To give a numerical example, here are the dates of *all* umbral lunar eclipses taking place between February 5 and March 7, per periods of 100 years :

from 1800 to 1899 :	Feb. 6–8	Feb. 15–17	Feb. 26–28
from 1900 to 1999 :	Feb. 8–11	Feb. 19–21	Mar. 2–3
from 2000 to 2099 :	Feb. 11–13	Feb. 21–22	Mar. 3–5
from 2100 to 2199 :	Feb. 13–14	Feb. 24–26	Mar. 7

The periods are gradually shifting toward 'later' dates in the year, corresponding to the slight inclination of the 'columns' in the panorama. If we now compare these values with the dates mentioned in the last column of Table 22.B, we see that occultations of Regulus by the eclipsed Moon were possible in the 19th century, but that they are no longer possible between the years 2000 and 2199. After several more centuries, the preceding 'column' has arrived at the place of the first one, and occultations of the star by the eclipsed Moon are possible again.

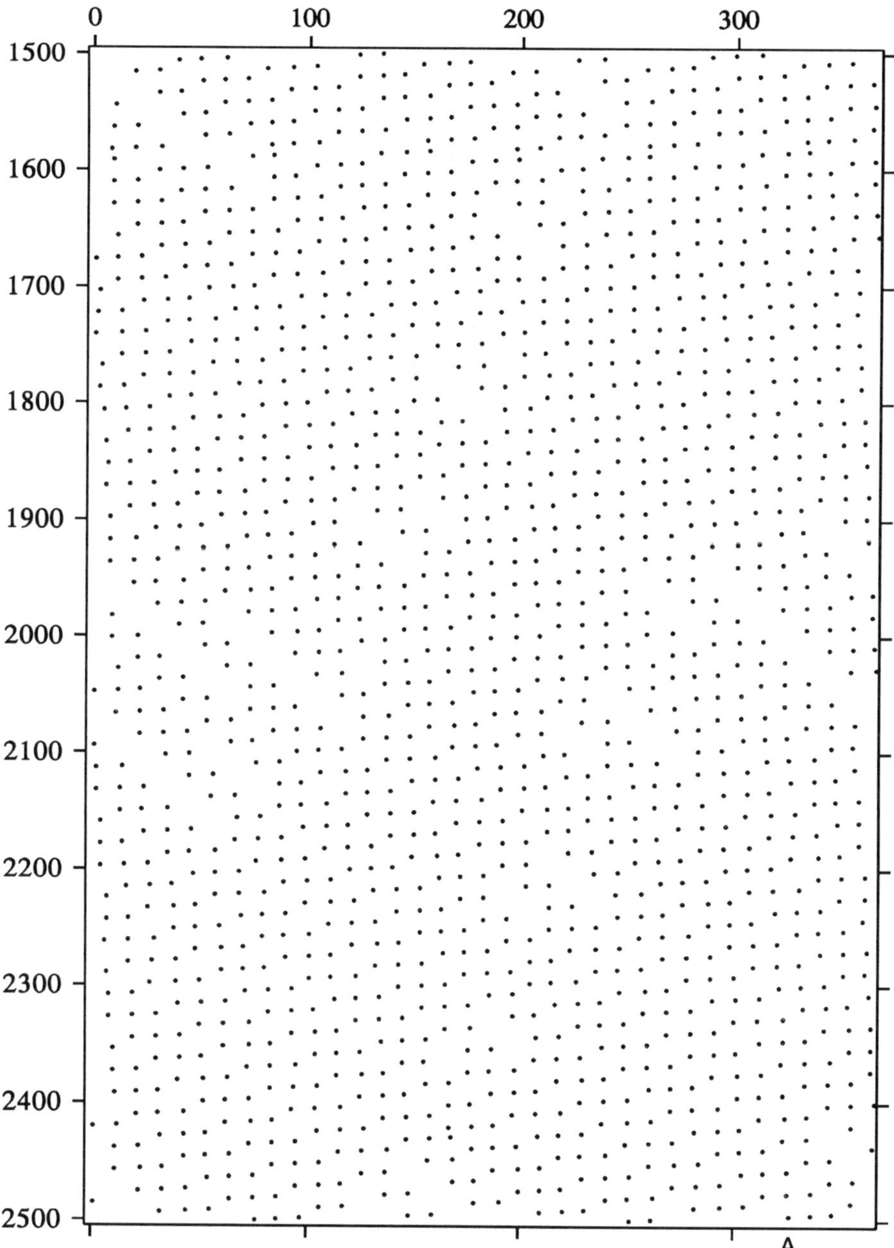

Fig. _22.b_ : _Panorama of the umbral lunar eclipses, from A.D. 1500 to 2500._
Vertically : the year.   _Horizontally : the day of the year._

The Praesepe cluster

Dr David W. Dunham, the American grazing occultation expert, wrote to one of the authors (J. M.):

> "Although the stars in the Praesepe cluster are fainter [than those of the Pleiades], the cluster is more compact, so that when conditions are favorable, passages of the Moon across it are quite spectacular. Epsilon Cancri, Z.C. 1299, is the brightest star, near the center of the cluster.
>
> "On January 30, 1972, the Moon occulted some of the stars of the Praesepe cluster as seen from Antarctica, only a few hours before a total lunar eclipse. It seems to me that, at least for the southern hemisphere, it would be possible for the totally eclipsed Moon to cross the southern part of the Praesepe cluster. It would be interesting to know when was the last time this occurred, and when it will next occur."

A photograph of the eclipsed Moon near the Praesepe cluster appeared in *Sky and Telescope*, May 1972, page 330. ε Cancri is a star of magnitude 6.3, and the cluster has a diameter of $1\,^1/_2$ degrees. As the star's latitude is greater than $0°.766$ (see Table 22.B), no occultation by the *totally* eclipsed Moon is visible from the Earth's center, nor *a fortiori* from the northern hemisphere. But the star's latitude is less than the limit of $1°.792$, so occultations of ε Cnc by the totally eclipsed Moon can actually be visible from a part of the southern hemisphere. Between the years 1600 and 2150 only the following cases occur:

> 1628 January 20: in a part of the southern hemisphere, occultation during the final *partial* phase of the total eclipse;
>
> 1647 January 20, 1674 January 22, 1777 January 23 and 1804 January 26: the star was occulted by the partially eclipsed Moon;
>
> 1823 January 26: for some regions of the southern hemisphere, occultation by the totally eclipsed Moon. This is the last case in the period 1600–2150.

As additional cases can be mentioned the following:

> 1953 January 29: ε Cancri was occulted in the southern hemisphere a few hours before the beginning of the eclipse;
>
> 1972 January 30: no occultation of ε Cnc; the geocentric conjunction took place at 6^h UT, three hours before the first contact with the umbra. A part of the Praesepe cluster was actually occulted;
>
> 2037 January 31: occultation of the star in the southern hemisphere, but once again a few hours before the first contact with the umbra.

Finally, during the *penumbral* lunar eclipse of 1991 January 30, the Moon was not far from ε Cnc, but nowhere on the Earth was the star occulted, the Moon being well south of the ecliptic. This will be the same situation too at the penumbral eclipse of 1999 January 31.

Occultations of other stars which can be occulted by the totally eclipsed Moon

β Scorpii. For some places this star was occulted during the second half of the partial eclipse of 1742 May 19, but at the limb which was in the penumbra. During the eclipses of 1826 May 21, 1845 May 21 and 1891 May 23, the star was occulted by the totally eclipsed Moon for some regions in the southern hemisphere.

During the small partial eclipse of 1872 May 22, β Scorpii was occulted. For some observers, the star disappeared *and* reappeared at the eclipsed limb; thus it was an occultation of type *P*. But for many other places these phenomena took place at the limb which was in the penumbra.

On 1975 May 25, occultation by the totally eclipsed Moon for places in the southern hemisphere. During the small partial eclipse of 1994 May 25, β Sco was occulted, for some observers at the eclipsed limb, for others at the illuminated one.

2021 May 26: for some observers, the star will be occulted during the beginning of this eclipse, but only emersions at the still non-eclipsed limb will be visible, hence type *(t')*.

On 2040 May 26, there will be an occultation in a part of the southern hemisphere shortly *before* the first contact with the umbra.

These are the only cases for β Sco in the period 1600–2150.

α Librae. There are 11 cases in the period 1600–2150, namely:

1836 May 1: occultation during the last $^1/_4$ hour of the eclipse, the disappearance taking place at the illuminated limb;

1855 May 2 and 1920 May 3: occultation during the final partial phase of these total eclipses;

1939 May 3: as for 1836;

1985 May 4, 2004 May 4, 2050 May 6 and 2069 May 6: occultation by the totally eclipsed Moon;

2088 May 5 and 2115 May 8: occultation by the partially eclipsed Moon;

2134 May 8: occultation by the totally eclipsed Moon.

On 1958 May 3, there was an occultation $1^1/_2$ hour *after* the last contact of the Moon with the umbra. On 1966 May 4, α Librae was occulted during a *penumbral* eclipse; this was visible from western Europe.

The other bright stars. For μ Geminorum, there was only one case between 1900 and 2050, namely on 1917 December 28, when for some places the emersion took place at the very beginning of the partial phase preceding the total eclipse, and at the bright limb.

δ Geminorum: For some regions on the Earth's surface, occultation by the *totally* eclipsed Moon on 1917 January 8, 1936 January 8, 1982 January 9 and 2001 January 9. Moreover, on 1955 January 8 the star was occulted during the penumbral eclipse.

ξ *Sagittarii* : Two cases in the period 1900–2050, namely : on 1936 July 4, occultation in the southern hemisphere, but for all places immersion and emersion of the star both took place at the illuminated limb; on 1963 July 6, in a part of the southern hemisphere emersion took place at the very beginning of the umbral eclipse, but at the illuminated limb.

π *Sagittarii* : No case between 1900 and 2050. On 1963 July 6 there was occultation in the southern hemisphere shortly *after* the last contact with the umbra.

λ *Aquarii* : Three cases between 1900 and 2050. On 1914 September 4, occultation in the northern hemisphere by the partially eclipsed Moon. On 1960 September 5 and on 1979 September 6, for some regions there was occultation during the initial partial phase of the total eclipse, but at the non-eclipsed limb.

Between A.D. 1900 and 2050, η *Geminorum*, ρ *Leonis* and β *Virginis* are *not* occulted by the eclipsed Moon.

Occultations of other stars by the partially eclipsed Moon

The greatest value of the Moon's apparent diameter is $0°.56$. Consequently, a star whose latitude is between $1°.79$ and $1°.79 + 0°.56 = 2°.35$ (north or south) may still be occulted by the partially eclipsed Moon, though not at the eclipsed limb. The situation is illustrated in Figure 22.c. Thus, for these stars only occultations of types *(p)* and *(p')* are possible. In Table 22.D, the data are given for the six brightest stars whose absolute latitude is between $1°.79$ and $2°.35$. In the last column, the date of conjunction with the anti-sun is again given.

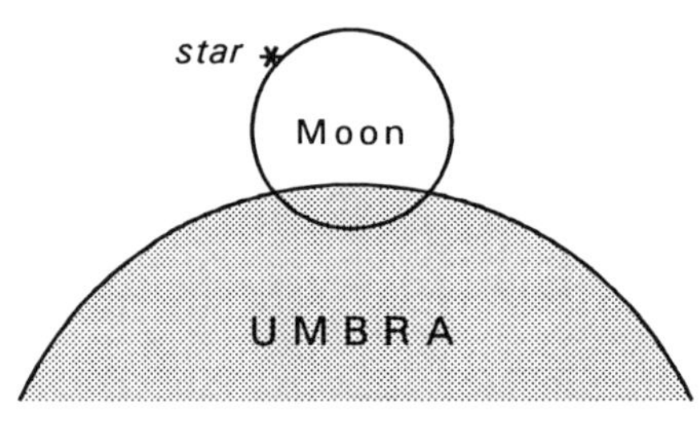

Fig. 22.c

TABLE 22.D

Star	Magnitude	Long. and Latit., 1950		Date, 1950
		°	°	
ζ Tau	3.0	84.08	−2.20	Dec. 16.6
ε Gem	3.2	99.24	+2.06	Dec. 31.5
α Vir	1.2	203.14	−2.05	Apr. 13.7
δ Sco	2.5	241.87	−1.98	May 23.6
θ Oph	3.4	260.70	−1.84	June 12.2
λ Sgr	2.9	275.62	−2.13	June 27.9

α *Virginis* (*Spica*) : There is *no* case between A.D. 1600 and 2100. On 1708 April 5, Spica was occulted about 6 hours after the end of the partial eclipse.

Spica was occulted in a part of the northern hemisphere during the *penumbral* eclipse of 1987 April 14. There was also an occultation of the star in a part of North America during the penumbral phase preceding the partial eclipse of 1995 April 15.

On 1949 April 13, the totally eclipsed Moon was in conjunction with Spica, but the star itself was not occulted. This was also the case on 1968 April 13; see the picture in *Sky and Telescope* of June 1968, page 351. It will occur again at the total eclipses of 2014 April 15 and 2033 April 14.

ζ *Tauri* and λ *Sagittarii* : No case between the years 1900 and 2050.

ε *Geminorum* : In the period 1900–2050, there is only one case, namely on 2009 December 31. During the small eclipse of that date (magnitude 0.08), the southern limb of the Moon will be eclipsed, while the northern limb will occult the star for observers in the southern hemisphere. Maximum eclipse will occur at 19^h23^m UT, about half an hour after the geocentric Moon-star conjunction in longitude.

On 1963 December 30 and on 1982 December 30, the totally eclipsed Moon was close to the star, but there was no occultation.

δ *Scorpii* : No case in the period A.D. 1900–2050. During the nearly total eclipse of 1956 May 24 the Moon was not far from the star, but there was no occultation.

θ *Ophiuchi* : No case in the period 1900 to 2050. A few hours before the partial eclipse of 1992 June 15, the star was occulted for a part of the northern hemisphere (Europe, northern Africa).

23. *Occultations of planets by the eclipsed Moon*

This is a part of a paper by Jean Meeus, J. van Maanen, and G. P. Können, which was published in the Journal *of the British Astronomical Association, Vol. 87, No. 2, pages 135–145 (February 1977).*

Due to the brightness of Mars, Jupiter and Saturn, occultations of these planets by the eclipsed Moon and also close conjunctions are spectacular phenomena, especially if the eclipse is total and Jupiter is involved. Therefore, a probability exists that these phenomena have been recorded ; such old observations may in principle be used for the study of the rotation of the Earth, fixing both the Dynamical Time and the hour angle at the place of observation. For this reason we traced these cases for the period −100 to +3000. For the faint planets Uranus and Neptune, however, we limited the study to the present time.

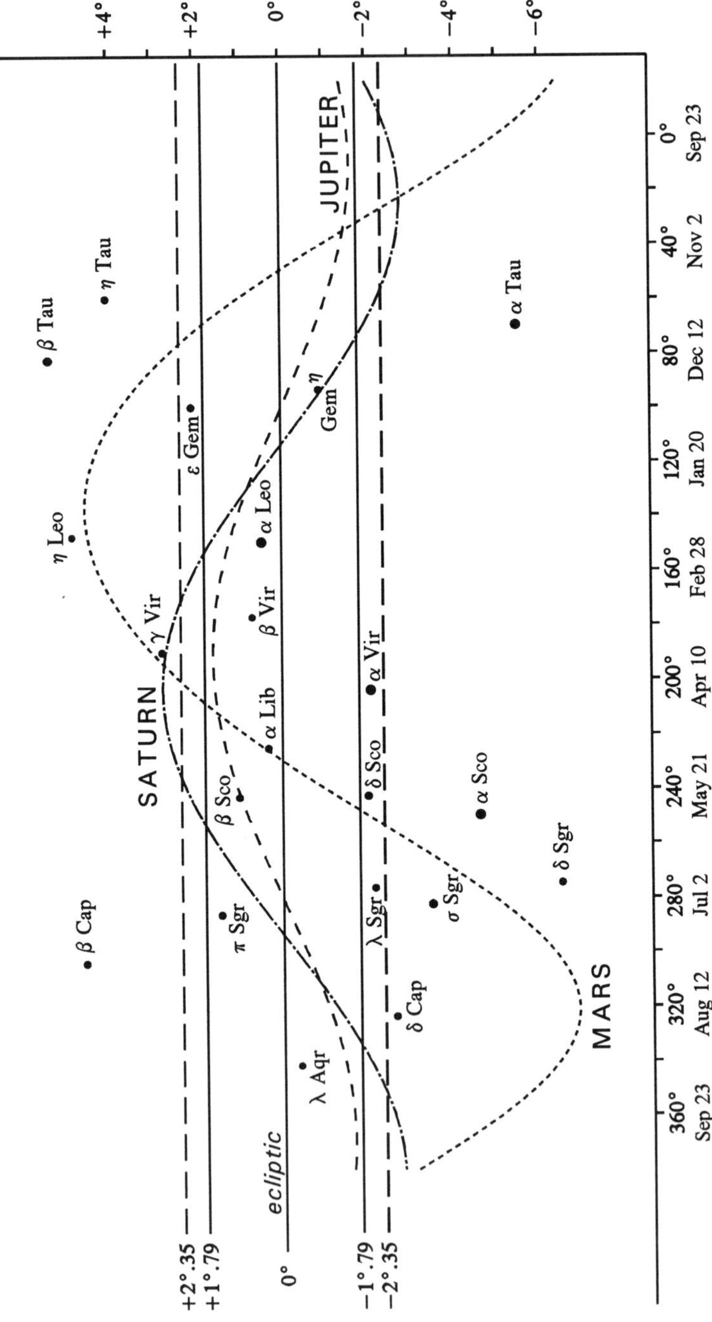

Figure 23.a : *Geocentric positions of the planets at opposition as a function of the opposition date and the longitude for the epoch 2000.0. The limits of 1°.79 and 2°.35, below which occultations by respectively the totally and the partially eclipsed Moon are possible, are also indicated. Some bright stars are added. For reason of clarity, the vertical scale (latitude) has been exaggerated. Horizontal scale : longitude and opposition date.*

Obviously, an occultation of a celestial object by the eclipsed Moon can take place only when the object is near its opposition with the Sun, while its absolute geocentric latitude has to be below a certain limit. Unlike the stars, however, a planet with a given orbital inclination has a variable latitude at opposition, depending on its position in orbit. In Figure 23.*a* the geocentric position of the bright planets and of some stars is given as a function of the opposition date and the longitude for the equinox and epoch 2000.0. Although these opposition dates and the longitudes are slowly varying with time, the position of the orbits with respect to the stars in the drawing remains almost unchanged during centuries.

In the drawing, the limiting latitudes of 1°.79 and 2°.35 have also been indicated, below which occultations by respectively the totally and the partially eclipsed Moon are possible (see Chapter 22). From the drawing it is clear that for Mars and Saturn these occultations are possible only in certain regions of the ecliptic, while Jupiter, thanks to its smaller orbital inclination, may be occulted by the eclipsed Moon at any longitude. A further consequence of the variable latitude of the planets near opposition is that this kind of occultation may be visible anywhere on Earth. This is not so for the stars, which have a fixed latitude.

In Table 23.A, my results are given for the three bright planets for the period 100 B.C. to A.D. 3000. The second column mentions the type of the event, for which I used the same symbols as in Chapter 22. The instants of maximum eclipse, in the third column, are in Dynamical Time.

The Jupiter case of A.D. 755 is of particular interest, since this occultation has been observed and recorded by Simeon of Durham in England (*). I have found that, for Durham, the immersion of Jupiter took place about 20 minutes after the end of the totality (†), while the emersion occurred shortly after the last contact of the Moon with the umbra. Using the value +50 minutes for the difference ΔT between Dynamical Time and Universal Time, the following results are obtained:

maximum eclipse	$18^h 43^m$ UT
end of totality	19 28
immersion of Jupiter	19 48
end of partial phase	20 37
emersion of Jupiter	20 39

Even for a value of ΔT as small as +35 minutes, the results are similar. Jupiter was occulted by the northern part of the eclipsed Moon.

The remarkable fact appearing from Table 23.A is the irregular distribution of the events involving Jupiter. While there were four cases during the 15th century, there is none between the years 1531 and 2932. The explanation for this is not easy, and can be found in the original article.

(*) Ashbrook, J., *Sky and Telescope*, Vol. 44, No. 2, page 85 (August 1972).
Newton, R. R., *Medieval Chronicles and the Rotation of the Earth*, Baltimore, 1972.

(†) In the original article, it was stated that immersion took place *during* totality. This now seems to be incorrect.

TABLE 23.A

*Occultations of bright planets by the eclipsed Moon
for the period −100 to +3000*

SH, NH, Eq stand for southern hemisphere, northern
hemisphere, and equatorial regions, respectively

Date	Type	Time of maximum eclipse (TD)	Magnitude of eclipse	Visibility	Notes
		h m			
M a r s					
2 Nov. 8	P	1 26	0.45	SH	
412 Nov. 4	T	21 55	1.60	SH	
916 Oct. 13	(p)	23 44	0.15	Eq	
2488 Apr. 26	T	9 35	1.38	SH	a
J u p i t e r					
103 Dec. 1	(p′)	15 14	0.21	Eq	
158 June 29	(p′)	12 00	0.63	NH	
400 Dec. 17	T	20 29	1.06	Eq	
458 Nov. 7	P	0 18	0.80	NH	
524 May 3	T	19 37	1.65	SH	
755 Nov. 23	T	19 33	1.40	NH	b
799 July 21	T	16 23	1.55	NH	
810 June 20	T	20 34	1.84	Eq	c
821 May 20	(t)	21 04	1.41	Eq	
879 Apr. 10	T	11 46	1.36	SH	
995 Jan. 19	(t′)	15 53	1.25	Eq	
1052 Dec. 8	(t′)	22 42	1.65	NH	
1176 Apr. 25	P	19 26	0.67	SH	
1234 Mar. 17	P	3 38	0.65	SH	
1407 Nov. 15	t′	12 49	1.19	NH	
1418 Oct. 14	T	22 09	1.12	NH	
1462 June 12	P	1 58	0.59	SH	
1473 May 12	P	7 28	0.37	Eq	
1531 Apr. 1	(p)	18 47	0.11	SH	
2932 June 10	P	0 01	0.20	Eq	
2990 May 1	P	1 40	0.09	SH	

(a) In Antarctica

(b) Observed in Europe

(c) Eclipse and occultation both nearly central

TABLE 23.A (Cont.)

Date	Type	Time of maximum eclipse (TD)	Magnitude of eclipse	Visibility	Notes
		h m			
Saturn					
195 July 10	t	3 53	1.70	Eq	
354 Dec. 16	(t)	15 59	1.34	Eq	
502 Dec. 29	T	16 11	1.65	Eq	
771 Feb. 4	(p)	10 49	0.93	SH	
959 June 23	P	8 37	0.94	SH	
1312 June 19	P	19 42	0.76	Eq	
1580 July 26	T	11 09	1.26	NH	
1591 Dec. 30	t	4 00	1.57	NH	
1796 Dec. 14	P	14 17	0.49	NH	
2344 July 26	T	12 41	1.33	NH	
2429 June 17	P	11 11	0.02	SH	
2829 Jan. 11	(t)	4 25	1.81	NH	
2977 Jan. 26	T	10 01	1.65	Eq	

For Uranus, I limited the study to the period 1850–2050. Four cases were found, namely 1930 October 7 (type *P*, Eq); 1938 November 7 (type *T*, NH); 2014 October 8 (type *T*, SH); and 2022 November 8 (type *T*, NH). It is remarkable that the 1938 case, though visible from the east coast of North America and the Atlantic, was mentioned neither in the *Nautical Almanac* nor in the *American Ephemeris* of that year.

For Neptune, I found the following cases between A.D. 1900 and 2050. On 1925 February 8-9, there was an event of type (p), NH. On 1999 July 28, in a part of the U.S.A. the emersion of Neptune occurs during the penumbral phase preceding the partial lunar eclipse of that date. And on 2008 August 16, in a part of Asia the planet will be occulted during the initial penumbral phase of the partial lunar eclipse of that date.

In addition to Table 23.A, for the present time the investigation has been extended to penumbral eclipses. In the period 1900–2050, only two occultations of a bright planet during a penumbral eclipse have been found, both being of Saturn. The first case (1933 August 5) is of minor importance since it concerned a small penumbral eclipse which was absolutely invisible without sophisticated instruments (magnitude in the penumbra 0.23). But in the second case (1944 December 29) the eclipse was a *total* penumbral eclipse with magnitude 1.02 (see Chapter 17). The magnitude in the umbra was -0.02, which indicates that the Moon did not touch the umbra, but passed very near to it. Therefore the eclipse could easily be seen with the naked eye; in fact, there is little difference in the appearance of such an eclipse and a small partial one of magnitude +0.02 like, for instance, the 2429 case in Table 23.A. The 1944 phenomenon was visible in the Pacific Ocean; at mid-eclipse the occultation could be seen near Samoa.

24. *Occultations of planets by the eclipsed Sun*

The original article on this subject, by Jean Meeus and Edwin Goffin, was published in The Journal of the Astronomical Society of Victoria *(Australia), Vol. 33, No. 1, pages 2-4 (February 1980). However, for this chapter all data have been recalculated from scratch by the first author.*

During the solar eclipse of 1954 June 30, which was total in North America, southern Greenland, Scandinavia and the Soviet Union, the planet Jupiter was hidden by the Sun. This was announced by H. O. Grönstrand (*Stockholms Observatoriums Annaler*, Band 16, No. 2, page 22), while we found by means of the data from the *American Ephemeris* for 1954 that Jupiter would be behind the Sun's disk from June 30 at $9^h 30^m$ UT until July 1 at $2^h 15^m$ UT. These instants are geocentric and refer to the center of the disk of Jupiter. Because the eclipse of 1954 June 30 lasted from $10^h 01^m$ to $15^h 03^m$ UT, Jupiter was indeed hidden by the Sun during the whole duration of the eclipse.

At that time, we had no idea about the mean frequency of phenomena of this type, although it was intuitively felt that they must be very rare. Indeed, to be occulted by the eclipsed Sun, a planet must be very close to its conjunction with the Sun, *and* be close enough to a node of its orbit, at the time of an eclipse of the Sun.

It may be noted here that the occultation of a bright planet by the Sun *outside* an eclipse is not an unduly rare event. For the five naked-eye planets, there are fourteen such events from 1991 to 2010, namely:

Date	Planet	Date	Planet
1991 Jan. 18	Saturn	2000 May 9	Mercury
1991 Nov. 8	Mars	2000 June 11	Venus
1992 June 13	Venus	2002 Nov. 14	Mercury
1993 May 16	Mercury	2007 May 3	Mercury
1994 Apr. 30	Mercury	2007 Dec. 23	Jupiter
1995 Dec. 18	Jupiter	2008 June 9	Venus
1998 May 12	Mars	2009 Nov. 5	Mercury

At the request of J.B. Trainor, of the Astronomical Society of Victoria, we have examined the problem carefully. Our aim was to find all occultations of naked-eye planets by the Sun during an eclipse over a period of thirty centuries. Finally we found fifteen cases for the period 0–3050, namely four involving Mercury, four with Venus, four with Mars, three with Jupiter, and *none* with Saturn. Our results are given in Table 24.A, of which the columns give the date of the solar eclipse, the planet which is occulted by the Sun, and finally the type of the eclipse.

TABLE 24.A

Occultations of planets by the eclipsed Sun, 0 to 3050

Date of eclipse	Planet	Type of eclipse
60 Apr. 19	Mercury	total *
420 Nov. 21	Venus	total
682 Nov. 5	Mars	annular
729 Oct. 27	Mars	total *
757 Apr. 23	Mercury	annular *
914 Nov. 20	Venus	total
1139 Oct. 24	Mars	partial
1302 June 26	Jupiter	partial *
1408 Apr. 26	Mercury	annular-total
1741 June 13	Venus	total
1954 June 30	Jupiter	total
2105 May 14	Mercury	partial *
2235 June 16	Venus	partial
2903 July 16	Jupiter	total
3008 May 27	Mars	total

* Indicates that the planet is behind the Sun
during only part of the duration of the eclipse

All four cases involving Mercury occur at the planet's ascending node. Moreover, the eclipses of the years 60 and 757 are separated by 697 years, as are those of 1408 and 2105.

For Venus, we find two intervals of 494 years (between 420 and 914, and between 1741 and 2235). A near-miss will occur at the partial eclipse of 2607 December 11.

The eclipse of 1139 October 24 was a small partial one, visible only in the north of North America. On 1302 June 26, Jupiter was occulted by the Sun during the second half of the eclipse, which was visible in the northern hemisphere.

For Saturn, there was a near-miss at the partial eclipse of 1106 December 27. Another near-miss will occur on 2550 January 19.

Of course, the occultation of a planet during a solar eclipse is *not* an observable event. We made these calculations for our own pleasure and to satisfy our curiosity — and that of the reader. It should be noted that only occultations by the *Sun* during a solar eclipse have been calculated. Of course, during an eclipse a planet could be occulted by the Moon without being occulted by the Sun, but such events have not been investigated.

Further, our investigation has shown that for the years under review no eclipse of the Sun occurs during a transit of Mercury or Venus over the Sun's disk. The total solar eclipse of 1769 June 4 occurred a few hours *after* the end of a transit of Venus.

Planets near the eclipsed Sun

Obviously, observers have more interest for the cases when a bright planet is seen *near* the eclipsed Sun, rather than being occulted by it. Table 24.B lists all these cases when one of the planets Mercury to Saturn is within two degrees of the center of the Sun's disk (that is, within $1^3/_4$ degrees from its limb) during a total or an annular eclipse, from A.D. 1900 to 2100.

For the eclipse type (second column), we used the abbreviations mentioned on page 43. For two instants (Universal Time) enclosing the total or the annular phase of the eclipse, we give the planet's position angle P with respect to the center of the solar disk, and its angular distance d to that center in minutes of arc. These values are geocentric and refer to the center of the planet's disk. The position angles are measured in the usual way, so for instance a value of $0°$ means that the planet is exactly north of the Sun's center (in the celestial equatorial system), $90°$ means to the east, and so on.

It appears from the table that during the total eclipse of 1965 May 30, the planet Jupiter was situated less than half a degree from the Sun's limb.

During the total eclipse of 2057 December 26, Venus will be only $1/_4$ degree from the Sun's limb. Unfortunately, the path of totality of that eclipse will run only over Antarctica.

During the annular eclipse of 2070 October 4, both Mercury and Venus will be within 2 degrees of the center of the solar disk.

On 1967 November 2, Mercury was near its inferior conjunction with the Sun. At all other cases involving Mercury and Venus, the planet is near its *superior* conjunction.

TABLE 24.B

Planets near the eclipsed Sun (total or annular eclipses), 1900 to 2100

Date	Eclipse type	Planet	UT	P	d
			h m	°	′
1908 Dec. 23	A-T	Mercury	10 10	196	95
			13 20	193	94
1911 Oct. 22	A	Mercury	2 20	326	88
			6 00	328	83
1923 Sep. 10	T	Venus	19 10	27	85
			22 20	28	85
1943 Aug. 1	A	Jupiter	3 00	308	76
			5 30	307	80
1954 Dec. 25	A	Mercury	5 40	184	95
			9 30	181	95
1957 Oct. 23	(T)	Mercury	4 30	340	59
			5 30	341	57
1965 May 30	T	Jupiter	19 40	205	41
			23 00	211	44
1966 Nov. 12	T	Venus	12 40	71	64
			16 10	73	65
1967 Nov. 2	(T)	Mercury	5 00	273	80
			6 00	275	86
1976 Apr. 29	A	Jupiter	8 30	211	89
			12 20	214	94
1979 Aug. 22	A	Venus	16 40	350	92
			18 00	351	92
2002 June 10 −11	A	Saturn	21 50	218	105
			1 40	221	111
2009 Jan. 26	A	Jupiter	6 00	240	98
			10 00	241	105
2010 Jan. 15	A	Venus	5 10	129	73
			9 00	128	75
2019 Dec. 26	A	Jupiter	3 30	84	77
			7 10	83	70

TABLE 24.B (Cont.)

Date	Eclipse type	Planet	UT	P	d
			$h \quad m$	°	′
2024 Oct. 2	A	Mercury	16 50 20 40	74 77	109 114
2030 June 1	A	Mars	4 40 8 10	273 273	106 108
2053 Mar. 20	A	Venus	5 20 9 00	122 121	100 102
2057 Dec. 26	T	Venus	0 30 2 00	215 213	31 31
2059 May 11	T	Mercury	17 40 21 00	250 250	97 87
2070 Oct. 4	A	Mercury	5 20 9 00	79 81	117 123
2070 Oct. 4	A	Venus	5 20 9 00	342 343	105 103
2077 May 22	T	Mars	1 10 4 20	265 265	92 94
2077 Nov. 15	A	Jupiter	15 20 19 00	65 61	75 69
2096 May 21 −22	T	Venus	23 50 1 50	87 87	112 114

25. *Occultations of bright stars by planets*

Occultations of bright stars by planets, though relatively rare phenomena, have been observed on several occasions. Perhaps the most famous cases were the Venus–Regulus event of 1959 July 7 which was observed in Europe in full daylight (the author saw it easily with a 6-inch refractor), the occultation of β Sco by Jupiter and its satellite Io in 1971, and the occultation of ε Gem by Mars on 1976 April 8.

During the period 1900–2100, there are only fourteen occultations of stars brighter than visual magnitude 3.5 by a planet which are visible somewhere on the Earth's surface. They are listed in Table 25.A. The visual magnitudes of the stars (fifth column) are those of the Revised Harvard Photometry. The last column mentions the elongation, that is, the angular distance from the Sun; E = east from the Sun = visible in the evening sky; W = west from the Sun = visible in the morning. This list has been calculated, independently, by G. P. Können and Jan van Maanen (*Journal of the British Astronomical Association*, Vol. 91, No. 2, p. 156; February 1981) and by Edwin Goffin (*L'Astronomie*, Vol. 95, p. 332; July-August 1981).

TABLE 25.A

Occultations of stars brighter than magnitude 3.5
by a planet, 1900 to 2100

Date	UT	Planet	Star	Visual magn.	Elonga-tion
1906 Dec. 9	18h	Venus	β Sco	2.9	15°W
1910 July 27	3	Venus	η Gem	3.2–4.0	31°W
1940 June 10	2	Mercury	ε Gem	3.2	20°E
1947 Oct. 25	2	Venus	α_2 Lib	2.9	14°E
1953 June 11	11	Mercury	ε Gem	3.2	19°E
1959 July 7	14	Venus	α Leo	1.3	45°E
1971 May 13	19	Jupiter	β Sco	2.9	170°W
1976 Apr. 8	1	Mars	ε Gem	3.2	81°E
1981 Nov. 17	16	Venus	σ Sgr	2.1	47°E
1984 Nov. 19	2	Venus	λ Sgr	2.9	39°E
2035 Feb. 17	15	Venus	π Sgr	3.0	42°W
2044 Oct. 1	22	Venus	α Leo	1.3	39°W
2052 Nov. 10	7	Mercury	α_2 Lib	2.9	3°W
2078 Oct. 3	22	Mars	θ Oph	3.4	71°E

Obviously, occultations of first-magnitude stars are the best events. For these stars, the three above-mentioned authors extended their calculations to the period A.D. 1000–3000, and the following occultations were found:

Regulus by Mercury: on 2253 August 1 and 2608 August 6;

Spica by Mercury: on 1080 September 2;

Regulus by Venus: 1128 September 11, 1959 July 7, 2044 October 1, and 2271 October 6, the latter case being a graze;

Spica by Venus: 1783 November 10 and 2197 September 2;

Antares by Venus: 1201 October 30 and 2400 November 17.

Moreover, Goffin found that on 1876 February 28 Jupiter occulted the star β Scorpii, and that Saturn occulted δ Geminorum on 1857 June 30. However, this latter event was not observable, as it occurred only 8 degrees from the Sun.

On 1990 July 29, Mercury passed 2′ north of Regulus, without occulting this star. Mercury will pass 3′ south of Regulus on 2004 September 10, and 3′ north of it on 2036 July 29.

Occultations of stars by Uranus and Neptune

Occultations of relatively bright stars by Uranus and Neptune are very rare events, because these two planets have a small angular diameter and because they are moving slowly on the starry background. Moreover, few relatively bright stars *can* be occulted by these planets: as seen from the Earth, Uranus and Neptune always remain in a small band situated near the ecliptic. For Uranus, this belt is only 0.08 degree wide; for Neptune, the width is 0.12 degree.

Of all stars brighter than visual magnitude 6.3, only the following *can* be occulted by Uranus and Neptune:

by Uranus: 44 Psc, λ Vir, α_1 Lib, α_2 Lib, 28 Lib, 41 Lib, 11 Sgr, SAO 187080, and SAO 187468. (The latter two stars are in Sagittarius);

by Neptune: 114 Tau, ψ Leo, η Vir, 82 Vir, ν Sco, ψ Oph, o Sgr and φ Aqr.

Moreover, close conjunctions are possible between Uranus and the following stars: β Vir, λ Lib, ω_1 and ω_2 Sco, θ Cap, 44 Cap, μ Cap and 96 Aqr. For Neptune, close approaches are possible with o Psc, α Leo (Regulus), ν Cap and 20 Psc. However, actual occultations of these stars by the mentioned planets are not possible.

It should be noted that all this is the *present* situation. In the course of the centuries this picture slowly varies by reason of the slowly variable planetary orbits (inclinations, longitudes of the nodes) *and* by the proper motions of the stars. To consider an extreme case: if we wait long enough (several millions of years?), the Pole Star will have arrived in the vicinity of the ecliptic, so even occultations of this star by one of the planets will be possible!

Uranus had triple conjunctions with β Vir in 1967–1968, with λ Vir in 1976, and with the couple α_1–α_2 Lib in 1977–1978, but every time there was no occultation. (A triple conjunction is a series of three conjunctions of a planet with a given star in the course of several months, due to the change of the planet's apparent motion: direct, then retrograde, then direct again).

In 1980–1981 there was a triple conjunction between Uranus and the star 41 Lib, but here too that did not result in an occultation. But one revolution of Uranus earlier, on 1897 September 8, 41 Librae *was* occulted. I don't know whether this event has been observed; the occultation was not visible in Europe.

Neptune had a *quintuple* conjunction with Regulus in 1927–1929, a triple conjunction with ν Sco in 1972, and a quintuple one with ψ Oph in 1973–1974. At none of these conjunctions were the stars occulted.

Goffin made a systematic search for the period A.D. 1800 to 2100. For the above-mentioned stars he found only *two* occultations, namely that of 41 Librae (magnitude 5.5) by Uranus on 1897 September 8, which I already mentioned, and that of ν Scorpii (magnitude 4.3) by Neptune on 1808 July 29 — that is, 38 years before that planet was discovered. Goffin also found the following very close conjunctions (without an occultation):

1974 September 24	Neptune 11″ south of ψ Ophiuchi,
1976 January 8	Uranus 8″ south of λ Virginis,
1981 September 12	Uranus 7″ north of 41 Librae,
2060 January 10	Uranus 9″ south of λ Virginis,
2065 September 14	Uranus 9″ north of 41 Librae.

For Uranus, note the periodicity of 84 years, the planet's revolution period.

Performing the calculation for fainter stars, up to magnitude 7.0, Goffin found only one more case during the period 1800–2100, namely the occultation of 81 Aquarii (magnitude 6.4) by Uranus on 1839 February 16. So, the three cases found by Goffin, for stars of magnitude 7.0 and brighter, all took place during the 19th century.

*   *   *
* *   * *   * *

And what about the occultations of a *planet* by another planet? The interested reader can consult the article by Steven C. Albers in *Sky and Telescope*, Vol. 57, No. 3, pages 220-222 (March 1979), with a correction on page 137 of the issue of August 1979. Albers found 21 planet-planet occultations from A.D. 1570 to 2223. But through bad luck we are now living in a period of drought. The last event took place on 1818 January 3, when Venus occulted Jupiter. The next mutual planetary occultation is due on 2065 November 22, when Venus again will pass in front of Jupiter.

PLANETARY MOTIONS

26. *The barycenter of the solar system*

The masses of the planets are small in comparison with the mass of the Sun. The 2×10^{30} kilograms of the solar body is 745 times as large as the sum of the masses of all planets. However, it is evident that the center of mass — the *barycenter* — of the total system "Sun + planets" does not coincid exactly with the center of the solar globe. Depending on the positions of the planets — and principally those with the greatest masses (Jupiter, Saturn, Uranus, Neptune) — on their orbits around the Sun, the barycenter of the solar system moves inside of the solar globe, and even can remain outside of this body during several years. The figures on the next two pages show this motion for the years 1940–2000 and 2000–2060, respectively.

Let us design by G the distance between the barycenter of the solar system and the center of the Sun, the radius R_0 of the solar globe being taken as unit of length (1 R_0 = 0.004 6524 astronomical unit = 695 990 kilometers). Table 26.A gives, for the period 1940–2061, the epochs when $G = 1$ (the barycenter is then moving through the surface of the Sun), and those when G is a maximum or a minimum.

TABLE 26.A

Epochs when G is a minimum, a maximum, or equal to 1

Epoch	G	Epoch	G	Epoch	G
1943 Nov.	1.759	1977 June	1.000	2013 Nov.	0.531
1948 Sep.	1.000	1983 Mar.	2.098	2016 July	1.000
1951 May	0.084	1988 Jan.	1.000	2022 Feb.	1.982
1953 Nov.	1.000	1990 Apr.	0.066	2027 Mar.	1.000
1958 Oct.	1.900	1992 Oct.	1.000	2030 Jan.	0.133
1964 Mar.	1.000	1997 Aug.	1.880	2033 Feb.	1.000
1966 Apr.	0.739	2003 July	1.000	2037 Oct.	1.667
1969 Oct.	1.000	2004 Oct.	0.924	2046 July	1.000
1970 July	1.018	2006 Sep.	1.000	2052 Feb.	0.638
1971 Apr.	1.000	2008 Sep.	1.078	2055 July	1.000
1975 Mar.	0.593	2010 Apr.	1.000	2061 Apr.	1.806

From the table and the drawings, it appears that the barycenter is more often outside of the Sun's globe than inside. During the years 1940–2060, G is larger than 1 during 62 percent of the time. On the other hand, it is possible for the barycenter to stay inside of the Sun during more than 12 successive years. For example, we will have $G < 1$ from December 2182 to June 2195.

In May 1951, the barycenter passed close to the center of the Sun; the distance G reached a minimum value of 0.084. In April 1990 the barycenter came even closer to the Sun's center, the least distance being 0.000 305 987 astronomical unit, or 45775 kilometers ($G = 0.066$) on April 23. Not before A.D. 2130 will the barycenter come still closer to the center of the solar globe than in 1990, with $G = 0.021$ on 2130 March 10. Another close approach will take place on 2169 April 5 ($G = 0.038$).

The greatest possible value for G is 2.26. This happens when all planets have the same heliocentric longitude (that is, when they are aligned on the same side of the Sun) *and* when they are at the *aphelion* of their orbits. In fact, the latter never can happen, because the aphelia of the planets have different longitudes. If we make use of the *mean* distances to the Sun, we find that the greatest possible value for G is 2.17, again when all planets are aligned, or nearly so.

This maximum value was almost reached in March 1983 ($G = 2.098$), when

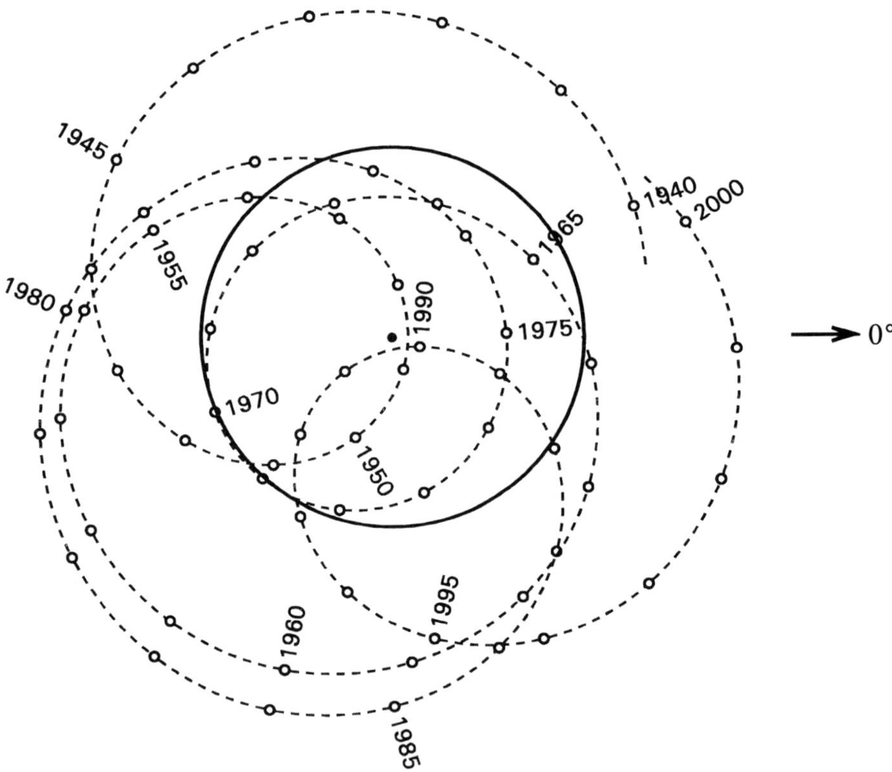

Fig. 26.a : *The displacement of the barycenter of the solar system with respect to the solar globe, from 1940 to 2000. In reality, it's the Sun which oscillates around the barycenter ! The positions are those for the beginning of the years indicated. The plane of the drawing is the plane of the ecliptic. The arrow gives the direction of the vernal equinox, longitude 0°. The solid circle represents the limb of the Sun. The central dot is the center of the Sun's globe.*

the heliocentric longitudes of the giant planets were: Jupiter 241°, Saturn 210°, Uranus 246°, and Neptune 267°. Not before A.D. 2161–2163 will G again be larger than 2, reaching a maximum value of 2.053 in February 2162. In April 1804, G reached a maximum value of 2.108. The great maxima of G (in the years 1958, 1983, 1997, etc.) occur at mean intervals of twenty years, near the times of the heliocentric conjunctions Jupiter–Saturn.

The contribution of each planet to the displacement of the barycenter away from the Sun's center is proportional to

planet's mass × distance from the Sun.

Table 26.B gives, for each planet:

— the reciprocal mass, or the ratio of the Sun's mass to that of the planet (including, for the latter, the masses of the atmosphere and satellites). For instance, the mass of Uranus is equal to 1/22869 times that of the Sun;

— the 'contribution' of the planet when it is in the aphelion of its orbit. Here, the contribution of Jupiter at its mean distance to the Sun is taken as unity.

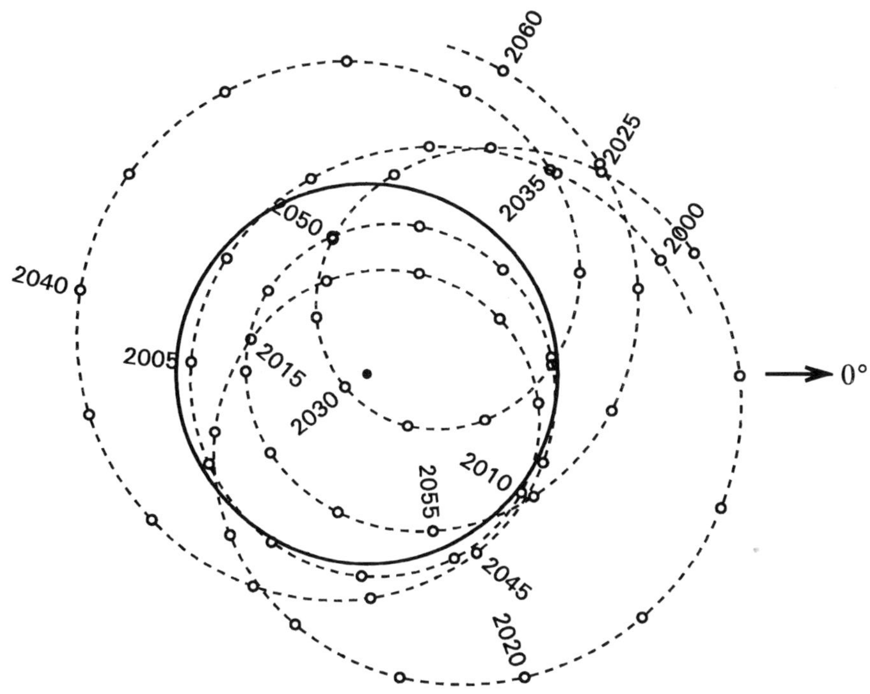

Fig. 26.b : *The displacement of the barycenter of the solar system with respect to the solar globe, from 2000 to 2060*

TABLE 26.B

Planet	$1/m$	Contribution in the aphelion
Mercury	6 023 600	0.00002
Venus	408 523.5	0.0004
Earth	328 900.5	0.0006
Mars	3 098 710	0.0001
Jupiter	1 047.355	1.0485
Saturn	3 498.5	0.5794
Uranus	22 869	0.1768
Neptune	19 314	0.3160
Pluto	130 000 000	0.0001

It appears that the contribution of Jupiter is almost equal to the sum of the contributions of the other three giant planets Saturn, Uranus and Neptune. Hence, when these three planets have nearly the same heliocentric longitude, and Jupiter is situated at the other side of the Sun, the whole system will be almost in 'equilibrium', and the barycenter of the solar system will be close to the center of the Sun. (The other planets do not play a significant role). On 1990 April 23, the heliocentric longitudes of the giant planets were : Jupiter 106°, Saturn 290°, Uranus 277°, and Neptune 283°.

The position of the barycenter of the solar system with respect to the center of the Sun can be calculated as follows. For a given instant, let l be the planet's heliocentric longitude, b its heliocentric latitude, and r its radius vector (the distance of the planet to the center of the Sun). The rectangular heliocentric ecliptical coordinates of the planet are then given by

$$x = r \cos b \cos l$$
$$y = r \cos b \sin l$$
$$z = r \sin b$$

Then the coordinates of the barycenter are

$$X = \frac{\Sigma mx}{M} \qquad Y = \frac{\Sigma my}{M} \qquad Z = \frac{\Sigma mz}{M}$$

where Σmx is the sum of the products mx of all the planets, etc.; m is the mass of the planet in units of the mass of the Sun; table 26.B gives the value of $1/m$ for each planet; M is the total mass of the Sun + the planets, or 1.001 342 111.

Then, for the given instant, the distance of the barycenter to the center of the Sun is given by $G^2 = X^2 + Y^2 + Z^2$. Because the distances r, and hence the coordinates x, y, z too, are expressed in astronomical units, G is obtained in the same units. Division by 0.004 6524 gives the distance in units of the Sun's radius.

27. *On the passages of Earth in perihelion*

The following text was first published in L'Astronomie, *Vol. 97, pages 294–297 (June 1983).*

It is well-known that the distance of the Earth from the Sun reaches a minimum in early January, and a maximum in the first days of July. But if we look at the instants when, in the course of the years, the distance between the centers of Earth and Sun reaches its least value, we observe that the time intervals between the successive passages of the Earth in perihelion are not constant — see Table 27.A.

TABLE 27.A

Instants of passage of Earth in perihelion, 1993 to 2000

Year	Date	UT	Difference
1993	January 4	3 h	
			363 days 3 hours
1994	January 2	6 h	
			367 days 5 hours
1995	January 4	11 h	
			364 days 20 hours
1996	January 4	7 h	
			363 days 16 hours
1997	January 1	23 h	
			367 days 22 hours
1998	January 4	21 h	
			363 days 16 hours
1999	January 3	13 h	
			364 days 16 hours
2000	January 3	5 h	

According to the equations of the Keplerian motion (Kepler's laws), the time intervals between the successive passages of the Earth in the perihelion of its orbit should be rigorously constant, because the Earth always needs the same time to describe its elliptic orbit. So what is the reason for the large differences we found?

The first possible cause which comes in mind is the perturbation of the Earth's motion due to the gravitational attraction by the other planets. However, the action of Venus on the heliocentric longitude of the Earth is never larger than 12″, that of Mars 5″, that of Jupiter 13″, and that of Saturn only 1″. The action of the other planets is much smaller still. From this it results that the perturbations of the Earth in longitude are never larger than 31″. The Earth travels over this distance in less than 15 minutes, because its speed is approximately one degree per day.

So, due to the planetary perturbations the Earth reaches a given point of its orbit (for instance, a given heliocentric longitude) less than 15 minutes earlier or later than 'on average'. So, it's not these perturbations which can explain the large variations noticed in Table 27.A.

TABLE 27.B

*Instants of passage of the Earth–Moon barycenter
in perihelion, 1993 to 2000*

Year	Date	UT	Difference
1993	January 3	3 h	
1994	January 3	8 h	365 days 5 hours
1995	January 3	16 h	365 days 8 hours
1996	January 3	23 h	365 days 7 hours
1997	January 3	5 h	365 days 6 hours
1998	January 3	14 h	365 days 9 hours
1999	January 3	22 h	365 days 8 hours
2000	January 3	24 h	365 days 2 hours

Let us now give the answer to the riddle: the large observed deviations are mainly due to the perturbing action of the *Moon*. If we calculate the times when the *center of mass* (the barycenter) of the Earth–Moon system is nearest to the Sun, then we find that the large differences almost vanish — see Table 27.B. This proves that the Moon is indeed the principal cause of the effect.

One might make the objection that the maximum distance between the center of the Earth and the barycenter of the Earth–Moon system is only 4942 kilometers. This distance is obtained as follows. The greatest possible distance between the centers of Earth and Moon is 406720 kilometers. The ratio of the mass of the Earth to that of the Moon is $m = 81.30$. The maximum distance mentioned above is obtained by dividing 406720 kilometers by $(m + 1)$.

But as the speed of the Earth in its orbit around the Sun is 30 kilometers per second, that distance of 4942 kilometers is covered by the Earth *in less than three minutes*. So, it is not *this* that explains the differences of several days which we found in Table 27.A.

In fact, while the Moon is indeed the cause of the observed deviations, its effect does not cause the Earth arriving much earlier or later than 'normally' at a given point of its orbit, for instance at its mean perihelion. The true reason is that the Moon *deforms* the path of the Earth's center with respect to a rigorously elliptical orbit. This is illustrated in Figure 27.*a*.

Due to the presence of the Moon, the Earth's center does not describe the path of the barycenter (dotted in the drawing), but it moves along the dashed line. Let us suppose that the barycenter G reaches perihelion at A. At that instant, in the case illustrated, the Earth's center T is still approaching the Sun, and it's only at B that the Earth will be nearest to the Sun. The distance AB is much larger than the distance TG, and corresponds to a time interval which can be larger than one day.

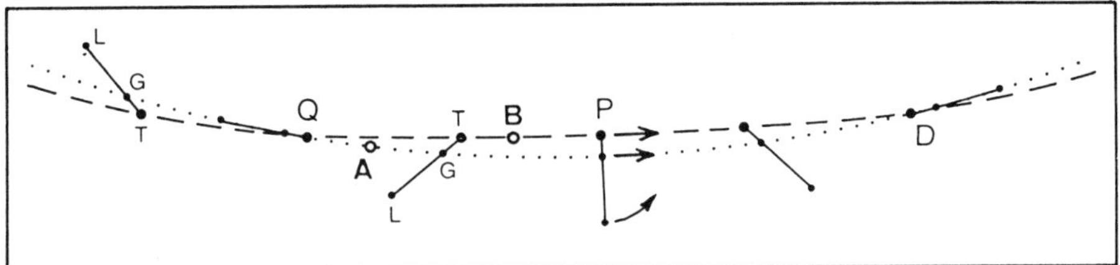

Fig. 27.a : *Paths of the Earth, the Moon, and their center of mass (barycenter). The Sun is upward, outside of the drawing. The barycenter G of the Earth–Moon system moves along the dotted line. Meanwhile, the Moon revolves around the Earth T. Here, T is actually the center of the Earth, and point G is always inside of the Earth's globe. In Q it is First Quarter, in P it is Full Moon, and in D we have Last Quarter. For clarity, the distances are not to scale. In fact, the distance QP is approximately 4000 times as large as the distance TG.*

Of course, it is also possible that point B precedes point A. If, at the time of passage of G in perihelion, the Moon is near First Quarter, then B occurs later than A (as in the drawing), and the Earth passes in perihelion *later* than the barycenter does. That was the case in 1993, when First Quarter took place on January 1. If the Moon is near Last Quarter, as in 1994 and in 1997, the passage of Earth in perihelion precedes that of the barycenter. Finally, if it's near New or Full Moon, then the difference of the times will be small.

During the period 1976–2000, the largest time difference is 32 hours. This occurred in 1990, when the Earth–Moon barycenter reached perihelion on January 3 at 9^h Universal Time, while the center of the Earth was nearest to the Sun on January 4 at 17^h.

During the period 1980–2020, the extreme times of passage of Earth in perihelion are January 1 at 22^h UT (in 1989), and January 5 at 8^h UT (in 2020). Of course, the jumps due to the leap (bissextile) days play a role too. Moreover, the perihelion of the Earth's orbit progresses by $12''$ per year with respect to the stars, while the vernal equinox regresses by $50''$ annually. Consequently, the tropical longitude (that is, measured from the moving equinox) of the perihelion increases by $+62$ arcseconds per year, with the consequence that the passages of the Earth in perihelion occur gradually at later dates in the course of the centuries. Near A.D. 1600, the extreme dates were December 26–28. If we express the instants in UT, the last time the Earth reached perihelion in December was in 1898 (on December 31, at 22^h). Near the year 2500, the extreme dates will be January 10–13.

The gravitational attractions of the *planets*, too, somewhat deform the path of the Earth–Moon barycenter itself. This explains the small remaining differences seen in the last column of Table 27.B.

In this chapter we considered only the passages of the Earth in the perihelion. What happens at the *aphelion* we leave as in exercise to the reader.

28. *Pe/ riheloids and apheloids*

Summary. — *An oddity in the motion of Neptune is described: the radius vector of this planet often reaches* two *minima near the perihelion, and* two *maxima near the aphelion.*

Neptune is a distant and slow planet. With an eccentricity of 0.009, its orbit around the Sun is almost circular. The period of revolution is 164 years. Simon *e.a.* [1] give the following values, referred to the mean equinox of the date, and for the epoch 2000 January 1 at 12^h Dynamical Time:

	Value	Variation per day
L = mean longitude of Neptune	$304° 20' 55''$	$+21''.67228$
π = longitude of perihelion	$48° 07' 13''$	$+ 0''.14058$

By definition, the difference $L - \pi$ is the planet's mean anomaly M. This is the angular distance from the perihelion to the 'mean Neptune'. For the epoch 2000 January 1 at 12^h TD we find $M = 256° 13' 42''$, the daily variation of M being $+21.53170$ arcseconds. According to the formulae for the elliptic motion, we have $M = 180°$ in the aphelion, when the planet is at its greatest distance from the Sun. At that instant, the 'mean' planet coincides with the true planet. From the numbers just given, we find that M was equal to $180°$ on 1965 February 8.

But if you think that the distance of Neptune to the Sun was indeed a maximum on 1965 February 8, you are wrong! You might object that the planets attract each other, and hence perturb each other's motion, and that for this reason Neptune passed through aphelion not exactly on 1965 February 8.

However, if you look in the astronomical almanacs of the years around 1965, you will find at your astonishment that, for instance in 1957, the distance of Neptune to the Sun increased, that it decreased in 1962, *but that it increased again in 1967*, to decrease — and now definitively — from 1969 on. Surely an uncommon behaviour for a planet! Was it an error in the almanacs, or did something special come into play?

Intermezzo

Let us consider the case of a triple star, consisting of a close double around which a third star is in revolution (Figure 28.*a*/1). The close binary consists of the components A and B which have almost identical masses and which revolve in nearly circular orbits around their common center of mass P. The third star, C, moves around the couple A–B in an elliptical orbit having a moderate eccentricity. The distance between A and B is supposed to be so small with respect to the distance to star C, that we can say with a very good approximation that C revolves around the *barycenter P* of the couple A–B.

Once per revolution, star C is nearest to P, in the perihelion of its elliptic orbit. Half a revolution later, C is in its apastron, farthest from P. Drawing 2 of Figure 28.*a* shows the variation of the distance C–P during a part of the revolution period of C. The points p and a correspond to the passages in periastron and in apastron, respectively.

But now suppose that we are interested in the distance of C, not to P, but for instance to component B of the close binary. Because B moves rapidly arount the fixed point P, the distance from C to B will show a variation of short period, represented by the undulating curve in drawing 3. In such a case, in the vicinity of the 'true' periastron, the distance CB may reach *two* minima, indicated by the small arrows. Between these two minima there is a *maximum* which, of course, has nothing to do with the apastron.

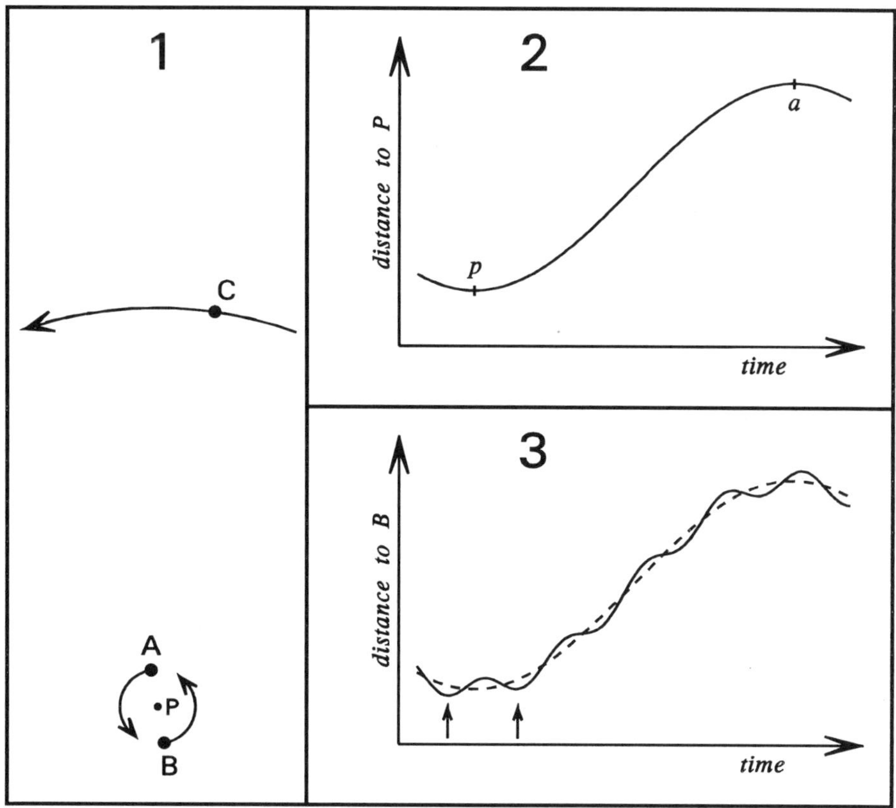

Fig 28.a

Similarly, near the apastron it is possible that the distance *CB* reaches two maxima instead of one, between which the distance reaches a least value which has nothing to do with the periastron.

This is rather evident, you will say...

Back to Neptune

Something similar occurs with Neptune. We can say that, in a first approximation, Neptune does not revolve around the Sun, but around the barycenter of the system Sun–Jupiter. Indeed, Jupiter is the planet with the largest mass in the solar system, and the distance Sun–Jupiter (5 astronomical units) is small in comparison to the distance Sun–Neptune (30 AU).

Of course, contrarily to the fictive example of Figure 28.*a*, where *A* and *B* had almost the same mass, the mass of the Sun is much greater than that of Jupiter. Nevertheless, under influence of this giant planet, the Sun is subject to an oscillating motion around the center of mass of the solar system. The other giant planets play a role too, but the biggest effect is due to Jupiter — see Chapter 26.

As a consequence, the distance of Neptune to the Sun will not vary regularly in the course of the years. Instead, just as in drawing 3 of Figure 28.*a*, it will exhibit periodic oscillations. Hence, these oscillations are not 'real perturbations'. They merely are the consequence of the fact that we consider the distance of Neptune to the *Sun*, not to the barycenter of the solar system. So, in fact it is a purely 'geometric' effect.

Of course, there *are* true perturbations in the motion of Neptune, due to the attraction of the other planets. But the geometric effect described above is predominant, and the distance from Neptune to the Sun will exhibit a variation with a period of about 13 years, the synodic revolution period of Jupiter with respect to Neptune. Hence, this 13-year variation is superponed on the 'big' curve which has a period of 164 years.

Figure 28.*b* shows the variation of the distance between the centers of Neptune and the Sun from A.D. 1954 to 1972. This distance reached a maximum on 1959 July 13 (arrow *1*), namely 30.33174 AU. Then the distance decreased until a minimum (*2*) of 30.32272 AU was reached on 1965 October 6. Then Neptune again moved away from the Sun, and on 1968 November 21 *a second maximum* was reached, 30.32407 AU (arrow *3*).

Of course, the minimum of 1965 *was not a perihelion*, as it took place in the vicinity of the planet's aphelion. For such a minimum that is not a perihelion, I coined the new term *periheloid* (= 'which resembles a perihelion') [2].

Half a revolution after the 1959-1965-1968 case, we have the situation illustrated in Figure 28.*c*. This will be almost a 'limiting case' : the principal perihelion (*1'*) will occur in 2042, while in 2049-2050 the distance of Neptune to the Sun will decrease only very slightly from the *apheloid* (*2'*) to the secondary perihelion (*3'*) — see also reference [3].

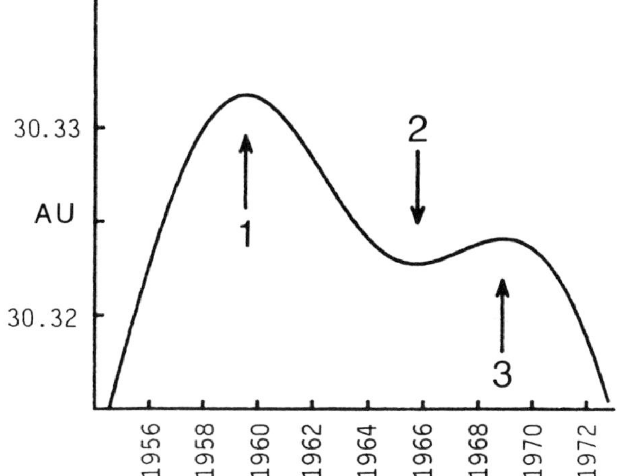

*Fig. **28.b** : The variation of the distance of Neptune to the Sun, 1954 to 1972.*

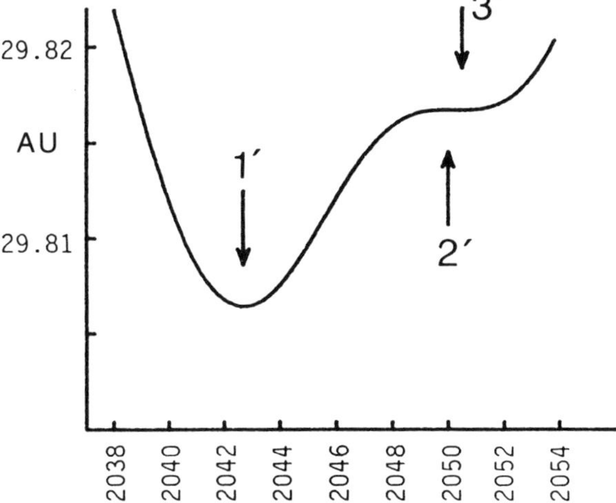

*Fig. **28.c** : The variation of the distance of Neptune to the Sun, 2038 to 2054.*

Table 28.A mentions all passages of Neptune through the perihelion and aphelion between A.D. 1500 and 2750. Beside the years, the succession of the extreme values of the distance of Neptune to the Sun is indicated. *P* means the main perihelion, *A* the main aphelion; *p* and *a* are the secondary perihelion and aphelion, respectively; *(p)* is a periheloid, and *(a)* an apheloid. For instance, in 1876 first the principal perihelion was reached; then came an apheloid in 1881, and finally in 1886 there was a secondary minimum of the distance Neptune–Sun.

TABLE 28.A

Perihelia and aphelia of Neptune,
1500 to 2750

1545	− 1548 −	1554	p	(a)	P
1628	− 1631 −	1637	a	(p)	A
1712	− 1714 −	1720	p	(a)	P
1794	− 1797 −	1803	a	(p)	A
1876	− 1881 −	1886	P	(a)	p
1959	− 1965 −	1968	A	(p)	a
2042	− 2049 −	2050	P	(a)	p
2125	− 2132 −	2134	A	(p)	a
2208	− 2215 −	2217	P	(a)	p
2291	− 2299 −	2300	A	(p)	a
	2374			P	
	2457			A	
	2540			P	
	2623			A	
	2705			P	

It is a remarkable fact that from A.D. 1500 to 2300 *all* the perihelia and aphelia of Neptune are double, while the five next cases are all *single*. This is probably due to the fact that one revolution period of Neptune is nearly equal to 14 (more exactly 13.89) revolutions of Jupiter, so that similar situations repeat for several revolutions of Neptune. Our successors of the middle of the third millennium will have to live without periheloids and without apheloids!

The peculiarity in the motion of Neptune I described in this chapter cannot occur for the planets Saturn and Uranus: these planets have a larger orbital speed than Neptune, and the eccentricity of their orbit is larger. Consequently, near the perihelion or near the aphelion the distance to the Sun varies much faster than is the case with Neptune, and there is simply no possibility for a double perihelion or a double aphelion to occur.

Using the osculating orbital elements

The peculiar motion of Neptune can also be described by means of the so-called *osculating elements* of its orbit.

"Osculating elements at a particular epoch are defined as the elements of an unperturbed elliptical orbit, referred to as the *osculating orbit*, in which the position and velocity of the planet at the epoch are identical with the actual position and velocity of the planet in its perturbed orbit at the same instant. The osculating elements therefore contain the effects of the perturbations due to the other planets, so that, unlike the mean elements, they are subject to periodic variations." [4]

Osculating (*not* oscillating!) elements are nothing else than the orbital elements of the 'instantaneous' orbit of the planet with respect to the Sun. Table 28.B mentions the extreme values of four of the osculating elements of Neptune during the period A.D. 1960–2010: the semimajor axis a in astronomical units, the eccentricity e, the longitude Ω of the ascending node, and the longitude π of the perihelion. These two longitudes are referred to the standard equinox of 2000.0.

TABLE 28.B

*Extreme values of osculating elements
of Neptune, 1960 to 2010*

a	min.	1964 Oct. 13	29.95096
	max.	1971 Apr. 8	30.22326
	min.	1977 Jan. 31	29.94848
	max.	1984 Apr. 1	30.28585
	min.	1990 Jan. 27	29.99977
	max.	1996 Aug. 6	30.27756
	min.	2003 Jan. 19	29.94310
	max.	2009 Aug. 15	30.23929
e	max.	1964 Oct. 8	0.012447
	min.	1970 Aug. 8	0.003925
	max.	1976 Dec. 20	0.012225
	min.	1982 Apr. 18	0.003909
	max.	1989 Apr. 5	0.010513
	min.	1993 Nov. 24	0.006182
	max.	2000 Dec. 23	0.011293
	min.	2006 Apr. 7	0.006751
	max.	2010 Oct. 15	0.011677
Ω	max.	1963 June	131° 51′ 23″
	min.	1969 Mar.	131° 43′ 57″
	max.	1975 Feb.	131° 52′ 05″
	min.	1981 Jan.	131° 44′ 08″
	max.	1986 Oct.	131° 49′ 21″
	min.	1992 Sep.	131° 45′ 09″
	max.	1998 Aug.	131° 47′ 48″
	min.	2010 Sep.	131° 46′ 01″
π	min.	1960 Oct. 2	22° 47′
	max.	1968 Jan. 29	58° 19′
	min.	1972 June 9	18° 38′
	max.	1979 Feb. 14	62° 17′
	min.	1984 Mar. 19	352° 32′
	max.	1991 Mar. 3	58° 01′
	min.	1996 June 16	1° 37′
	max.	2003 Sep. 1	69° 26′
	min.	2008 Oct. 30	14° 48′

The reader may be horrified by the large variations in the orbital eccentricity and especially in the longitude of the perihelion. The instantaneous orbital eccentricity varies between 0.004 and 0.012. From February 1979 to March 1984, the longitude of the perihelion decreased by *seventy degrees*. During the next seven years, it increased again by 65 degrees!

However, it is important to note that, as we have seen in Chapter 26, the Sun is oscillating around the barycenter of the solar system, and that the osculating elements are in fact referred to this moving Sun. If, instead, we referred the orbital elements to the barycenter, one would observe much smaller periodic variations. Moreover, the eccentricity of Neptune's orbit is so small, that a slight variation in this orbit results in a large variation in the orientation of its major axis.

It is interesting to look at what happened in 1965; λ is the true heliocentric longitude of Neptune, π the longitude of the perihelion of the osculating orbit, both referred to the equinox 2000.0, and M is the planet's mean anomaly, again in the osculating orbit. The difference $\lambda - \pi$ is the *true* anomaly. In the case of Neptune, the true anomaly differs little from the mean anomaly because the orbit is nearly circular.

	$\lambda =$	$\pi =$	$M =$
1965 Sep. 6.0	229° 49′ 38″	49° 33′ 36″	180° 16′ 12″
1965 Oct. 6.0	230° 00′ 12″	49° 58′ 56″	180° 01′ 03″
1965 Nov. 5.0	230° 10′ 46″	50° 28′ 07″	179° 42′ 00″

So, it appears that in 1965 *the motion of the instantaneous perihelion was larger than that of the planet itself.* As a consequence, the mean anomaly of Neptune was (temporarily) *decreasing* — a strange fact indeed for a 'normal' planet! The value $M = 180°\,00'\,00''$ was reached very near the time the planet passed at the periheloid, 1965 October 6. And so, paradoxically, in 1965 the distance of Neptune to the Sun was a *minimum* at the time the planet passed at the *aphelion* of its osculating orbit!

Let us now consider the values of the semimajor axis a and of the orbital eccentricity e during the second half of 1965:

Date, 0^h	a	e	$a\,(1 + e)$
1965 July 8.0	29.960910	0.01207784	30.32277
Aug. 7.0	29.963073	0.01200390	30.32275
Sept. 6.0	29.964985	0.01193877	30.32273
Oct. 6.0	29.966566	0.01188517	30.32272
Nov. 5.0	29.968348	0.01182522	30.32273
Dec. 5.0	29.970995	0.01173673	30.32276

We see that a was increasing, and that e was decreasing. But if we calculate the aphelion distance $a\,(1 + e)$, we find that this value passed through a minimum in early October 1965, when Neptune passed at the periheloid, and that this minimum value was precisely equal to the minimum value of Neptune's distance to the Sun.

The numerical values given in this chapter have been calculated by means of the planetary theory VSOP87, which was constructed in 1987 at the Bureau des Longitudes of Paris by P. Bretagnon and G. Francou. VSOP means 'Variations Séculaires des Orbites Planétaires'.

REFERENCES

1. J. L. Simon *e.a.*, 'Numerical expressions for precession formulae and mean elements for the Moon and the planets', *Astronomy & Astrophysics*, Vol. 282, pages 663–683 (1994).

2. J. Meeus, 'Le centre de gravité du système solaire et le mouvement de Neptune', *Ciel et Terre* (Belgium), Vol. 68, pages 288–292 (November-December 1952).

3. J. Meeus, *Astronomical Tables of the Sun, Moon and Planets*, 2nd ed., page 33 (Willmann-Bell, ed.; 1995).

4. *Explanatory Supplement to The Astronomical Ephemeris*, page 114 (London, 1961).

5. J. Meeus, 'Remarques sur le mouvement de Neptune', *L'Astronomie* (France), Vol. 79, pages 276–280 (July-August 1965).

29. A periodicity of 179 years ?

The American spacecraft Voyager 2, launched on 1977 August 20, arrived close to the planet Jupiter on 1979 July 9. Its path was chosen in such a manner that, after being deflected by the gravity of the giant planet, Voyager 2 was directed to Saturn. The craft arrived at this planet on 1981 August 25. Later, it had close encounters with Uranus (1986 January 24) and Neptune (1989 August 25) — see the drawing below.

This 'Grand Tour' to Neptune, with relatively little fuel on board, and using the gravitational assist of successively Jupiter, Saturn and Uranus, was possible thanks to the lucky relative positions of the four giant planets. It should be noted that a visit to all four planets, in a manner as shown in Figure 29.*a*, and using a *small* amount of fuel, is possible when the positions of the four planets (at the time of launch from Earth) are within well-defined limits; the planets don't need to be situated *precisely* as on the launch date, 1977 August 20. But of course, if *much* fuel were available, then the Grand Tour is possible with *any* initial positions of the planets.

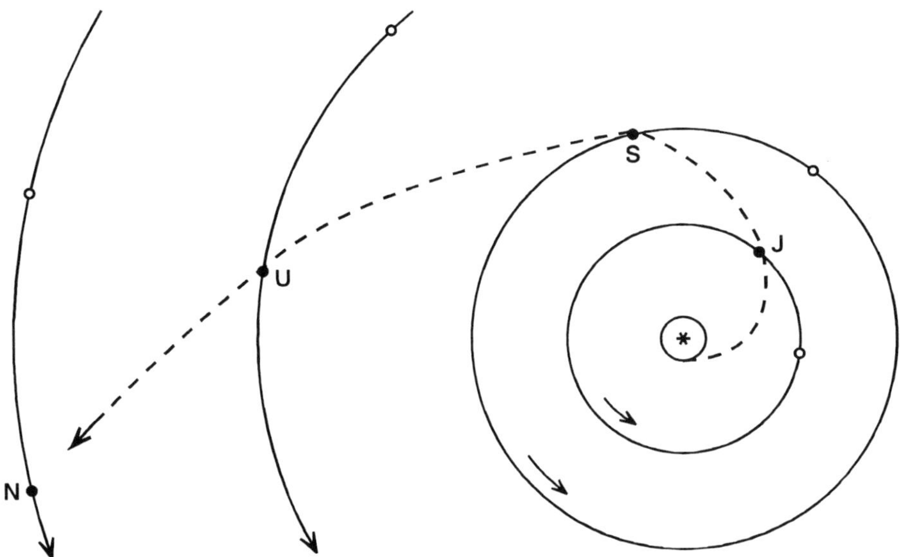

Fig. 29.a : *The path (dashed line) of the spacecraft Voyager 2 through the solar system. The central star represents the Sun. The smallest circle is the orbit of the Earth. J = Jupiter, S = Saturn, U = Uranus, N = Neptune. For each of these planets, the open circle is the position on 1977 August 20, when Voyager 2 left the Earth, while the black dots are the positions at the time of the flyby of Voyager 2. Consequently, the open circles are simultaneous positions, while the black dots are not.*

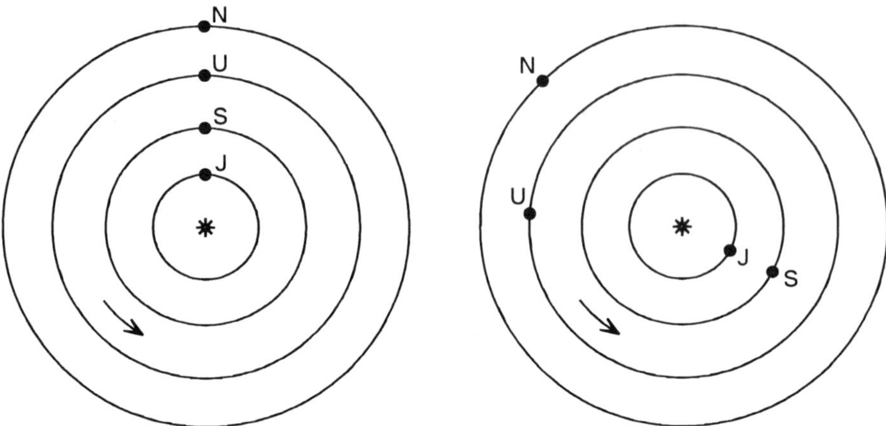

Fig. 29.b : *The orbits of the four giant planets Jupiter (J), Saturn (S), Uranus (U) and Neptune (N) around the Sun.* Left : *the 'start', when the four planets are supposed to be exactly aligned with the Sun.* Right: *the positions 7253 days later; Jupiter and Saturn are again in heliocentric conjunction, but now this takes place far from Uranus and Neptune.*

With reference to the favorable relative positions of the four giant planets in 1977, mention is sometimes made of a periodicity of 179 years after which these relative positions repeat. This gives the *wrong* impression that the 'favorable situation' will ever repeat, during centuries and centuries, every 179 years. However, as we will see, this 'periodicity' of 179 years should not be taken too literally. Certainly it is *not an exact periodicity*, and after a few periods of 179 years nothing is left from the original favorable situation!

The following table gives the sidereal revolution periods P and the mean daily motions n of the four giant planets. The value of n is obtained by dividing 360° by the revolution period.

Planet	*P (days)*	*n (degrees/day)*
Jupiter	4332.589	0.083 091 20
Saturn	10759.23	0.033 459 64
Uranus	30688.48	0.011 730 79
Neptune	60182.3	0.005 981 83

In one day, Jupiter moves over $0.083 091 20 - 0.033 459 64$, or $0.049 631 56$ degree, more than Saturn does. So, after $360 / 0.049 631 56 = 7253.45$ days (a little less than 20 years), Jupiter has again overtaken Saturn. This period can also be calculated as follows :

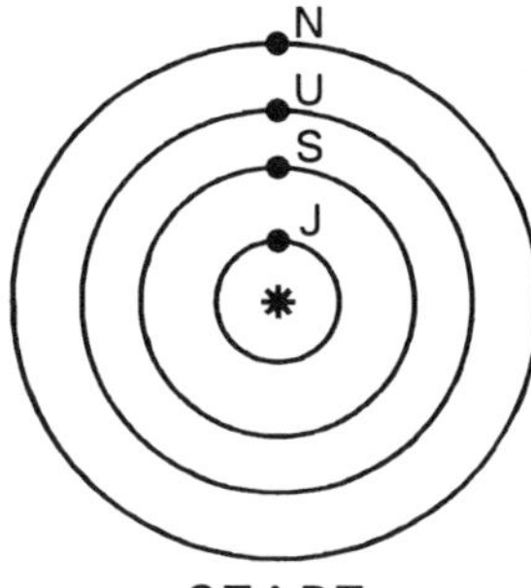

START

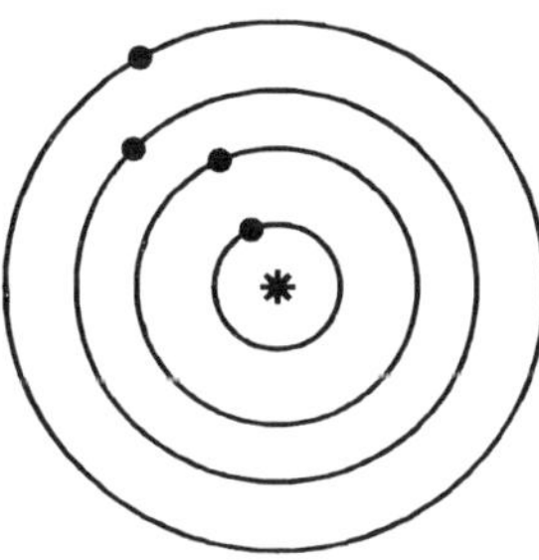

After 179 years

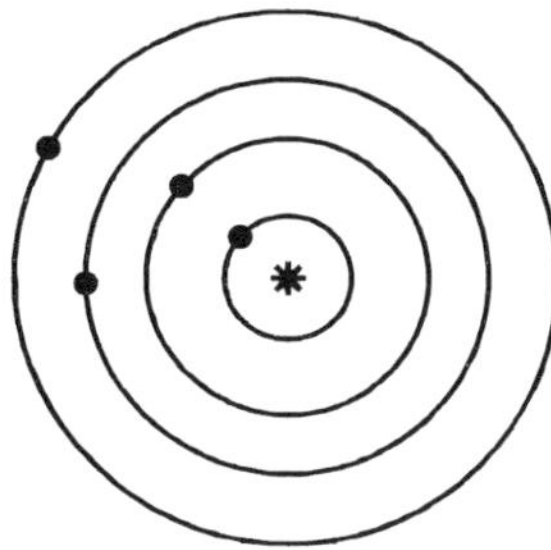

After 2 × 179 years

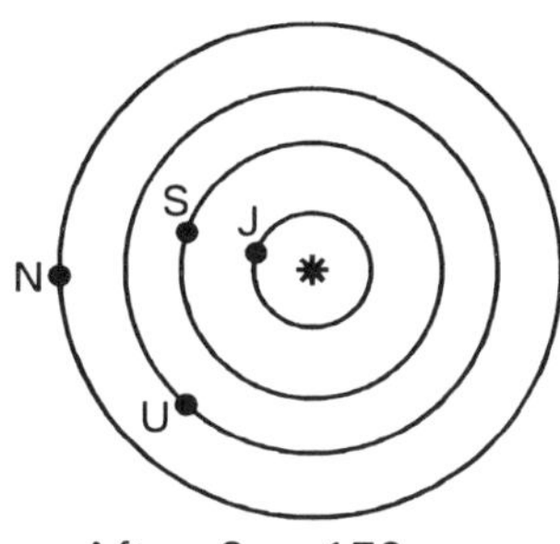

After 3 × 179 years

$$x = \frac{1}{4332.589} - \frac{1}{10759.23}$$

$$\frac{1}{x} = 7253.45 \text{ days}$$

So, after 7253 days, Jupiter is again in heliocentric conjunction with Saturn. (*Heliocentric* means 'as seen from the Sun'). However, if at the first Jupiter–Saturn conjunction Uranus and Neptune too were in that direction, then the second Jupiter–Saturn conjunction occurs far from them; see Figure 29.*b*. Here we should note that during a period of 7253 days, Jupiter and Saturn move over 603 and 243 degrees, respectively, in their orbits around the Sun; in other words: Jupiter makes one complete revolution + approximately 2/3 of a revolution, while Saturn performs only 2/3 of a revolution.

But let us now consider 9 times the period of 7253.45 days. This is 65281 days, or 178.73 years. During this period the planets move around the Sun over the following angles:

Jupiter	15 revolutions	+ 24.28 degrees
Saturn	6 revolutions	+ 24.28 degrees
Uranus	2 revolutions	+ 45.80 degrees
Neptune	1 revolution	+ 30.50 degrees

So, when Jupiter and Saturn are, for the 9th time since the 'start', again in heliocentric conjunction, Neptune is 30.50 − 24.28, or a little more than 6° farther. But for Uranus the deviation from the Jupiter–Saturn position is as large as 45.80 − 24.28, or 21.5 degrees; after another 179 years, the longitude difference will be 43 degrees. So, after a few times 179 years, nothing is left from a nice periodicity — see Figure 29.*c*.

Fig. 29.c : *The not-exact 'periodicity' of 179 years. For clarity, the orbits are not to scale. J = Jupiter, S = Saturn, U = Uranus, and N = Neptune.*

30. *Planetary quadrants and planetary sectors*

On 1982 March 10, the nine planets Mercury to Pluto were situated inside a heliocentric sector 95 degrees wide, the narrowest of the 20th century (Figure 30.*a*). How rare are such planetary sectors? When did a narrower one occur and when will the next take place?

For the entire time interval between the years 0 and 4000, I have investigated when planetary *quadrants* occur, that is, when all the planets lie within a heliocentric sector of 90 degrees. (*Heliocentric* means 'centered on the Sun'). I did not include Pluto for two reasons. First, several astronomers don't consider Pluto as being a 'true' planet, because his mass is so small and his orbit is rather highly inclined. Secondly, its mean orbital elements have not been calculated; thus, it is not easy to obtain accurate positions of Pluto for the remote past or future.

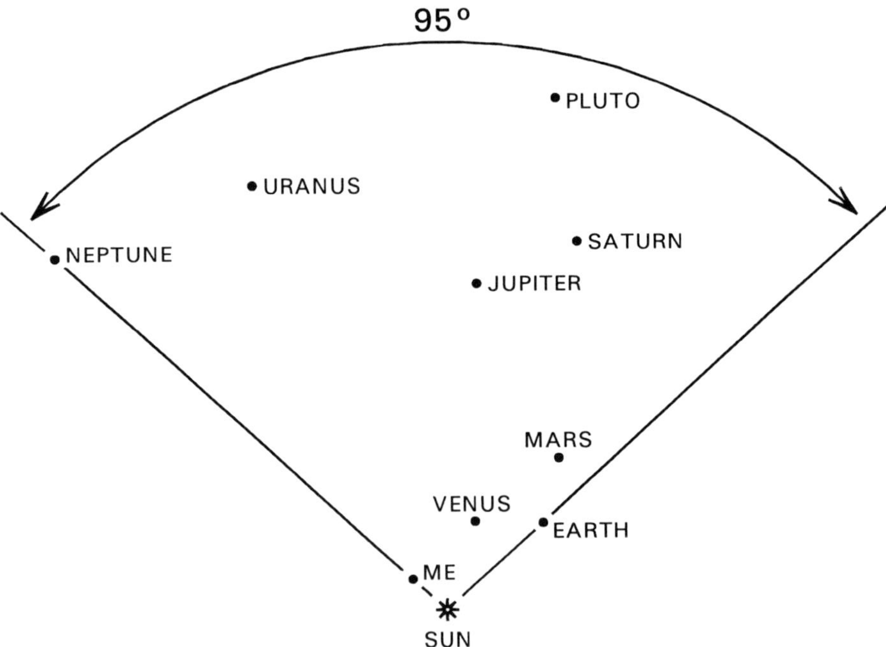

Fig. 30.a : *The smallest heliocentric planetary sector of the 20th century. It took place on 1982 March 10. In this drawing, the positions of the nine planets have been projected on the plane of the ecliptic. For clarity, the distances to the Sun are not to scale.*

TABLE 30.A

Planetary quadrants and minimum sectors, 0 to 4000

Year	Quadrant			Minimum sector			
	Start	*End*	*Dur. (days)*	*Date*	*Angle*	*Limiting planets*	
117	Nov. 26	Dec. 4	8	Dec. 1	82°	VE	ME-UR
310	Mar. 6	Mar. 13	7	Mar. 8	87	EA	VE-UR
410	Sep. 11	Sep. 24	13	Sep. 22	87	UR	VE-NE
449	Jan. 16	Feb. 13	28	Jan. 22	57	ME-JU	VE
626	Feb. 2	Feb. 14	12	Feb. 11	85	JU	MA-UR
628	Jan. 2	Feb. 5	34	Jan. 23	65	EA	ME-SA
768	Nov. 16	Nov. 26	10	Nov. 17	86	ME-UR	MA
949	Jan. 20	Feb. 4	16	Feb. 1	80	MA	ME-JU
987	May 28	July 7	40	June 28	66	MA	ME-UR
989	May 7	June 12	35	June 8	76	MA	ME-JU
1126	May 1	May 17	16	May 9	83	VE-JU	EA
1128	Mar. 30	May 10	41	Apr. 11	40	ME-UR	EA
1130	Mar. 16	Mar. 29	13	Mar. 18	84	ME-UR	JU
1166	Aug. 21	Sep. 13	23	Aug. 31	72	ME-NE	EA
1307	Mar. 18	May 9	51	Apr. 14	46	VE	ME-JU
1666	Sep. 14	Oct. 3	19	Sep. 19	85	VE-MA	EA
1817	June 4	June 22	17	June 9	83	JU	MA-SA
2161	Apr. 29	June 2	33	May 19	69	SA	VE-NE
2176	Nov. 4	Nov. 24	20	Nov. 7	78	ME-NE	SA
2492	Apr. 29	May 7	7	May 6	90	JU	ME-SA
2520	Oct. 4	Oct. 23	19	Oct. 22	90	SA	EA-UR
2851	Sep. 15	Oct. 1	16	Sep. 30	82	MA	ME-SA
2892	Jan. 1	Jan. 20	19	Jan. 6	82	EA-NE	UR
2992	July 8	Aug. 6	30	July 21	73	MA	VE-NE
3030	Sep. 16	Sep. 20	4	Sep. 19	85	VE	ME-UR
3030	Dec. 6	Dec. 8	2	Dec. 7	88	ME-JU	VE
3171	July 3	Aug. 10	38	July 31	50	EA	ME-NE
3209	Nov. 21	Dec. 11	20	Nov. 26	65	ME-JU	MA
3211	Oct. 27	Nov. 12	15	Nov. 3	85	EA	VE-MA
3350	Oct. 6	Oct. 26	19	Oct. 10	79	ME-UR	MA
3530	Dec. 3	Dec. 23	20	Dec. 14	83	MA	EA-JU
3569	Apr. 16	May 12	26	May 2	70	MA	EA-UR
3571	Mar. 22	Apr. 17	26	Apr. 5	72	MA-NE	UR
3671	Oct. 11	Oct. 29	18	Oct. 12	87	ME-UR	JU
3710	Feb. 27	Mar. 18	19	Mar. 2	72	ME-SA	EA
3712	Feb. 5	Feb. 13	7	Feb. 7	81	ME-NE	VE
3850	Oct. 12	Oct. 27	15	Oct. 20	78	VE-UR	JU
3889	Feb. 9	Mar. 7	26	Feb. 25	52	VE	ME-MA
3891	Jan. 23	Feb. 14	22	Feb. 9	74	EA	ME-JU

The results of the calculations are given in Table 30.A. A similar list, for the period 0–3000, had been published earlier (*Sky & Telescope*, Vol. 63, No. 1, page 6; January 1982). But now I have calculated the data again from scratch, the list is extended to A.D. 4000, and my new calculations are based on the VSOP87 planetary theory of Bretagnon and Francou (Bureau des Longitudes, Paris).

During the period 0–4000, there are 39 cases when the eight major planets come inside a 90° heliocentric sector. The table gives the beginning and ending dates and the duration of the quadrant. The date and the angle of the narrowest sector are also given. The dates are based on the Dynamical Time, which differs less than three hours from Universal Time during the period considered. Values in the fourth and sixth columns have been rounded to the nearest integer. In 2492 and 2520, the angle of the minimum sector will be between 89°30′ and 90°00′.

For example, the 18th line of the table should be read as follows: in the year 2161 the eight planets (Mercury to Neptune) will be inside a heliocentric sector of 90° during 33 days, namely from April 29 to June 2; on 2161 May 19 these planets will form the smallest sector, namely 69 degrees.

When a planetary sector is a minimum, it is limited at one side by *two* planets (at the moment one is overtaking the other), and at the other side by *one* planet. For each sector in the table, the last two columns contain the abbreviated names of the three 'limiting' planets: ME = Mercury, EA = Earth, etc. It should

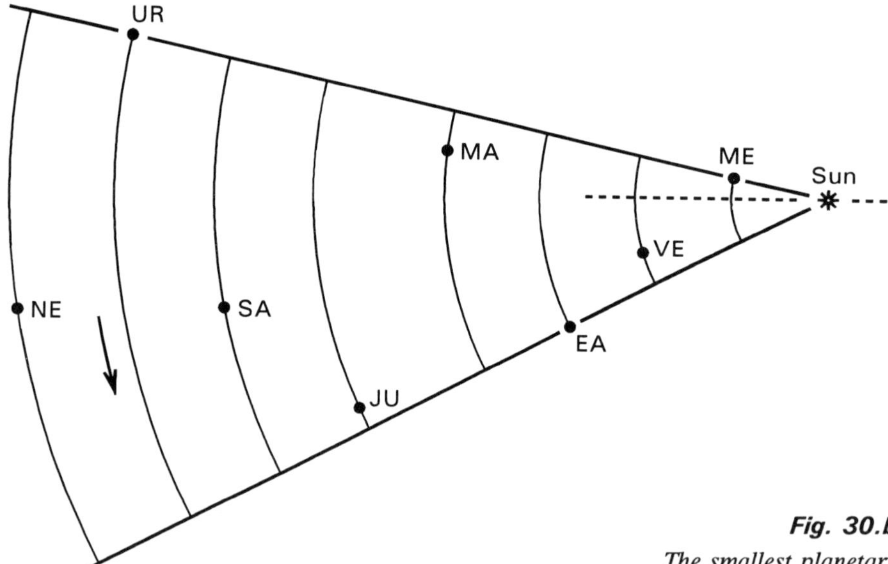

Fig. 30.b

*The smallest planetary
sector in the four millennia studied
by the author: that of 1128 April 11, when all planets (except Pluto)
were situated inside a heliocentric sector of 40 degrees. Abbreviations (the first two
letters of each planet's name) are the same as in the table. The orbits are not to scale.
The dashed line gives the direction (to the right) of the vernal equinox, longitude zero.*

be further noted that, at the time of the smallest sector, neither Mercury (the fastest planet) nor Neptune (the slowest) can be alone at one end of the sector.

We see from the table that sometimes there is no planetary quadrant during more than 300 years, for example between 1307 and 1666, and presently from 1817 until 2161. On the other hand, there may be more than one quadrant in the same century, such as the four events in 1126, 1128, 1130 and 1166.

As mentioned at the beginning of this chapter, the smallest heliocentric sector of the 20th century was 95°, on 1982 March 10; it included Pluto. The smallest sector of the 21st century will be 102° wide, on 2024 November 22; Pluto will *not* be situated in that sector.

The smallest sector in the whole period 0–4000 occurred on 1128 April 11, when the eight planets were within a heliocentric sector of only 40° (Figure 30.*b*). Another very small sector (46°) occurred on 1307 April 14; the corresponding planetary quadrant had a duration of 51 days. The longest possible is 55 days (with the present values of the eccentricities of the planetary orbits), and it takes place when all planets are aligned at the longitude of Mercury's aphelion.

In the table we find three times an interval of 100 years (310–410, 2892–2992, and 3571–3671), four times an interval of 39 years, and eight times an interval of 2 years. The latter interval is easily explained: the exterior planets Jupiter to Neptune move rather slowly; on the other hand, after two years the Earth and Mars come again near the same places in their orbits.

The giant planets

The gathering of the four giant planets (Jupiter, Saturn, Uranus, Neptune) into a heliocentric sector of 60° is not too rare. This situation occurs seven times between A.D. 1400 and 2200:

> October 1483 to October 1487
> August 1625 to April 1626
> May 1663 to July 1666
> August 1803 to February 1806
> July 1844 to January 1845
> December 1982 to November 1984
> May 2162 to July 2163

Between the years 0 and 4000, the smallest sector of these planets was 7°06', on 1306 September 4. And this was the *only* sector of these planets smaller than 10 degrees during these four millennia.

And during these 40 centuries, the *three* outermost and slowest giants (Saturn, Uranus, Neptune) come seven times into a heliocentric sector of 10°. The minimum sectors are: 4°49' on 625 October 6, 1°05' on 1307 June 17, 7°29' on 1989 July 18 (Figure 30.*c*), 7°31' on 2169 February 12, 3°08' on 2850 November 22, 9°20' on 3030 February 13, and only 0°58' on 3711 July 1.

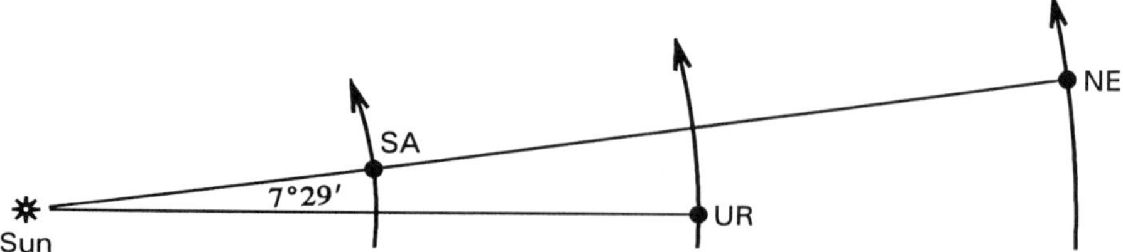

Fig. 30.c : *On 1989 July 18 the three outermost giant planets Saturn,*
Uranus and Neptune formed a minimum heliocentric sector of 7°29'.
But on 3711 July 1 the same planets will form a much smaller sector.

31. How often are the planets aligned ?

The following text was originally published in Heelal *(Belgium),*
Vol. 40, No. 10, pages 270-272 (October 1995).

Sometimes the following question is posed to an astronomer: suppose that
at some instant all nine planets (Mercury to Pluto) are exactly aligned with the Sun;
after what time interval will they again be situated on a straight line?

More precisely stated: if all planets have the same heliocentric longitude,
when will they again have the same heliocentric longitude (not necessarily the *same*
as at the first alignment, of course). In this prospect, the heliocentric *latitudes* of the
planets are not taken into account; only their longitudes are considered here.

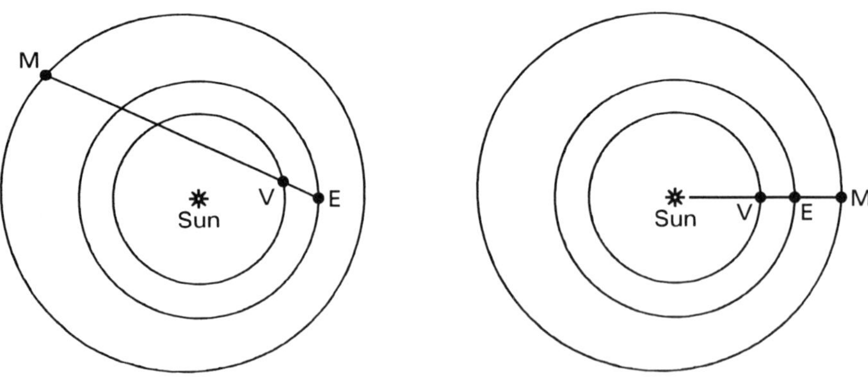

Fig. 31.a

Actually, I could restrict myself to the answer *"never"* and end this chapter here. But presumably the reader would want some more explanation.

Why *never*? For the evident reason that even only *three* planets never *can* have the same heliocentric longitude. And when I say 'exactly', then I really mean *exactly*, to an accuracy of one arcsecond, and even to an accuracy of a thousandth of an arcsecond. You might call this an exaggerated form of hairsplitting, but such a precision is needed to state what is really meant.

When a planet catches another one in its revolution around the Sun, then both bodies have again the same heliocentric longitude. Let us suppose that both are then at longitude $123°57'49''.08$. At that very instant, a third planet never can have *exactly* that same longitude; its longitude will always differ by at least a bit from the value mentioned (and in most cases much more than 'a bit') — see the 'mathematical approach' further on.

Here we should point out that it occurs often enough that three planets are exactly aligned. This occurs, for instance, when as seen from the Earth there is a conjunction between Venus and Mars. At that instant Mars, Venus and the Earth are situated on a straight line, but this line does not pass through the Sun (see the left drawing of Figure 31.*a*). A situation such as that shown in the drawing at right in the figure (and it is *this* what we mean) *never* can occur because, as we said, three planets never can have *exactly* the same heliocentric longitude. Although such an alignment is indeed possible *theoretically*, it never takes place in actual practice because the probability of this occurrence is equal to *zero*.

If we still want to be accurate, then there is already a problem in the case of only *two* planets. Consider for instance the case of Jupiter and Saturn. During the period 1940–2020, these planets are five times in heliocentric conjunction, namely on the following dates; for each case, we mention the common heliocentric longitude of the two planets, referred to the mean equinox of the date:

1940 November 15	$41°43'$
1961 April 16	$293°41'$
1981 April 16	$187°08'$
2000 June 22	$52°01'$
2020 November 2	$301°50'$

We see that the time intervals are not equal. Between the conjunctions of 1981 and 2000 the interval is 19 years and 2 months, while those of 1940 and 1961 are separated by 20 years and 5 months. These differences are easily explained. The orbits of the planets are not exactly circular; they are ellipses, and a planet describes its elliptical orbit at a variable speed: it is fastest at perihelion, slowest at aphelion. So it is not surprising that the time differences between successive heliocentric conjunctions of two planets are not equal.

For this reason, it is not possible to give an accurate answer to the question "After what time are Jupiter and Saturn in heliocentric conjunction again?" except, of course, in a well-defined case. It *is* possible to calculate the exact date of the

Jupiter-Saturn conjunction following that of A.D. 2020. But *generally* speaking only a *mean* value can be given.

To complicate the matter even more: the eccentricities of the planetary orbits slowly vary with time. The longitudes of their perihelia are also subject to variations. Presently, the eccentricity of the orbit of Jupiter is increasing, while that of Saturn decreases. The longitude of Saturn's perihelion increases faster than that of Jupiter by 21 arcminutes per century.

And that is not yet the end of the story! Planets do attract each other, so their motions are somewhat perturbed. For instance, there exists a so-called *long-period inequality* in the motions of Jupiter and Saturn, with a period of 883 years. During half this period, the speed of Jupiter in its orbit is a bit faster than its mean value, while on the contrary Saturn is somewhat slower than normally. During the other half of the 883-year period, the opposite takes place. It is a sort of periodic exchange of energy between the two giant planets: when one of them accelerates, the other is slowing down. Due to this long-period perturbation, Jupiter can be up to 20′, and Saturn up to 49′, ahead or behind in its orbit. There is a similar inequality of long period between the motions of Uranus and Neptune; here the period is 4233 years.

At this point, it will be clear to the reader why it is hopeless to give an *accurate* answer to some questions, such as the one posed at the beginning of this chapter.

Simplifying the problem

Well, under these circumstances, let us simplify the problem by considering imaginary planets moving around the Sun on *circular* orbits with *constant* speeds. From what has been said above, it is evident that even only *three* planets never can have *exactly* the same heliocentric longitude. Let us repeat again that it could indeed happen theoretically, but that in practice the probability of the event is zero.

The reader now will reply: "When I ask after what time interval all planets arrive back again on a straight line with the Sun, then I do *not* mean *exactly* on a straight line". — Well, then this is a very different problem! But we should know precisely what the real question is.

So, let us suppose that it is asked after what time period all the planets are again inside *a very narrow sector*, for instance in a heliocentric sector of one degree.

The probability for *this* occurrence is no longer zero, although it remains a very rare event. As we have seen in the preceding chapter, between the years 0 and 4000 the three outermost giant planets Saturn, Uranus and Neptune come only seven times into a heliocentric sector of 10°. On 1306 September 4 the *four* giant planets were in a sector of 7°06′, their smallest one during the period 0–4000. That is still a much larger sector than 1°, and it concerns only four planets.

Let us now make a calculation using the mean speeds of the planets, and for a heliocentric sector of one degree. The revolution period of Mercury, referred to the mean equinox of the date, is 87.9684336 days, so that Mercury's mean daily motion is 4.0923771 degrees. Similarly we find that the mean daily motion of Venus is 1.6021687 degrees. The difference between the two values is 2.4902084 degrees per day. The mean time interval between two successive heliocentric conjunctions Mercury–Venus is obtained by dividing 360 (degrees) by 2.4902084; we obtain 114.57 days, or 0.3958 year.

The probability that a third planet lies, together with Mercury and Venus, in a sector of one degree is 1/360. Therefore, the three planets will be situated inside a sector of $1°$ every 0.3958×360 years.

For all nine planets, that is, Mercury–Venus + 7 planets, the mean frequency is therefore once every 0.3958×360^7 years, or 3×10^{17} years.

This reasoning is not quite correct, however. Suppose that Mercury and Venus are in conjunction at longitude $50°00'$, and that at this instant Neptune is at longitude $51°01'$. Then the sector delimited by these three planets is $1°01'$, or slightly larger than the preset limit of one degree. After approximately six hours, Mercury has overtaken the slow Neptune, and is now too at longitude $51°01'$. But meanwhile Venus is arrived at longitude $50°24'$, so that the minimum sector of the three planets is actually $0°37'$, not $1°01'$.

Nevertheless, the reasoning gives an approximate order of magnitude : a 'period' much longer than the age of our solar system!

Problem : Can the reader find a more accurate value for the period? Use constant, mean values for the motions of the planets, and calculate the frequency of the occurrence of the nine planets in a heliocentric sector of one degree. What is the mean frequency for a sector of ten degrees?

Mathematical approach

Nevertheless, let us try to answer the question posed at the beginning of this chapter. Suppose that at a given instant all planets are exactly aligned with the Sun, so that they have the same heliocentric longitude. After what time will they be aligned again? To avoid the difficulties mentioned earlier, we will consider fictional planets having a constant speed and not subject to perturbations.

Because the new alignment should not occur necessarily at the same heliocentric longitude as the first one, we will work with the *synodic* periods of revolution, not with the sidereal periods. The synodic period of a planet is the (mean) interval of time between successive conjunctions of that planet and the Earth, as observed from the Sun. And for practical reasons we will not take Pluto into consideration. The synodic periods, rounded to the nearest integer number of days, are as follows :

Mercury	116 days
Venus	584
Mars	780
Jupiter	399
Saturn	378
Uranus	370
Neptune	367

Then the least common multiple of these numbers will give the answer. First, we decompose the numbers into their prime factors:

Mercury	$2^2 \times 29$
Venus	$2^3 \times 73$
Mars	$2^2 \times 3 \times 5 \times 13$
Jupiter	$3 \times 7 \times 19$
Saturn	$2 \times 3^3 \times 7$
Uranus	$2 \times 5 \times 37$
Neptune	367 (prime)

The least common multiple is $2^3 \times 3^3 \times 5 \times 7 \times 13 \times 19 \times 29 \times 37 \times 73 \times 367$ $= 5.37 \times 10^{13}$ days, or 147 billion years. This is ten times the age of the universe, which is estimated to be 15 billion years!

But the reader will protest, and rightly so: the calculation is incorrect, because the synodic periods of the planets are not integer numbers of days. For instance, that of Mercury is equal to 115.877 4771 days, not exactly 116. I admit that the calculation above was really very naive. So, let us repeat the calculation with more accurate values for the synodic revolution periods; we now will round them to the nearest hundredth of a day. But in order to be able to work with integer numbers, we shall now take 0.01 day as the unit of time. So we obtain the following values:

Mercury	11588
Venus	58392
Mars	77994
Jupiter	39888
Saturn	37809
Uranus	36966
Neptune	36749

I now leave it as an exercise to the reader. Decompose these new numbers into their prime factors (attention: 36749 is a prime number!), and calculate the least common multiple. Don't forget that the unit of time is now 0.01 day. My result is 3.86×10^{24} days, or 10^{22} years, a much longer time period than that obtained by the previous calculation.

But the new values just used are not 'correct' either. If you wish, repeat the calculation with still more accurate values. The more significant digits you use for the revolution periods, the larger the final result will be. This result tends to 'infinitely large'. This is in accordance with the earlier mentioned value for the probability of the event : *zero*.

32. On 'remarkable' relations between the mean motions of the planets

In 1876, in the eminent journal *Astronomische Nachrichten* (No. 2093, pages 77-78), there appeared an article entitled "*On some remarkable Relations between the mean Motions of the Primary Planets*". The author was Daniel Kirkwood, *the* Kirkwood who is known in connection with the gaps in the distribution of the major axes of the orbits of the minor planets : the so-called Kirkwood gaps. The article of Kirkwood begins as follows :

"I have found the following remarkable relations between the mean motions of the primary planets :

$$13\,n^{\mathrm{I}} + 93\,n^{\mathrm{II}} - 98\,n^{\mathrm{III}} - 238\,n^{\mathrm{IV}} + 227\,n^{\mathrm{V}} + 8\,n^{\mathrm{VI}} + 2\,n^{\mathrm{VII}} - 7\,n^{\mathrm{VIII}} = 0 \qquad (1)$$

$$68\,n^{\mathrm{V}} - 145\,n^{\mathrm{VI}} - 219\,n^{\mathrm{VII}} + 296\,n^{\mathrm{VIII}} = 0 \qquad (2)$$

where n^{I}, n^{II} etc. represent the mean motions of Mercury, Venus etc. in a julian year. In each equation, it will be observed, the algebraic sum of the coefficients is equal to zero. With the received values of the mean motions the second member of (1) is $-0''.309$. The equation will therefore be verified if we diminish Newcomb's value of n^{VIII} by $^1/_{23}$ of a second. Equation (2) is rendered exact by diminishing n^{VIII} by $^1/_{30}$ of a second."

Well, if they are true, the relations (1) and (2) are really remarkable. They remind us of the *exact* relation between the mean motions of the satellites I, II and III of Jupiter :

$$n_1 - 3\,n_2 + 2\,n_3 = 0 \qquad (A)$$

This you can verify for yourself. According to Jay H. Lieske, the mean motions of the three satellites, in degrees per day and referred to the mean equinox of the date, are

satellite I	Io	$n_1 = 203.488\,955\,432$
satellite II	Europe	$n_2 = 101.374\,724\,550$
satellite III	Ganymede	$n_3 = 50.317\,609\,110$

Similar *exact* relations do not exist for the *planets*, although for these bodies there are several *approximate* relations, so for instance:

$$8\,n^{II} \approx 13\,n^{III} \qquad (B) \qquad\qquad 7\,n^{III} \approx 83\,n^{V}$$

$$42\,n^{III} \approx 79\,n^{IV} \qquad (C) \qquad\qquad 360\,n^{V} \approx 894\,n^{VI}$$

Relation (B) is the cause of the near-periodicity of 8 years in the phenomena of Venus, while (C) results in the near-periodicity of 79 years in the oppositions of Mars — see Chapter 38.

There are also relations between the mean motions of more than two planets, so for instance

$$3\,n^{V} - 5\,n^{VI} - 7\,n^{VII} \approx 0 \qquad\qquad 4\,n^{VI} - 16\,n^{VII} + 9\,n^{VIII} \approx 0$$

You can check this for yourself. Table 32.A gives the mean sidereal motions, in *seconds of arc per day*, according to the VSOP82 theory of Bretagnon (Bureau des Longitudes, Paris).

But all these relations are only *approximate*. For example, 13 revolution periods of Venus are only *approximately* equal to 8 periods of the Earth. Kirkwood, however, believed his formulae (1) and (2) to be exact. He had found the first member of expression (1) to be equal to -0.309 arcsecond (when the values of the n are expressed in arcseconds per *year*). But if one uses the modern values given in Table 32.A, then the first member of (1) becomes -0.012136 arcsecond per day, or -4.43 arcseconds per year, a much larger value than that found by Kirkwood. Nevertheless, it still remains a very small value; bear in mind that the mean annual motion of Neptune, the slowest planet, is $7866''$.

Nor is the first member of (2) equal to zero. Using the modern values from the table, one finds 0.466 arcsecond per day, or $170''$ per year.

Does Kirkwood's formula (1) point to a physical relation? Very probably not. In the case of the first three satellites of Jupiter, relation (A) originates from the mutual attractions (tides) in the course of millions of years. But in the case of the planets the effect of the mutual tides is next to nil. Moreover, while (A) is an *exact* relation, (1) is only approximate. And finally, in formula (A) the coefficients of the n-values are very small integers (1, 2, and 3), while in (1) they range from 2 to 238.

This suggests that rela-

TABLE 32.A

Mean sidereal motions of the planets
in arcseconds per day (VSOP 82)

Mercury	n^{I}	14 732.419 676 63
Venus	n^{II}	5 767.669 717 55
Earth	n^{III}	3 548.192 806 75
Mars	n^{IV}	1 886.518 209 25
Jupiter	n^{V}	299.128 280 04
Saturn	n^{VI}	120.454 747 69
Uranus	n^{VII}	42.230 833 51
Neptune	n^{VIII}	21.534 574 15

tion (1) is rather far-fetched. The interested reader could search for other coefficients giving a still better result than that obtained by Kirkwood. Try to use no larger coefficient than 238, Kirkwood's largest. And, if possible, try to find a formula in which, as in (1), the algebraic sum of the eight coefficients is zero.

The preceding text appeared in the Belgian journal Heelal *(Vol. 33, No. 8, pages 179-180) of August 1988. The next issues contained solutions and comments from some readers :*

Erik Vackier, of Tielt (Flanders West, Belgium) found the following new relation which yields a much more accurate result than Kirkwood's formula (1):

$$-26\,n^{I} + 53\,n^{II} + 3\,n^{III} + 34\,n^{IV} + 28\,n^{V} - 60\,n^{VI} + 102\,n^{VII} - 134\,n^{VIII} = 0$$

The first member is equal to $+0.000\,034\,16$ arcsecond per day, or only $+0.012$ arcsecond per year! Moreover, this formula fulfils the conditions we posed: the algebraic sum of the coefficients should be zero, and no coefficient should be larger than 238 in absolute value.

In the issue of October 1988, Herman Sallé, of Bussum, Netherlands, wrote that formulae such as (1) have no physical meaning. It is possible to choose the coefficients such that in equation (1) the error is smaller than $1 / 10\,000\,000$ arcsecond per year. "The 'error', Sallé noted, can always be made smaller than

$$\sqrt{\frac{\pi\,m\,(k^{2} - 1)}{6}}\ (k - 1)\,\Sigma n\,/\,k^{m}$$

where m = the number of planets = 8, $k = 239$ = the largest allowed coefficient minus 1, and Σn = the sum of the mean motions. Therefore there is not the least reason to call his [Kirkwood's] equations remarkable."

On page 256 of the November issue, Luc Van den Brempt, of Kraainem (Belgium), gave the solution

$$7\,n^{I} - 16\,n^{II} - 10\,n^{III} + 11\,n^{IV} + 14\,n^{V} - 3\,n^{VI} + 6\,n^{VII} - 9\,n^{VIII}$$

With the n-values he had at his disposal, Kirkwood obtained for the first member of his formula (1) the value -0.309 arcsecond per year. But using the modern values of n as given in Table 32.A, the above expression by Van den Brempt takes the value $+0.000\,001\,0600$ arcsecond per day, or only $+0.000\,387$ arcsecond per year. This not only is very much better than Kirkwood's result, but is is also much better than that of Vackier. And the most surprising fact is that in the expression no coefficient is larger than 16.

Another expression found by Van den Brempt is

$$31\,n^{I} - 78\,n^{II} - 50\,n^{III} + 86\,n^{IV} + 16\,n^{V} + 18\,n^{VI} + 91\,n^{VII} - 114\,n^{VIII}$$

the value of which is equal to only $-0.000\,000\,015$ arcsecond per year. This confirms the statement of H. Sallé: it is possible to choose the coefficients such that the error is smaller than one ten-millionth arcsecond per year.

Not the slightest physical meaning should be attached to the several formulae mentioned above. If the n-values were different, then too similar expressions could be found with a result practically equal to zero.

On the other hand, with formula (A), relating to the motions of the satellites I, II and III of Jupiter, things are different. Not only is this relation truly *exact*, but moreover here the coefficients are *small* integers. These three satellites are rather close to each other. Their mutual attractions are far from negligible. The satellites are 'locked' to each other by an exact resonance built up in the course of many millennia. One well-known consequence of this resonance is that these three satellites never can be eclipsed simultaneously in the shadow of Jupiter — see also Chapter 43.

33. *Ceres and Pallas, and other couples*

Ceres and Pallas, the first two minor planets, were discovered in 1801 and in 1802, respectively. They describe their orbits at nearly the same mean distance from the Sun, 2.77 astronomical units, and so they have nearly the same period of revolution. Actually, Ceres is a little closer to the Sun than Pallas, its mean sidereal period of revolution being 1681.57 days, while that of Pallas is 1685.04 days. From these numbers one deduces that it takes about *22 centuries* for Ceres to perform one revolution more than Pallas. Table 33.A mentions, for some dates at intervals of 10 000 days, the difference between the mean longitudes of the two bodies, in the sense Ceres minus Pallas. We see that indeed this difference in now slowly increasing with time.

The values in the table have been deduced from numerical integrations performed by the Belgian calculator Edwin Goffin. From his osculating orbital elements the mean longitude L of a body at a given date is easily found: it is the sum of the longitude Ω of the ascending node, the argument ω of the perihelion, and the mean anomaly M:

$$L = \Omega + \omega + M.$$

Figure 33.*a*, due to Goffin, shows the variation of the distance in astronomical units between Ceres and Pallas from 1800 to 2050.

TABLE 33.A

*Difference between the
mean longitudes of
Ceres and Pallas*

	$^{\circ}$
1804 Feb. 14	$+12.23$
1831 July 2	$+16.44$
1858 Nov. 17	$+22.55$
1886 Apr. 4	$+25.52$
1913 Aug. 21	$+29.71$
1941 Jan. 6	$+35.95$
1968 May 24	$+38.98$
1995 Oct. 10	$+43.13$
2023 Feb. 25	$+49.31$

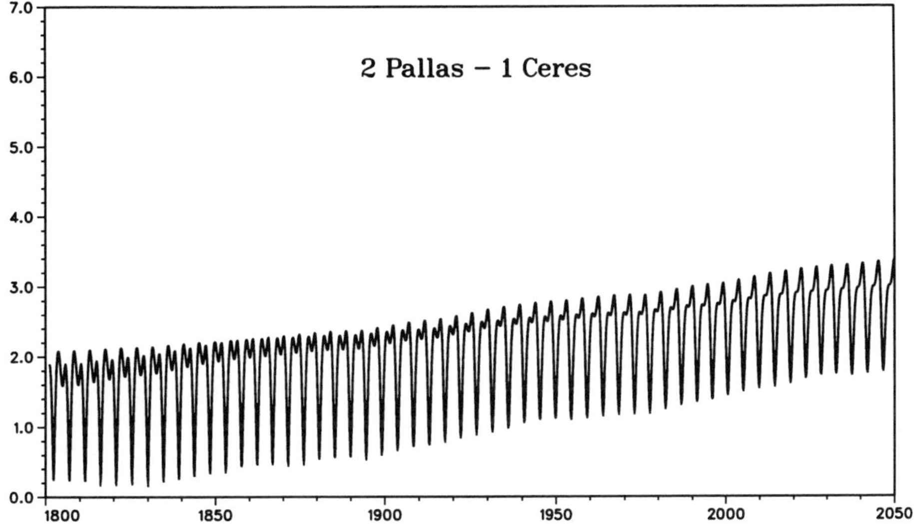

Fig. 33.a *: The distance in astronomical units
between Ceres and Pallas, from 1800 to 2050.*

The distance between the two asteroids is now gradually increasing, as the
difference between their mean longitudes increases. The curve presents an oscillation
with a period of 4.6 years, due to the fact that the orbits of the two objects are not
identical: because the orbital eccentricities and inclinations are different, Ceres
moves alternately faster and slower than Pallas in its orbit around the Sun, whence
the periodic variation in the mutual distance.

We see from the figure that during the first half of the 19th century the
distance between Ceres and Pallas dropped several times below 0.3 AU. But
nowadays, and for several more centuries, this is no longer possible. The closest
distance between the two asteroids was 0.1814 AU, on 1820 October 27. Since
February 1936, their mutual distance remains larger than 1 astronomical unit.

Turning our attention to other couples of minor planets, let us mention that
during the period 1800–2050 the least distance between Ceres and Vesta was 0.186
astronomical unit, on 1893 March 7.

During the same two-and-a-half centuries, the least distances between Pallas
and Vesta are

1825 April 28	0.360 AU
1995 December 16	0.157
2046 September 26	0.107

However, we find more interesting situations with minor planet 197 *Arete*.
Figure 33.*b*, due to Goffin (as are all other graphs in this chapter), shows the
distance in astronomical units between Ceres and Arete. Here too we see periodic

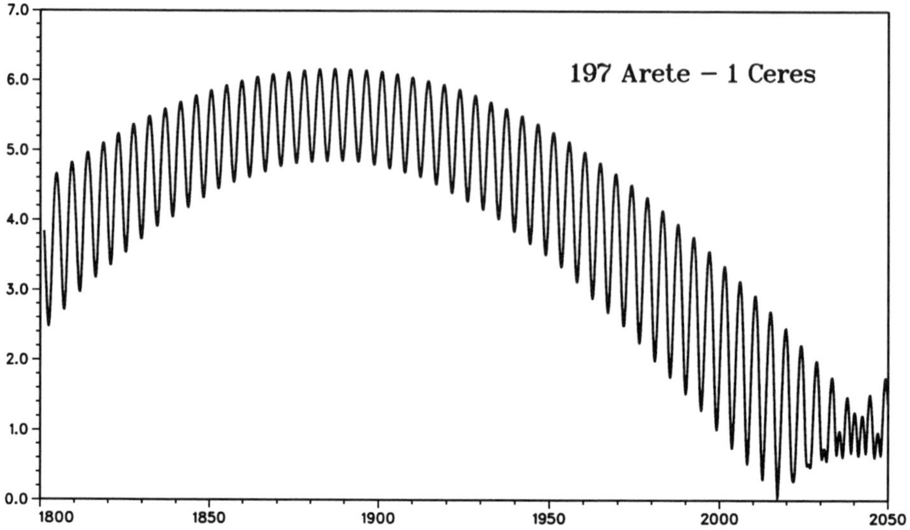

Fig. 33.b : *The distance in astronomical units
between Ceres and Arete, from 1800 to 2050.*

variations due to the fact that the orbits of the two bodies have different shapes and
inclinations.

From A.D. 1800 to September 1980 the distance between Ceres and Arete
was always larger than 2 astronomical units. A close approach will take place on
2017 May 2, when the distance between the two asteroids will be only 0.022 AU.

Figure 33.*c* shows the variation of the distance between Arete and Pallas.
The sidereal revolution period of Arete is 1657 days, which is not too different from
the periods of Ceres and Pallas, whence the long-period variation shown in Figures
33.*b* and 33.*c*.

A completely different picture appears with the couple Arete–Vesta. The
revolution period of Vesta is 1326 days, so this asteroid is in 5:4 resonance with
Arete: five revolution periods of Vesta have almost the same duration as four
periods of Arete. Whence it results that the two bodies come close to each other
every 18.14 years, as shown in Figure 33.*d*. During the period 1800–2050, the
deep minima of the distance Arete–Vesta (in AU) occur on the following dates:

1812 Oct. 26	0.0065	1903 July 7	0.0274	1994 Mar. 19	0.0420
1830 Dec. 15	0.0051	1921 Aug. 25	0.0290	2012 May 6	0.0396
1849 Feb. 4	0.0063	1939 Oct. 17	0.0319	2030 June 18	0.0488
1867 Mar. 25	0.0177	1957 Dec. 9	0.0354	2048 July 31	0.0548
1885 May 14	0.0180	1976 Jan. 27	0.0346		

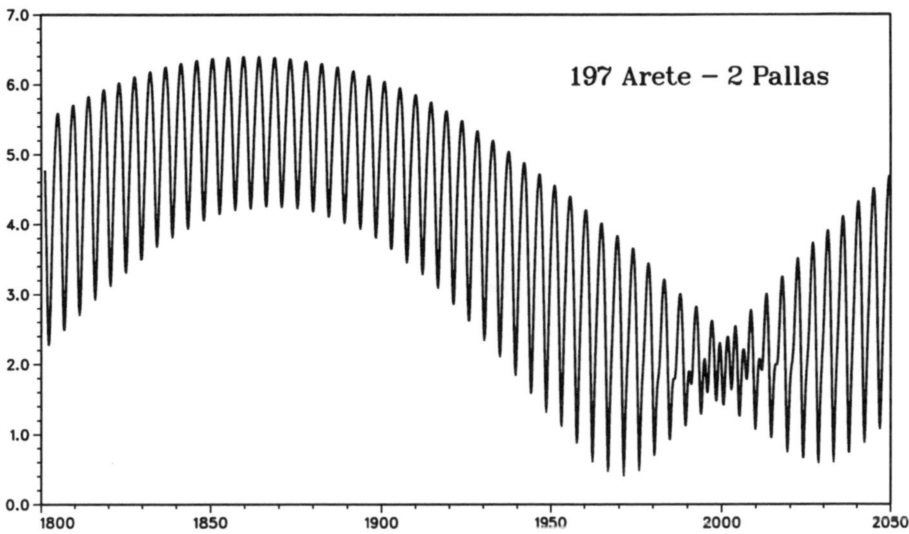

Fig. 33.c : *The distance between Pallas and Arete, 1800-2050.*

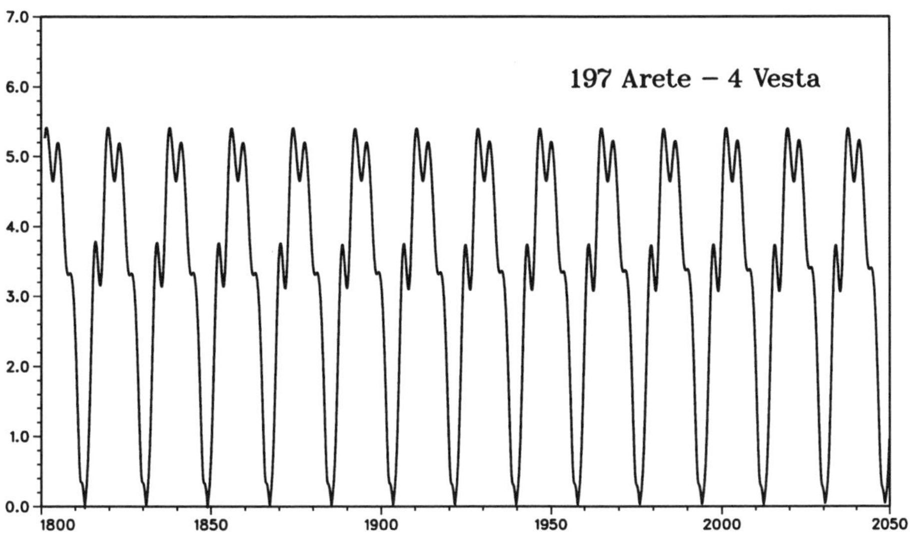

Fig. 33.d : *The distance between Vesta and Arete, 1800-2050.*

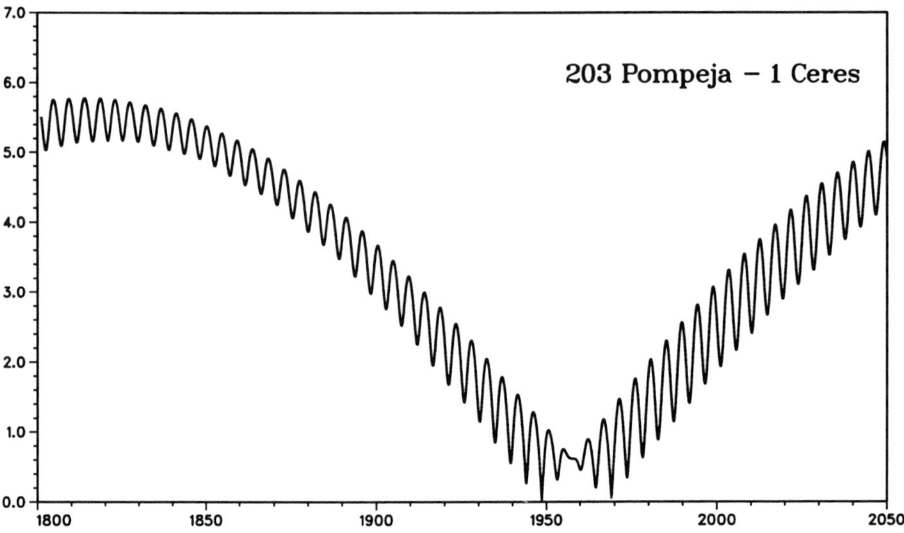

Fig. 33.e : The distance between Ceres and Pompeja, 1800-2050.

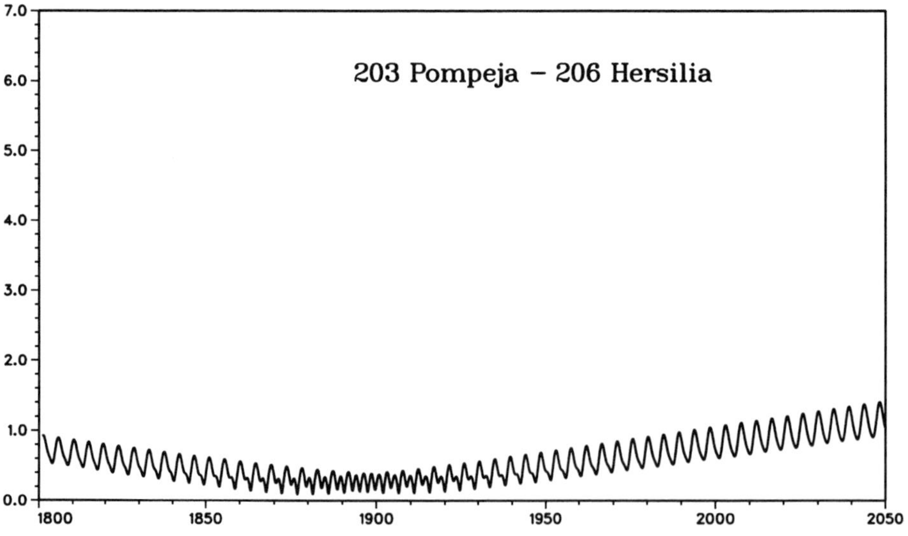

Fig. 33.f : *The distance between Pompeja and Hersilia, 1800-2050.*

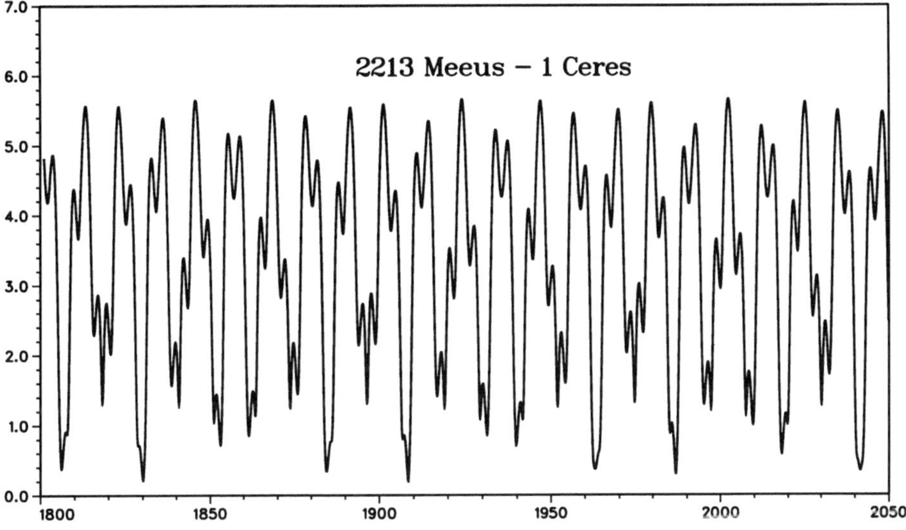

Fig. 33.g

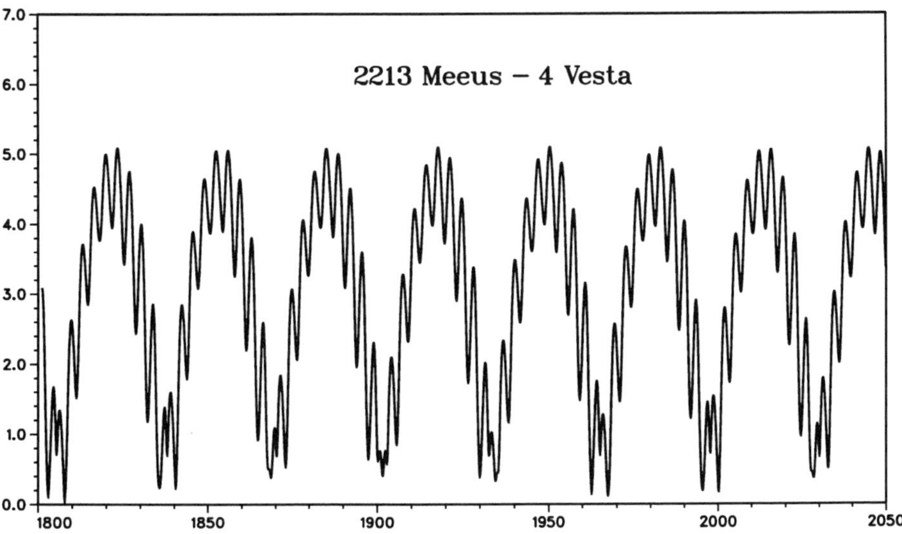

Fig. 33.h

Mutual perturbations of two minor planets caused by a close approach can be used for a determination of the masses of these bodies. The mass of Vesta was calculated by H. Hertz in 1968 using the close approaches to Arete.

E. Goffin has derived the mass of Ceres from its perturbations on the motion of 203 Pompeja (*Astronomy & Astrophysics*, Vol. 249, pages 563-568; 1991). Because the semimajor axis of Pompeja differs little from that of Ceres, both objects can approach each other only every 265 years. Figure 33.*e* shows the evolution of their mutual distance from 1800 to 2050. As can be seen, both asteroids stayed to within 1 AU of each other from 1947 to about 1970. Their closest approach (0.016 AU) took place on 1948 August 22, and a second minimum (0.078) occurred on 1969 February 15.

In his article, Goffin mentions the special case of 91 *Aegina*. This asteroid had a close approach (0.033 AU) to Ceres on 1973 September 12. After this approach, Aegina lost about $7''$ per 10 years with respect to a pre-1973 orbit. So, the gravitational action of Ceres cannot be neglected. Goffin insists on the importance of adding Ceres as a perturbing body in the determination of orbits of minor planets. Though its mass is 16 times smaller than that of Pluto, the fact that it describes its orbit amid the asteroid belt causes measurable perturbations on a number of other minor planets. On the other hand, it is superfluous to take Pluto into account in the calculation of perturbations due to its rather small mass combined with its large distance. So, for the calculator of minor-planet orbits, the 'ninth' planet clearly is Ceres, not Pluto.

Another interesting couple is Pompeja–Hersilia. The revolution period of the former is about 3 days shorter than that of the latter, so their mutual distance is subject to a variation of long period, as shown in Figure 33.*f*. Because both orbits have rather small inclinations ($3°$ and $4°$, respectively) and small eccentricities (0.06 and 0.04), the short-period variation in their mutual distance has a small amplitude. The two asteroids remained less than one astronomical unit from each other from the beginning of the 19th century till September 1993.

Finally, figures 33.*g* and 33.*h* show the variation of the mutual distances of two 'ordinary' couples of minor planets.

34. *Seneca, Orthos, and Quetzálcoatl*

On 1978 February 17 a minor planet was discovered by H.-E. Schuster at the European Southern Observatory at La Silla, Chile. The object, which received the provisional designation 1978 DA, passed 0.090 astronomical unit from the Earth on the next March 15. The perihelion distance was 1.025 AU, so the asteroid was a so-called 'Earth-grazer' of the Amor type. Four years later the definitive number 2608 was assigned to the object, which was named *Seneca* for the great Roman philosopher and statesman Lucius Annaeus Seneca.

As it is interesting to examine the past and the future of such an asteroid, the Belgian calculator Edwin Goffin performed a *numerical integration* of the equations of motion. This is a method of calculation in which, starting from the known position and velocity of the asteroid at a given instant (the *Epoch*), new positions into the past or the future are obtained step by step, taking into account the gravitational attraction of the planets. Successive positions and velocities can then be used to obtain the *osculating* orbital elements. These are the elements of the 'instantaneous' elliptic orbit of the body — see also page 176. If these elements are calculated for different times and the values are then plotted in a graph, we obtain a picture showing the evolution of the orbit with time.

Figure 34.*a* shows the variation of four orbital elements of 2608 Seneca during two and a half centuries, from A.D. 1801 to 2050. These graphs are based on the results of Goffin's numerical integration: the semimajor axis a of the elliptic orbit, in astronomical units; the orbital eccentricity e; the perihelion distance q in AU, calculated from $q = a(1 - e)$; and the orbit's inclination i on the ecliptic of 2000.0, in degrees.

It is remarkable that the four graphs show similar variations. Let us first have a close look at the first graph. The semimajor axis of the orbit varies periodically between 2.45 and 2.55 astronomical units, so the mean value of a is 2.50 AU. Then, by means of Kepler's third law $a^3 = T^2$ it is possible to calculate the corresponding period of revolution T (in years). This gives $T = 3.95$, which is *exactly 1/3 of the revolution period of Jupiter*. This so-called 3 : 1 commensurability gives rise to important perturbations with a period equal to the revolution period of Jupiter, a little less than twelve years.

Consider the situation near A.D. 1925. Each time Seneca reaches the aphelion of its orbit, at approximately 4 AU from the Sun, Jupiter is successively 60° behind, 60° ahead, and then at the other side of the Sun 180 degrees distant from Seneca. So the two bodies never come close together in such a situation, and the orbital elements are almost invariable. But as the revolution period of Seneca is a little shorter than 1/3 of that of Jupiter, the giant planet gradually drops behind. Near 1950, when Seneca reaches its aphelion Jupiter is successively 80° behind, 40° ahead, and 160° ahead. The mutual distances in the second case are smaller than at the others, and so the perturbations are stronger. At such times, the orbital elements are subject to larger variations, which are distinctly visible in the graphs as jumps every twelve years. These approaches occur gradually closer to Seneca's aphelion, and so the jumps become larger and larger.

However, the gradual increase of the semimajor axis results in an increase of the revolution period. Hence Jupiter's 'approach' towards the aphelion of Seneca will slow down. Near A.D. 2000 the 3 : 1 resonance between the revolution periods is reached; then the approaches occur at the same places on the orbits, Jupiter being a little ahead of Seneca's aphelion, and the perturbing action of the giant planet is greatest. From then on the mutual minimum distance increases again, and near A.D. 2080 (at right, outside of the drawing) the starting situation is reached again, but now the revolution period of Seneca is a little *larger* than 1/3 of that of Jupiter.

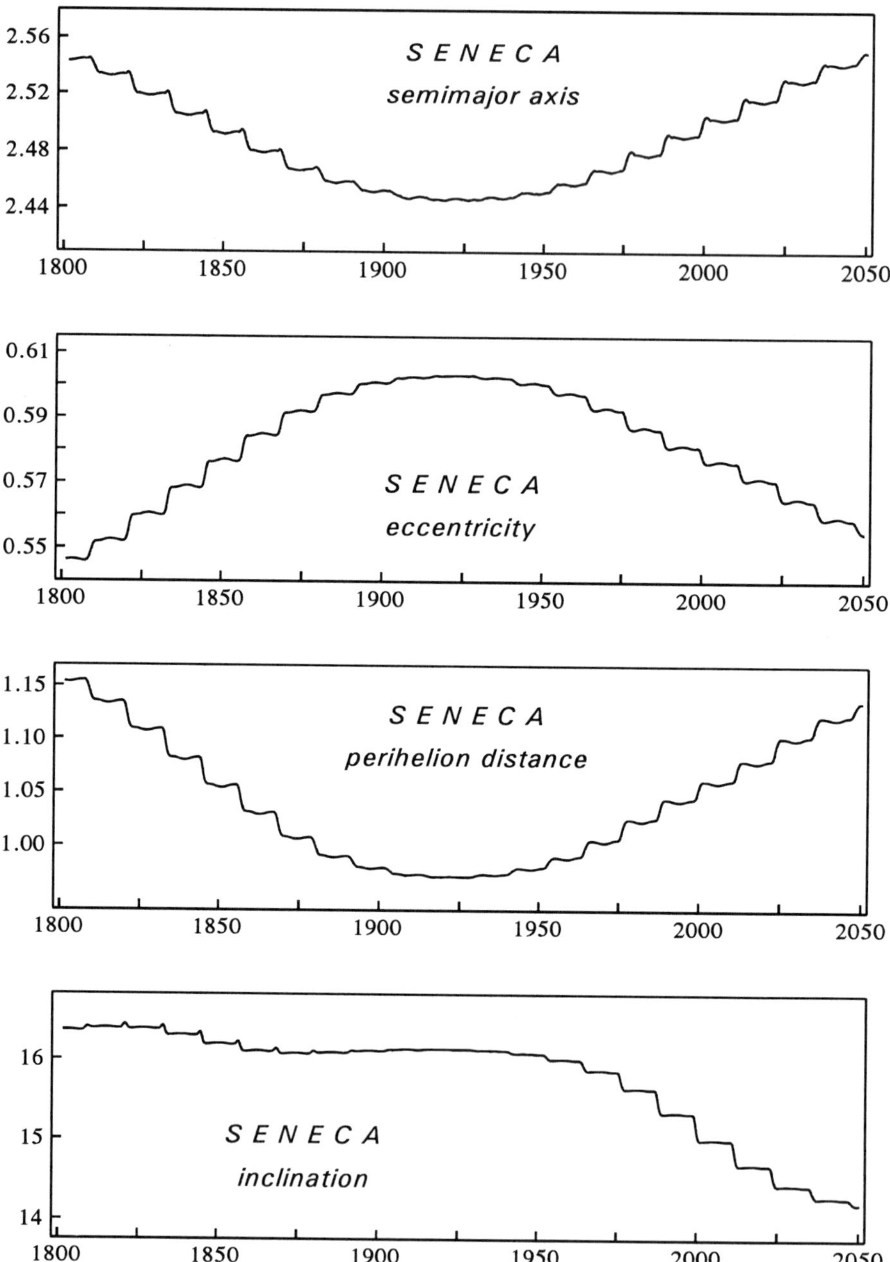

Fig. 34.a : *Evolution of the orbit of 2608 Seneca, 1801 to 2050.*

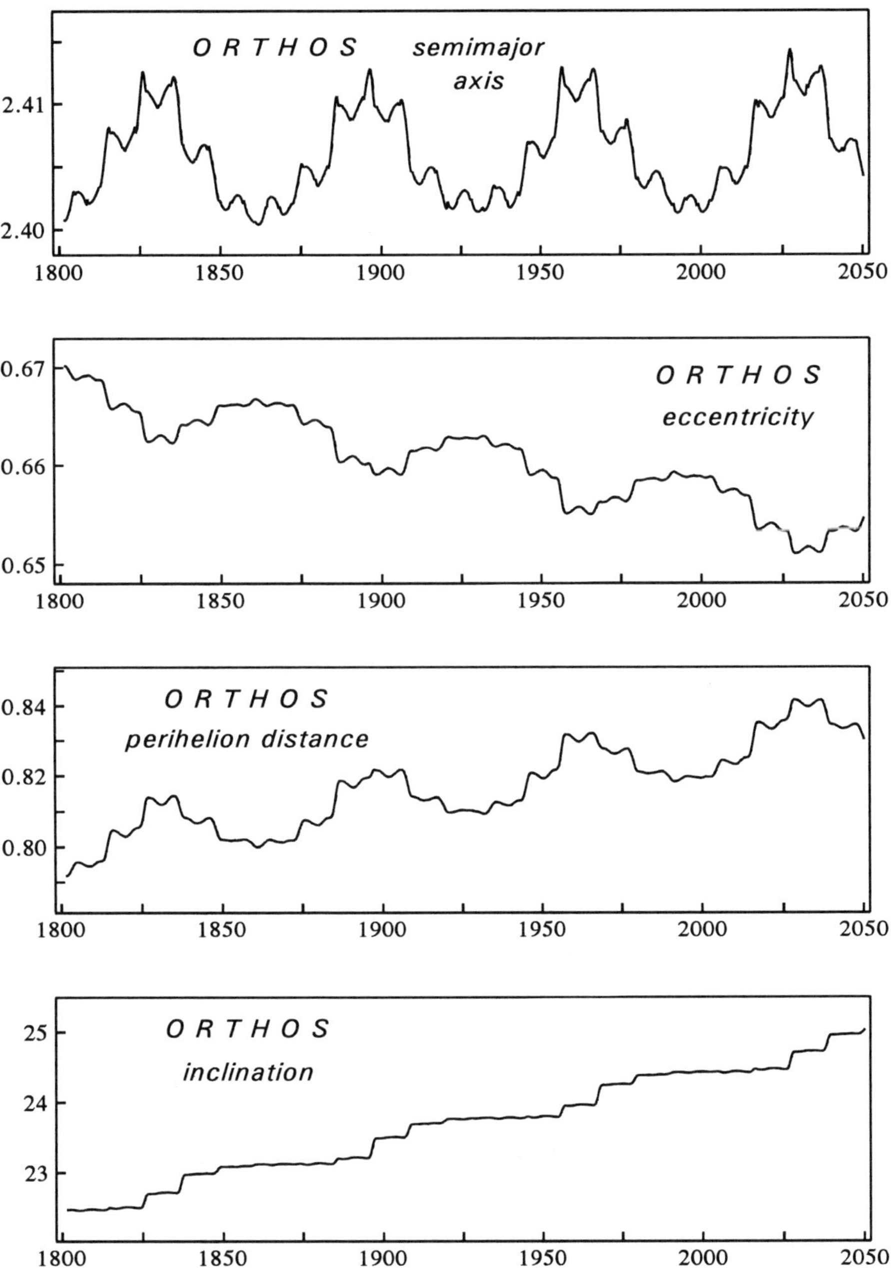

Fig. 34.b : *Evolution of the orbit of 2329 Orthos, 1801 to 2050.*

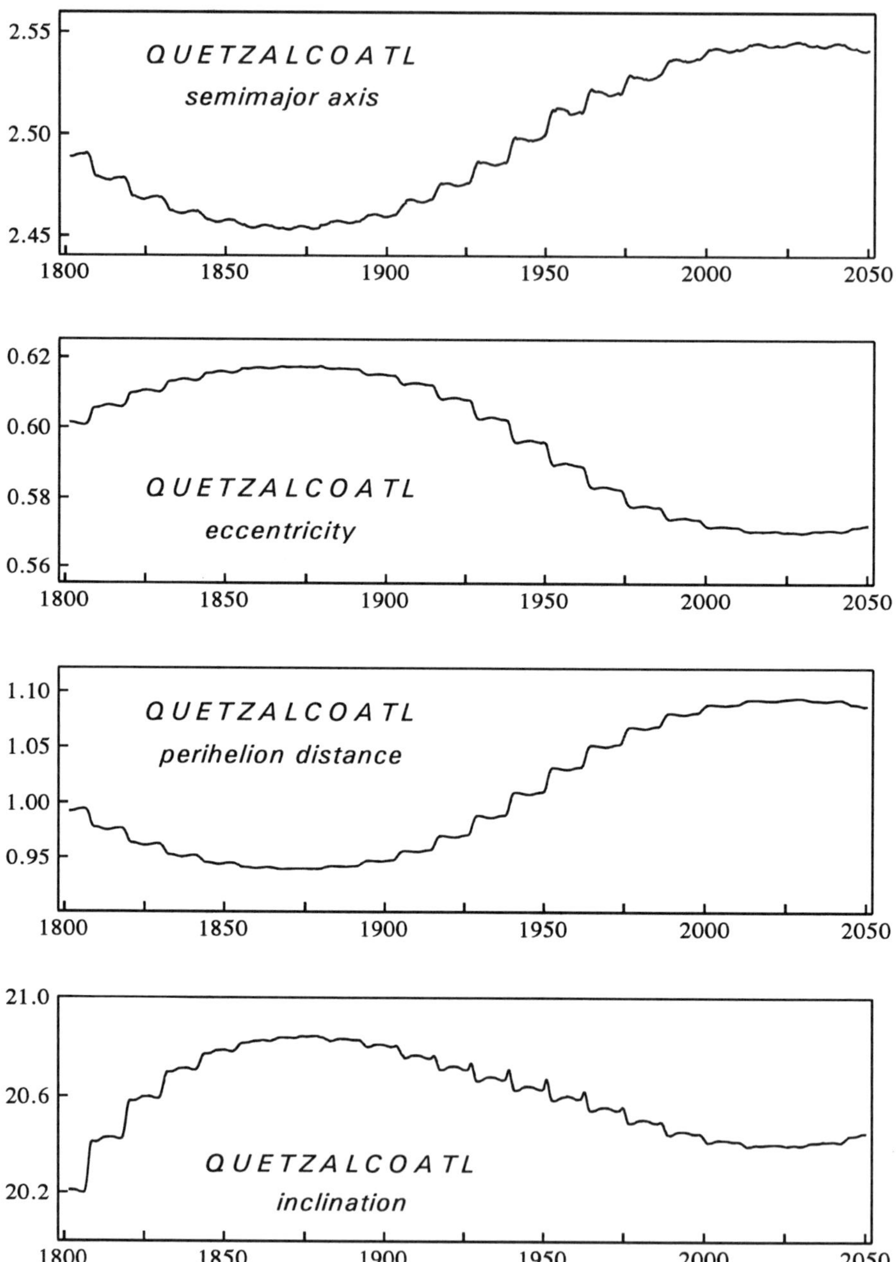

Fig. 34.c : *Evolution of the orbit of 1915 Quetzálcoatl, 1801 to 2050.*

From then on, the approaches of Jupiter *behind* Seneca's aphelion will have the greatest influence, and the orbital elements of the minor planet will vary in the opposite sense. So the above-described scenario will repeat as a mirror-image. In this manner, an orbit which at first sight is rather unstable actually is maintained thanks to a *libration* — an oscillation — around a resonance.

The perihelion distance (third graph of Figure 34.*a*) is generally larger than 1, and thus most of the time Seneca is an asteroid of the Amor type. However, during some years q is smaller than 1 AU, and then the asteroid becomes a member of the Apollo group; this was the case from A.D. 1879 to 1965. The smallest value of the perihelion distance of Seneca was 0.9710 AU, in 1925.

We also note that the jumps in the orbital inclination (lower graph) occur in phase with the other orbital elements.

Orthos

This minor planet was discovered on 1976 November 19 at the European Southern Observatory (Chile) by the same Schuster who also found Seneca. The object first received the provisional designation 1976 WA, and later the definitive number 2329 and the name *Orthos*, for a two-headed dog from Greek mythology.

The four curves of Figure 34.*b* show the evolution of the orbit of Orthos from 1801 to 2050, according to calculations by Goffin. The existence of two periods is obvious: one of approximately 67 years (maxima near A.D. 1830, 1895, 1960, 2030) and another of 10 years. The main maxima correspond to the 'steps' in the lowest curve, that of the orbital inclination.

Between 1801 and 2050 the semimajor axis of the orbit oscillates between the extreme values 2.4004 and 2.4143 astronomical units, so the mean value is close to $a = 2.4073$. This corresponds to a mean revolution period of 3.735 years. So, contrarily to Seneca, the (mean) revolution period of Orthos is *not* exactly 1/3 of that of Jupiter.

If O is the mean motion of Orthos, and J that of Jupiter, then we have, in degrees per year, $O = 360/3.735 = 96.39$, and $J = 30.35$. The above-mentioned periods of 10 and 67 years correspond to the arguments $O - 2J$ and $O - 3J$, respectively.

The orbital eccentricity (second curve) presents maxima which coincide with the *minima* of the semimajor axis. However, we note a slow gradual decrease of the eccentricity, from 0.670 in 1801 to 0.651 around A.D. 2030. As a consequence of this decrease of the orbital eccentricity, the perihelion distance of Orthos is slowly increasing (third curve), from 0.79 AU near A.D. 1801 to 0.84 around 2035. In any case, during the time period considered here the perihelion remains between the orbits of Venus and the Earth: Orthos is an asteroid of the Apollo type.

The orbit's inclination too is gradually increasing, from 22°27′ in 1805 to 25°02′ in 2050 (lowest curve). This increase does not occur at a constant rate, but it takes place as jumps. The principal jumps occur in the years 1825, 1836, 1896,

1907, 1955, 1967, 1978, 2027, and 2038. They clearly are due to the gravitational influence of Jupiter: the interval of 11 years (for instance from 1967 to 1978) corresponds to three revolutions of Orthos and one of Jupiter. The interval of 71 years (1825–1896–1967–2038) corresponds to 19 revolutions of Orthos and to 6 revolutions of Jupiter.

Quetzálcoatl

This minor planet was discovered on 1953 March 9 by A. G. Wilson at Palomar. It received the provisional designation 1953 EA, and later the definitive number 1915. It is named for the god of wisdom and culture who brought learning to the Toltec people.

Like Seneca, Quetzálcoatl is in 3:1 resonance with Jupiter, and so its orbit is subject to similar variations (Figure 34.*c*).

The perihelion distance of Quetzálcoatl was smaller than 1 AU from the beginning of the 19th century to A.D. 1939. The asteroid was of the Apollo type from 1807 to 1936. Since then, it is a member of the Amor family. See the definition of Apollo and Amor asteroids in the next chapter. The perihelion distance reached its smallest value (0.9390 AU) in 1867; it will have its largest value (1.0944) in 2028–2029.

Finally, note on the lowest curve the rapid increase of the orbital inclination, from 20°12′ to 20°25′, in 1806–1808.

The table below mentions the least distances of the three asteroids to the Earth during the period 1801–2050, in astronomical units.

2608 Seneca (< 0.3 AU)	1878 Apr. 10 1905 Mar. 1 1978 Mar. 15	0.1825 0.1723 0.0901
2329 Orthos (< 0.2 AU)	1864 Oct. 4 1920 Oct. 5 1976 Oct. 1 2017 Sep. 30 2032 Sep. 20	0.1660 0.1822 0.1627 0.1582 0.0950
1915 Quetzálcoatl (< 0.1 AU)	1879 Mar. 12 1906 Mar. 6 1953 Mar. 4 1957 Feb. 23 1981 Mar. 2	0.0777 0.0254 0.0534 0.0936 0.0832

35. Defining asteroids of the Apollo and Amor types

The following text was first published in the Journal *of the British Astronomical Association, Vol. 104, No. 5, page 214 (October 1994). It has been slightly edited.*

In the *Astronomy and Astrophysics Encyclopedia* [1], pages 32-33, Michael J. Gaffey writes that an asteroid is of the Apollo type when the semimajor axis a of its orbit is larger than 1 AU, and its perihelion distance q is smaller than 1.017 AU. This strange definition had already appeared in the first *Asteroids* book [2] but it should be considered incorrect. Indeed, it is not useful to call an asteroid of the Apollo type if no part of its orbit lies inside the orbit of the Earth.

The perihelion and aphelion distances of the Earth are 0.9833 and 1.0167 astronomical units, respectively. Consequently, if the perihelion distance q of an asteroid is less than 0.9833 AU, the object certainly is of the Apollo type. If q is larger than 1.0167, then the minor planet certainly is not an Apollo object.

But if q is between 0.9833 and 1.0167, closer examination is needed to find out whether the asteroid crosses or does not cross the orbit of the Earth. More precisely, we have to look at the longitude of the perihelion of the orbit, and find the radius vector of the Earth's orbit at this longitude. Let us consider, for instance, the case of minor planet 3752 *Camillo*. According to the Russian *Ephemerides of Minor Planets* for 1996, this asteroid has $a = 1.41362$ AU, and $e = 0.30238$. From these values we deduce $q = 0.9862$. Therefore, one might think that Camillo is an Apollo object, because its perihelion distance is smaller than 1 AU. However, the longitude of its ascending node is 148°, and its argument of perihelion is 312°, the sum of which is equal to the longitude of the perihelion, or 100°. This is close to the longitude of the perihelion of the Earth. At this longitude, the Earth's radius vector is 0.9833 AU, which is smaller than Camillo's perihelion distance. Consequently, the asteroid's orbit is completely outside the orbit of the Earth, so why would we call it an Apollo object? That would be quite misleading.

Michael Gaffey further writes: "An asteroid with a perihelion distance of 0.99 AU will not be Earth-crossing if its perihelion is near the ecliptic longitude of the Earth's perihelion (0.983 AU). However, the longitude of perihelion of such an orbit will precess rapidly, so that the shallow Apollo orbits ($q > 0.983$ AU) can vary between Earth-crossing and non-crossing many times during their dynamical lifetimes."

This is incorrect, however. The above argument would be valid only if the perihelion distance of the asteroid were invariable. This is not the case. Moreover, the longitude of the perihelion of an asteroid does *not* 'precess rapidly'. This precession is generally very slow, while the variation of the perihelion distance can be much larger and more rapid.

Consider, for instance, the case of 1915 *Quetzálcoatl*. In the preceding chapter we have seen that the perihelion distance of this minor planet varies from 0.939 to 1.094 astronomical units in less than two centuries. But let us now look at the longitude of the perihelion. Here are some values of this longitude according to E. Goffin; they are referred to the standard equinox 2000.0:

1801	May	20	150°06′
1851	Mar.	19	150°36′
1901	Jan.	16	150°18′
1950	Nov.	15	150°44′
2000	Sep.	13	150°53′
2050	July	13	150°30′

We see that during those 250 years the longitude of the perihelion of Quetzálcoatl *changes by less than one degree*, so that the 'precession of the perihelion' is almost negligible. (We note a slight trend to increase, but in the reference frame 2000.0 the longitude of the Earth's perihelion itself increases by 19 minutes of arc per century!).

Adopting the definition used by Gaffey (and others) thus does not at all guarantee 'long-term consistency'! When an Apollo object evolves to the Amor type, or vice versa, in practically all cases this is due to the fact that it is the perihelion distance which varies, not the longitude of the perihelion.

A similar criticism can be made concerning the definition of the Amor objects. Gaffey, after E. M. Shoemaker et al. (*Asteroids*, cited above, page 253), writes that Amor asteroids have $a > 1$ AU and q is between 1.017 and 1.3 AU. As we have seen, an orbit can have q as small as 0.984 AU, yet lie entirely outside the orbit of the Earth. Moreover, the semimajor axis of an orbit cannot be smaller than the perihelion distance. Hence, if q is 1.017 or larger, then a certainly is larger than 1.017. Consequently, the added condition $a > 1$ is superfluous!

REFERENCES

1. *The Astronomy and Astrophysics Encyclopedia*, edited by Stephen P. Maran, Cambridge University Press (UK), 1992.
2. Tom Gehrels (ed.), University of Arizona Press, Tucson (1979).

36. *Periodic comet Encke and Jupiter*

Of all known periodic comets, comet Encke has the shortest period of revolution. It describes its elliptic orbit around the Sun in only 3.30 years, or 3 years and a little less than 4 months. As the comet was discovered in 1786, it is the comet with the largest number of observed appearances.

Comet Encke was discovered by Méchain at Paris on the evening of 1786 January 17. By reason of bad weather no observation was possible the next day. The comet was seen again on January 19 by Méchain himself and Messier. These were the only observations, so the data were not sufficient to allow the calculation of an orbit. Nine years later, on 1795 November 7, at Slough, England, Caroline Herschel discovered a comet which was observed till November 27. Ten years later, on 1805 October 19, a comet was discovered independently by Pons (Marseille), Huth (Frankfurt/Oder) and Bouvard (Paris). On 1818 November 26 the comet was discovered for the fourth time, by Pons at Marseille — indeed, afterwards it appeared that each time it had been the same comet.

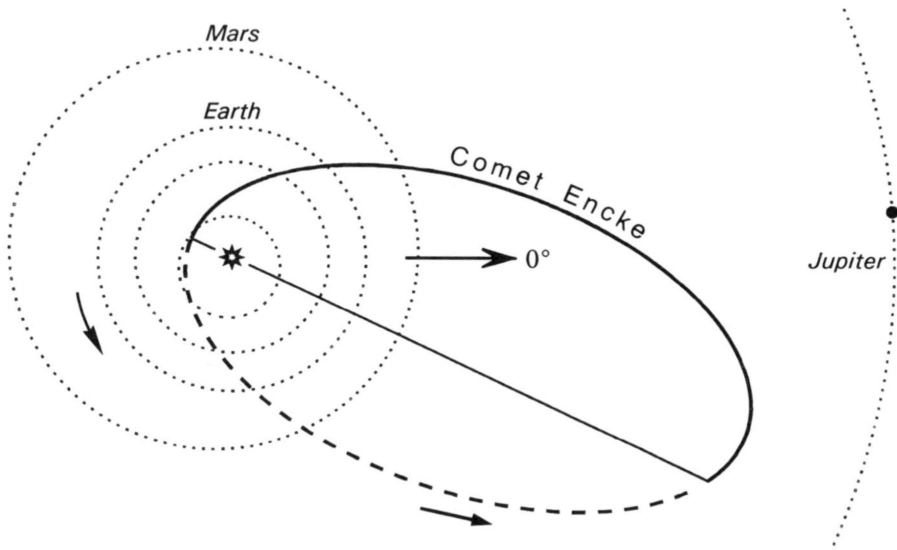

Fig. 36.a : The orbit of periodic comet Encke and the orbits (dotted) of the planets Mercury to Jupiter. The central star represents the Sun. The part of the comet's orbit that is drawn as a dashed line lies south of the plane of the ecliptic. The straight line passing through the Sun is the line of nodes, the intersection of the comet's orbital plane and the plane of the Earth's orbit. The descending node of the orbit is close to the perihelion of the comet. The long arrow indicates the direction of the vernal equinox, longitude zero. The position of Jupiter is plotted for 1999 January 15, when comet Encke reaches the aphelion of its orbit.

TABLE 36.A

Comet Encke: perihelion passages, orbital elements, and least distances to Earth

T	ΔT	q	e	i	t	Δ_{min}	J
1786 Jan. 31	—	0.3360	0.8484	13.69	1786 Jan. 23	0.618	24
* 1789 May 19	1203.90	0.3350	0.8488	13.71	1789 June 30	0.229	128
* 1792 Sep. 4	1204.46	0.3355	0.8486	13.70	1792 Aug. 20	1.295	220
1795 Dec. 21	1203.23	0.3355	0.8486	13.70	1795 Nov. 9	0.256	318
* 1799 Apr. 11	1206.57	0.3405	0.8467	13.60	1799 May 3	0.539	67
* 1802 Aug. 2	1207.86	0.3395	0.8471	13.62	1802 Aug. 26	1.120	166
1805 Nov. 21	1207.60	0.3406	0.8465	13.59	1805 Oct. 16	0.435	259
* 1809 Mar. 12	1206.38	0.3350	0.8487	13.65	1809 Mar. 16	0.652	3
* 1812 June 26	1202.47	0.3346	0.8488	13.66	1812 July 30	0.691	109
* 1815 Oct. 13	1203.30	0.3344	0.8489	13.66	1815 Sep. 16	0.954	203
1819 Jan. 27	1202.61	0.3351	0.8486	13.64	1819 Jan. 17	0.602	298
1822 May 24	1212.70	0.3459	0.8444	13.36	1822 July 4	0.269	47
1825 Sep. 16	1211.31	0.3448	0.8449	13.38	1825 Aug. 28	1.235	149
1829 Jan. 10	1211.47	0.3454	0.8446	13.37	1828 Dec. 12	0.472	242
1832 May 4	1210.24	0.3434	0.8455	13.39	1832 June 18	0.257	344
1835 Aug. 26	1209.38	0.3444	0.8451	13.38	1835 Sep. 10	1.314	92
1838 Dec. 19	1210.65	0.3440	0.8452	13.38	1838 Nov. 7	0.219	188
1842 Apr. 12	1210.01	0.3449	0.8449	13.36	1842 May 4	0.533	283
1845 Aug. 10	1215.60	0.3381	0.8474	13.15	1845 Aug. 31	1.184	31
1848 Nov. 26	1204.46	0.3370	0.8478	13.16	1848 Oct. 20	0.370	134
1852 Mar. 15	1204.62	0.3376	0.8476	13.15	1852 Mar. 20	0.646	226
1855 July 1	1203.34	0.3372	0.8477	13.15	1855 Aug. 2	0.737	325
1858 Oct. 18	1205.33	0.3408	0.8464	13.09	1858 Sep. 20	0.904	74
1862 Feb. 6	1206.88	0.3400	0.8467	13.10	1862 Jan. 31	0.625	172
1865 May 28	1206.68	0.3410	0.8463	13.08	1865 July 7	0.297	265
1868 Sep. 15	1205.70	0.3336	0.8492	13.13	1868 Aug. 27	1.233	9
1871 Dec. 29	1200.18	0.3330	0.8494	13.14	1871 Nov. 16	0.312	115
1875 Apr. 13	1201.19	0.3330	0.8494	13.14	1875 May 3	0.550	208
1878 July 26	1200.18	0.3335	0.8491	13.13	1878 Aug. 21	1.030	304
1881 Nov. 15	1208.14	0.3434	0.8453	12.90	1881 Oct. 11	0.540	53
1885 Mar. 8	1208.33	0.3424	0.8458	12.91	1885 Mar. 10	0.648	154
1888 June 28	1208.33	0.3431	0.8455	12.90	1888 July 31	0.713	246
1891 Oct. 18	1206.96	0.3405	0.8465	12.93	1891 Sep. 20	0.917	348
1895 Feb. 5	1205.82	0.3411	0.8462	12.92	1895 Jan. 28	0.618	96
1898 May 27	1207.12	0.3408	0.8464	12.92	1898 July 7	0.275	192

TABLE 36.A (Cont.)

T	ΔT	q	e	i	t	Δ_{min}	J
1901 Sep. 15	1206.60	0.3417	0.8460	12.90	1901 Aug. 28	1.252	286
1905 Jan. 12	1214.42	0.3389	0.8471	12.59	1904 Dec. 15	0.479	34
1908 May 1	1205.03	0.3378	0.8475	12.60	1908 June 16	0.317	137
1911 Aug. 19	1205.12	0.3384	0.8472	12.59	1911 Sep. 7	1.241	229
1914 Dec. 5	1203.90	0.3379	0.8475	12.59	1914 Oct. 27	0.288	329
1918 Mar. 24	1205.41	0.3408	0.8464	12.54	1918 Apr. 1	0.630	78
1921 July 13	1206.75	0.3400	0.8467	12.55	1921 Aug. 12	0.876	175
1924 Oct. 31	1206.35	0.3411	0.8463	12.53	1924 Sep. 30	0.756	268
1928 Feb. 19	1205.90	0.3326	0.8495	12.56	1928 Feb. 17	0.651	13
1931 June 3	1199.28	0.3319	0.8498	12.57	1931 July 12	0.312	119
1934 Sep. 15	1200.17	0.3319	0.8498	12.57	1934 Aug. 28	1.248	211
1937 Dec. 27	1199.47	0.3324	0.8496	12.55	1937 Nov. 14	0.271	307
1941 Apr. 17	1206.39	0.3414	0.8462	12.36	1941 May 12	0.514	56
* 1944 Aug. 6	1207.08	0.3403	0.8466	12.37	1944 Aug. 29	1.133	157
1947 Nov. 26	1207.10	0.3410	0.8463	12.36	1947 Oct. 20	0.420	249
1951 Mar. 16	1205.88	0.3380	0.8475	12.39	1951 Mar. 20	0.648	351
1954 July 2	1204.31	0.3384	0.8473	12.38	1954 Aug. 4	0.723	99
1957 Oct. 19	1205.33	0.3381	0.8474	12.38	1957 Sep. 21	0.912	194
1961 Feb. 5	1204.75	0.3390	0.8471	12.37	1961 Jan. 28	0.616	288
1964 June 3	1213.90	0.3393	0.8470	11.98	1964 July 13	0.330	37
1967 Sep. 22	1205.57	0.3382	0.8474	11.99	1967 Sep. 2	1.210	139
1971 Jan. 9	1205.92	0.3389	0.8472	11.98	1970 Dec. 1	0.425	232
1974 Apr. 28	1205.02	0.3381	0.8475	11.99	1974 June 13	0.363	332
1977 Aug. 17	1206.01	0.3407	0.8465	11.94	1977 Sep. 6	1.222	80
1980 Dec. 6	1207.57	0.3399	0.8468	11.95	1980 Oct. 28	0.278	177
1984 Mar. 27	1207.11	0.3410	0.8463	11.93	1984 Apr. 5	0.624	271
1987 July 17	1206.72	0.3317	0.8499	11.93	1987 Aug. 15	0.885	17
1990 Oct. 28	1199.17	0.3309	0.8502	11.95	1990 Sep. 28	0.797	121
1994 Feb. 9	1199.91	0.3309	0.8502	11.94	1994 Feb. 3	0.633	214
1997 May 23	1199.12	0.3314	0.8500	11.93	1997 July 4	0.190	310

T = date of passage at perihelion (UT date)
q = perihelion distance in astronomical units
e = eccentricity of the orbit
i = inclination of the orbit on the ecliptic of 2000.0
t = date of least distance to Earth (UT date)
Δ_{min} = least distance to Earth in astronomical units
J = heliocentric longitude (2000.0) of Jupiter at time T

It was the German astronomer J. F. Encke (1791–1865) who showed that the four apparitions concerned one and the same comet. He calculated an elliptic orbit and found a revolution period of 3.3 years. Contrary to the then general practice, the comet did not receive the name of its first discoverer, Méchain, but it was named after Encke who had investigated the comet's motion thoroughly.

Since 1819 comet Encke has been observed at each of its returns at the perihelion, except in 1944. Table 36.A lists all perihelion passages from 1786 to 1997. The orbital data have been taken from the tenth *Catalogue of Cometary Orbits* by Marsden and Williams [1]. For the unobserved returns, which are indicated by an asterisk in the first column, the data are from Marsden and Sekanina [2]. For the return of 1997, predicted elements are taken from *MPC* 23483 [3].

The first column of the table gives the date T of the passage at the perihelion. The second column gives the time difference ΔT, in days and decimals, since the previous perihelion time. The comet's orbit appears to be very stable. During the period 1786–1997, the quantity ΔT varies between the extreme values 1199 and 1216 days. The mean value of the period of revolution has been

$$
\begin{array}{lll}
1207.4 \text{ days} & \text{during} & 1786{-}1852 \\
1205.7 & - & 1852{-}1918 \\
1204.8 & - & 1918{-}1997
\end{array}
$$

The next three columns of the table list the perihelion distance q in astronomical units, the eccentricity e of the orbit, and the inclination i (in degrees and decimals) on the ecliptic of 2000.0.

Next, the table gives the date t of the comet's least distance to the Earth, and the value Δ_{min} of this least distance in AU, at each of the returns. Before 1997, the comet's least distance to Earth has been 0.219 AU (33 million kilometers), on 1838 November 7. However, on 1997 July 4 the comet approaches us to only 0.190 AU, or 28 million kilometers, its closest distance ever since its discovery in 1786.

It appears from the table that, for every return of the comet, the interval between the perihelion time T and the time t of the least distance to the Earth is at most 46 days.

Finally, in the last column of the table we find the heliocentric longitude J of Jupiter at the comet's perihelion time T. This longitude, which is referred to the equinox of 2000.0, will be used in a little while.

The comet's perihelion is just inside the orbit of Mercury. During the period 1786–1997, the extreme values of the perihelion distance q are 0.331 and 0.346 astronomical units. The aphelion distance is 4.1 AU, so the orbit of comet Encke lies completely inside that of Jupiter (Figure 36.*a*).

The orbital eccentricity e too remained between narrow limits during the period 1786–1997, namely between 0.844 and 0.850. This is not the case, however, for the orbit's inclination. Instead, we note a gradual decrease of this inclination, from 13.7 degrees near A.D. 1790 to 11.9 near 1990. Moreover, it appears from Table 36.A that this decrease took place in 'jumps'. So there occurred the following sudden decreases in i :

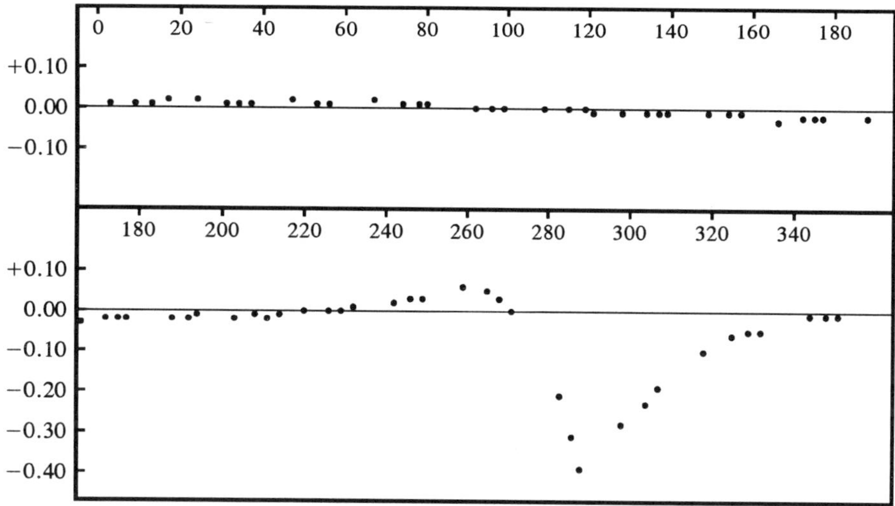

Fig. 36.b : *The variation of the orbital inclination of comet Encke at its next return (vertical scale, in degrees) as a function of J (horizontal scale). See the explanation in the text.*

0.10 degree	between the returns of	1795 and 1799
0.28	— —	1819 and 1822
0.21	— —	1842 and 1845
0.23	— —	1878 and 1881
0.31	— —	1901 and 1905
0.19	— —	1937 and 1941
0.39	— —	1961 and 1964

From the last column of the table, we see that these jumps always occur for approximately the same values of J. In Figure 36.b we have plotted, for each return of comet Encke, the variation Δi of the orbital inclination (in degrees) as a function of J. In other words, if J is the heliocentric longitude of Jupiter at the perihelion time of the comet, then the graph indicates the variation Δi of the orbital inclination at the *next* return. For example, for the perihelion passage of 1819 we have $J = 298°$ and $\Delta i = -0.28$ degree.

It is obvious that the plotted points are not distributed randomly. They are situated on a well-defined curve. This proves that indeed Δi depends on J. In other words, the variations of the comet's orbital inclination are caused principally by the gravitational action of Jupiter. When J is between 0° and 90°, or between 220° and 270°, then Δi is positive and small. That is, when at the time of the comet's perihelion passage Jupiter is situated in one of these two parts of its orbit, the inclination is a little larger at the comet's next return.

But when J is between 280° and 310°, then Δi has a large negative value. Only then does the inclination decrease considerably. The largest negative values of Δi occur when J is in the vicinity of 290°. What is the uniqueness of this particular value of J? When, at the time of the perihelion passage of the comet, Jupiter is at heliocentric longitude 290°, then half a revolution period of the comet later (that is, after about 603 days) Jupiter will be at longitude 340° approximately. This is precisely the longitude of the comet's *aphelion*. And, of course, the comet itself will then be at the aphelion of its orbit. Hence, the large variations in the inclination of the comet's orbit take place at *some* aphelion passages of the comet, namely only at those times when Jupiter is in the neighborhood.

Further, we note that the large jumps in i repeat after either 7 or 11 returns of the comet. This corresponds to 23 and 36 years, respectively, or 2 and 3 revolutions of Jupiter. From the value $J = 310°$ at the return of 1997, we thus may expect another negative jump in i at the next return in A.D. 2000. And, indeed, Yeomans and Wimberly [4] predict $i_{2000} = 11°.76$ for that return $(T = 2000$ September 9), or a variation of $\Delta i = -0.16$ degree with respect to the orbit of 1997. This jump occurs 11 revolutions of the comet after that of 1961–1964.

Finally, because the orbital inclination of the comet's orbit is gradually decreasing, the least separation between this orbit and that of the Earth is decreasing too: 0.210 AU in 1786, 0.196 AU in 1891, and 0.175 AU in 1997. Please note that these are the values of the least distance between the *orbits* of the comet and the Earth, *not* between these bodies themselves!

REFERENCES

1. B. G. Marsden and G. V. Williams, *Catalogue of Cometary Orbits*, tenth edition, 1995; International Astronomical Union, Central Bureau for Astronomical Telegrams / Minor Planet Center.

2. B. G. Marsden and Z. Sekanina, 'Periodic comet Encke 1786–1971', *Astronomical Journal*, Vol. 79, No. 3, pages 413–419 (March 1974).

3. *Minor Planet Circular* (*MPC*) No. 23483 (1994 May 25).

4. D. K. Yeomans and R. N. Wimberly, *Cometary Apparitions: 1990–2010*, Jet Propulsion Laboratory, Cometary Science Team, Preprint Series No. 129 (May 1990).

37. *The orbital inclinations of the four Galilean satellites*

In the literature, we find differing values for the inclinations of the orbits of the four great satellites of Jupiter on the equator of that planet. What are their *exact* values? It depends on the adopted definition for the 'inclination'. This requires some explanation, the subject of this chapter.

We first consider a simple case. Let i be the inclination of the orbit of a satellite on the equatorial plane of Jupiter (Figure 37.a). Let Ω be longitude of the ascending node N of the orbit, that is, the arc from the vernal equinox γ to N. At some given instant, the satellite is situated at S, and its longitude is $\lambda = \text{arc } \gamma N + \text{arc } NS$, so $\text{arc } NS = \lambda - \Omega$. Then the latitude β of the satellite, referred to the equatorial plane of Jupiter, is given by $\sin \beta = \sin i \sin (\lambda - \Omega)$. When the orbital inclination i is very small, as is the case for the great satellites of Jupiter, we may write

$$\beta = i \sin (\lambda - \Omega) \tag{1}$$

It is evident that the extreme values of β are $+i$ and $-i$. They are reached for $\lambda - \Omega = 90°$ and $\lambda - \Omega = 270°$, respectively, namely when the satellite is $90°$ from the node.

However, in the case of the Galilean satellites of Jupiter we should also take the mutual perturbations into account. The motion of each of the four satellites is somewhat perturbed by the attractions of the other three satellites, and also by that of the Sun and the flattening of Jupiter itself. For instance, the latitude of satellite III (with respect to the equator of Jupiter), expressed in degrees, is given by

$$\beta = 0.18543 \sin (\lambda - \Omega) + 0.09689 \sin (\lambda - A) + 0.03924 \sin (\lambda - B)$$
$$+ 0.01608 \sin (\lambda - C) + \text{terms with smaller coefficients} \tag{2}$$

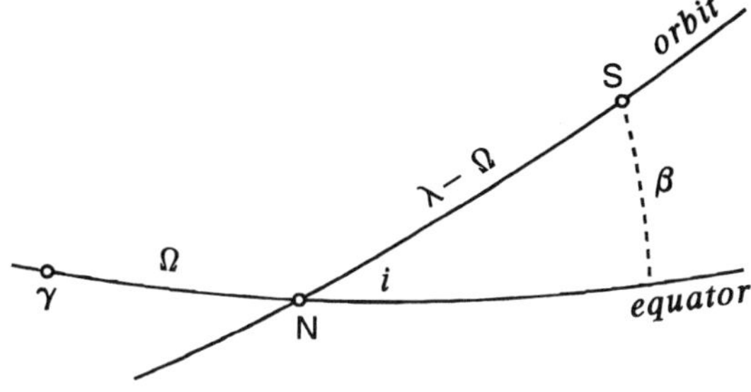

Fig. 37.a : *The two arcs are the intersections of the equatorial plane of Jupiter and the orbital plane of the satellite with the celestial sphere.*

where, as before, λ is the longitude of the satellite, and Ω the longitude of its ascending node on the equator of Jupiter, while A, B and C are angles whose definitions are irrelevant here. Ω as well as A, B and C are quantities *which very slowly vary with time.* For instance, for satellite III the angle Ω decreases by $0.007\,177\,04$ degree per day, so it needs 137 years to

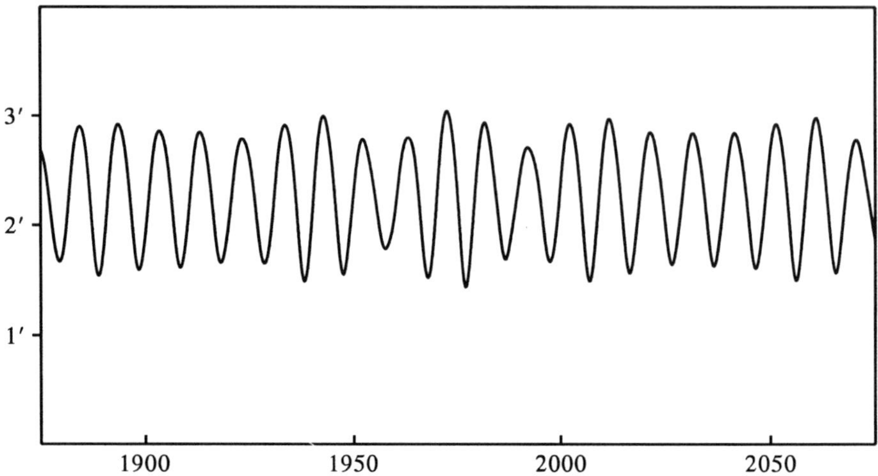

Fig. 37.b : *The effective inclination of the orbit of satellite I (Io)
on the equatorial plane of Jupiter, 1875 to 2075.*

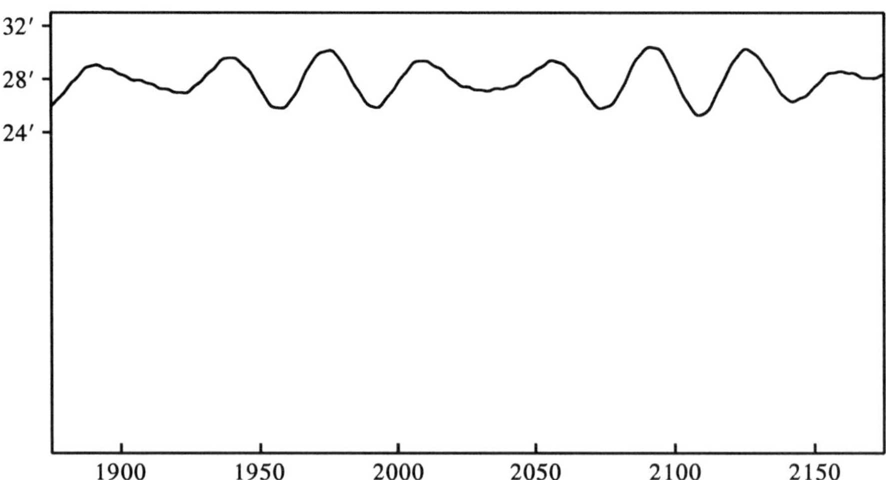

Fig. 37.c : *The effective inclination of the orbit of satellite II (Europa)
on the equatorial plane of Jupiter, 1875 to 2175.*

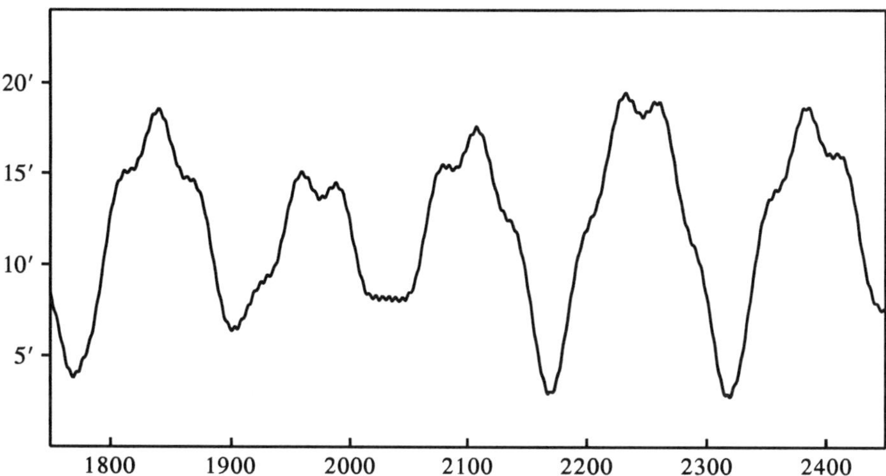

Fig. 37.d : *The effective inclination of the orbit of satellite III (Ganymede) on the equatorial plane of Jupiter, 1750 to 2450.*

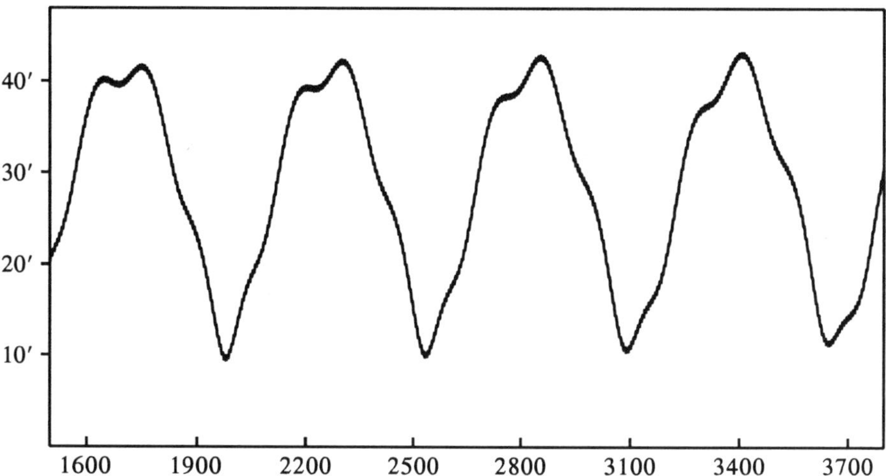

Fig. 37.e : *The effective inclination of the orbit of satellite IV (Callisto) on the equatorial plane of Jupiter, 1500 to 3800.*

change by 360 degrees. In other words, Ω, A, B and C are 'almost-constants'. In a first approximation, and to trace the behavior of the satellite, they may be considered as constants indeed. Of course, the longitude λ of the satellite itself varies very much faster. For instance, the revolution period of satellite III is 7.15 days, so for this satellite the quantity λ increases by 50 degrees per day.

By means of the classical formula $\sin(a - b) = \sin a \cos b - \cos a \sin b$, we may write expression (2) as

$$\begin{aligned} \beta = \ & 0.18543 \sin \lambda \cos \Omega - 0.18543 \cos \lambda \sin \Omega \\ &+0.09689 \sin \lambda \cos A - 0.09689 \cos \lambda \sin A \\ &+ \ . \ . \ . \ . \ . \end{aligned}$$

or as

$$\begin{aligned} \beta = \ & (0.18543 \cos \Omega + 0.09689 \cos A + \ . \ . \ . \ . \ .) \sin \lambda \\ &- (0.18543 \sin \Omega + 0.09689 \sin A + \ . \ . \ . \ . \ .) \cos \lambda \end{aligned}$$

Because the quantities Ω, A, B and C are almost constant during a short time period (a few weeks, say), so are the two expressions between parentheses. Let us call these expressions P and Q. So we have

$$\beta = P \sin \lambda - Q \cos \lambda \tag{3}$$

where P and Q slowly vary with time. During a period of a few weeks they may be considered as constants.

Actually, expression (3) is of the same form as (1). To see this, we can pose $\tan \xi = Q/P$, so that ξ too is a 'constant'. We then obtain

$$\beta = \sqrt{P^2 + Q^2} \ \sin(\lambda - \xi) \tag{4}$$

Prove this by way of exercise! In formula (2), Ω is the longitude of the ascending node of the orbit of the satellite on the equatorial plane of Jupiter. Therefore, $0°.18543$ is 'the' orbital inclination of the satellite. However, due to the other periodic terms in the expression for the satellite's latitude, this latitude actually varies according to formula (4). Hence, the satellite now no longer has an orbital inclination of 0.18543 degree, but it has an 'effective' inclination equal to

$$\sqrt{P^2 + Q^2} \tag{5}$$

which, of course, depends on the values of Ω, A, B and C, and which, just as these quantities, *slowly* varies with time. Considered over a short period (a few weeks), the satellite moves as if it had an inclination given by (5) instead of $0°.18543$. The latter value, hence the coefficient of $\sin(\lambda - \Omega)$, is called the *proper inclination* of the orbit. The 'apparent' inclination (5) can be called the *effective* inclination.

Said in words instead of with mathematics, each of the periodic terms such as $a \sin (\lambda - A)$ has as consequence that the satellite actually follows a path inclined by an angle a to the proper orbit which has inclination i (0°.18543 in the case of satellite III) on Jupiter's equatorial plane. The resulting effective orbital inclination will have a value somewhere between $i + a$ and $i - a$, depending on the position of the 'node' of the perturbation term $a \sin (\lambda - A)$ with respect to the node N of the proper orbit.

For satellite III, the effective orbital inclination can have the maximum value (in degrees)

$$0.18543 + 0.09689 + 0.03924 + 0.01608 + \text{ some smaller terms,}$$

that is 0°20′. The smallest possible effective orbital inclination for this satellite is

$$0.18543 - 0.09689 - 0.03924 - 0.01608 - \ldots$$

degrees, or 0°02′. So, the effective orbital inclination of satellite III can take all values between 0°02′ and 0°19′. The 'mean' value is 0°11′, which is the proper inclination.

In the case of satellite IV there is a peculiarity: one of the periodic terms has a *larger* coefficient (0°.43876) than the proper inclination (0°.25295), with the consequence that for this satellite the 'mean' inclination (0°26′) is *not* equal to the proper inclination (0°15′).

TABLE 37.A

Proper and effective orbital inclinations of the Galilean satellites

Satellite		*Proper orbital inclination*	*Extreme values of the effective orbital inclination*		*Period (years)*
I	Io	0° 02′ 14″	0° 01′	0° 03′	10
II	Europa	0° 27′ 56″	0° 25′	0° 31′	39
III	Ganymede	0° 11′ 08″	0° 02′	0° 20′	137
IV	Callisto	0° 15′ 11″	0° 09′	0° 44′	561

The figures on the previous pages show the variations of the effective orbital inclinations of the four satellites in the course of the years. Take notice of the different scales, vertically as well as horizontally. For satellite I the curve is given for two centuries, but it is given for 23 centuries in the case of satellite IV.

See how different the four curves are. In each case a well-defined period is readily apparent. This period results from the interference between the periods of the two largest periodic terms — in the case of satellite III, the first two terms of expression (2). The length of this period is given in the last column of Table 37.A.

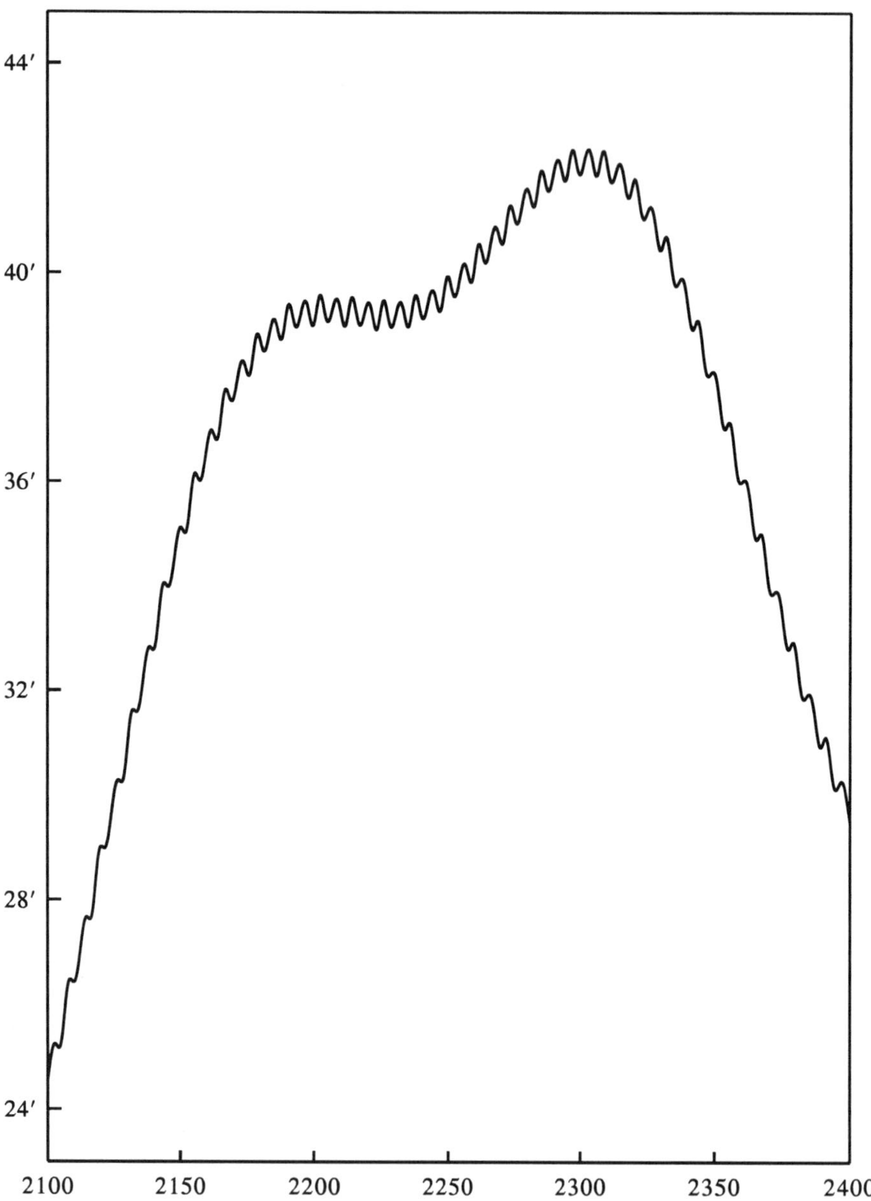

Figure 37.f : *The effective inclination of the orbit of satellite IV (Callisto)
for the shorter period 2100–2400 and at a larger scale than Figure 37.e,
to show more clearly the variations of short period.*

PLANETARY PHENOMENA

38. *Planetary motions: approximate periodicities*

Mathematical astronomy is an exact science. Astronomers involved in this branch of astronomy can calculate many years in advance, and with high accuracy, events such as solar and lunar eclipses, occultations by the Moon, or planetary conjunctions.

However, it is not necessary to perform *all* calculations with the highest possible accuracy. In many cases approximate results are sufficient, allowing the use of simplified and faster methods of calculation. For instance if one wishes to know, for a given instant, the position of Jupiter and if an accuracy of one degree is sufficient, we may consider an unperturbed elliptic orbit. But in order to obtain positions accurate to 1 arcsecond, one should take many periodic terms ('perturbations') into account, or execute the calculation by means of numerical integration.

Will bright moonlight interfere with observations of the Perseid meteor shower next August? My neighbor wants to look at Saturn through my telescope; when is this planet visible in the evening? To answer these and similar questions, approximate data are sufficient. Indeed, in such cases it is not necessary to know to the nearest minute the time of the Full Moon, nor when Saturn rises and sets.

In this respect, approximate periodicities of planetary motions can be of practical utility. Here are two fictive examples:

— Mr. X was born in 1949. He doesn't have any astronomical information for that year, but he has a collection of almanacs for recent years. How can he get an idea of the visibility of the planets in 1949 without making the actual calculations?

— The year is 1996. Dr. Nichtraucher writes a column for an astronomical magazine. He wants to write a text about the visibility of Mercury in September 1997. He already has the needed numerical information, but he is rather lazy and lacks inspiration about *what* to write. What help could he use?

So, let us first consider the case of *Mercury*. What we need is a period equal to an *integer* number of years, after which the planet is again in approximately the same position in its orbit. In this case, after such a time period the planet will be again in nearly the same position with respect to the perihelion and to the ascending node of its orbit, and so the equation of the center (the longitude difference between the true planet and the mean planet due to the orbital eccentricity) and the heliocentric latitude will again approximately repeat. Moreover, after an integer number of years, the Earth itself will of course have returned to nearly the same place in its orbit, so that the planetary phenomena will repeat on nearly the same date of the year. Finally, the ecliptic at sunrise or sunset will then be at almost exactly the same position with respect to the horizon, and this is of importance to infer the visibility of a planet which is situated close to the Sun in the sky.

Well, in the case of Mercury we have approximate periodicities of 6 and 7 years. But a more exact periodicity is that of 13 years, which is the sum of the two ones just mentioned. When using the period of 13 years, the errors of the periods of 6 and 7 years cancel each other for a large part). So we have, for instance, inferior conjunction of Mercury with the Sun on the following dates:

1993 July 15
1999 July 26   (11 days later than in 1993)
2000 July 6   (9 days earlier than in 1993)
2006 July 18   (3 days later than in 1993)

So, if you have astronomical almanacs for 13 successive years, you can deduce approximate data for the phenomena of Mercury during a long time! In one of the fictive examples cited above, Dr. Nichtraucher can take the text about Mercury's visibility written for September *1984*, and replace the old numerical data with those for September 1997 to have his copy for the September 1997 article.

As another example of the periodicity of 13 years, let us give the following instants (in UT) of the rise, transit through the southern meridian, and setting of Mercury at Paris, France:

	Rise	*Transit*	*Set*
1961 February 11	7^h32^m	13^h05^m	18^h38^m
1974 February 11	7^h39^m	13^h11^m	18^h45^m
1987 February 11	7^h43^m	13^h13^m	18^h45^m
2000 February 11	7^h45^m	13^h12^m	18^h40^m

Remarkable, isn't it?

There also exists a periodicity of 33 years. For example, the inferior conjunction of Mercury on 1973 July 20 will repeat on 2006 July 18. However, there is a more accurate periodicity of 46 years, which is the sum of those of 13 and 33 years. For example, on the following dates there was or will be a transit of Mercury over the solar disk:

1927 November 10
1973 November 10
2019 November 11
2065 November 11

Don't confuse periodicity and mean frequency. It is important to note that there are many other transits of Mercury, for instance on 1993 November 6 or on 2003 May 7, but they don't belong to *this* 46-year series.

As another example, let us mention the following greatest western (morning) elongations of Mercury, illustrating the periods of 33 and 46 years, respectively:

1939 Aug. 28	18° 16′	1913 Aug. 22	18° 26′
1972 Aug. 25	18° 20′	1959 Aug. 23	18° 25′
2005 Aug. 23	18° 24′	2005 Aug. 23	18° 24′
2038 Aug. 22	18° 29′	2051 Aug. 25	18° 23′

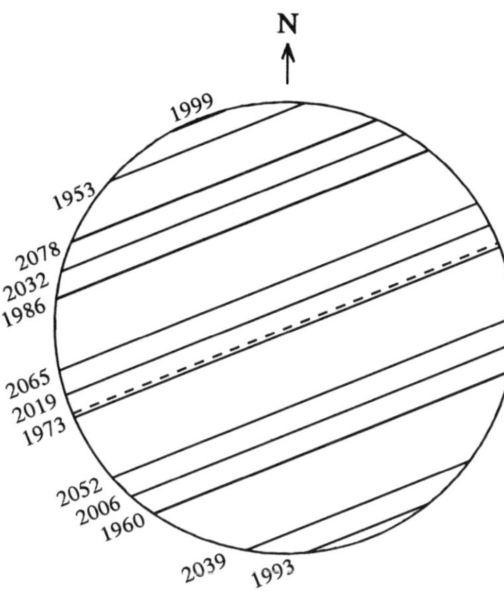

Fig. 38.a : *The paths of Mercury over the Sun's disk at the November transits from 1953 to 2078. North (N) is at the top. In each case, Mercury enters from the east (left) and leaves the Sun at the western limb (right). Note the periodicities of 6, 7, 13, 33 and 46 years, which are mentioned in the text. The dashed line is the chord described by Mercury at the transit of 2190 November 12, which occurs 217 years after that of 1973 November 10.*

For Mercury, there exists a still better periodicity of 217 years. The various periodicities of 6, 7, 13, 33, 46 and 217 years are illustrated in Figure 38.a. Look for instance at the chords described by Mercury over the solar disk in 1993–1999, in 1986–1993, 1960–1973–1986, 1953–1986–2019–2052, 1973–2019–2065, and in 1973–2190. There doesn't exist a truly *exact* periodicity, even of long period. Indeed, the orbital elements of the planetary orbits are subject to slow, 'secular' variations, which prevent the phenomena to repeat *exactly*.

The periods of 6, 7, 13, 33, 46 and 217 years can be found as follows. The sidereal revolution periods of the Earth are Mercury are 365.256363 and 87.969256 days, respectively. If we divide the first value by the second, we obtain 4.152091. If we now search for successive fractions with integer numerator and denominator, and which converge ever more closely to the given value 4.152091, we obtain

$$\frac{4}{1} \qquad \frac{25}{6} \qquad \frac{29}{7} \qquad \frac{54}{13} \qquad \frac{137}{33} \qquad \frac{191}{46} \qquad \frac{901}{217} \quad \text{etc} \ldots$$

The method how to find these fractions is described in the section 'Approximations of a decimal number by fractions' further in this chapter.

In the denominators of these fractions we find the periods mentioned before. For instance, the fraction 54/13 means that 13 sidereal periods of the Earth correspond to approximately 54 revolutions of Mercury.

The first fraction corresponds to 1 year, which in fact is a too inaccurate periodicity. For example, Mercury's greatest western elongation of 1997 May 22 (25° 22′) 'repeats', after 4 sidereal revolutions (or 3 synodic periods) of the planet, on 1998 May 4 (26° 44′ W), so this periodicity of one year is not very useful.

For *Venus* we have the well-known periodicity of 8 years, and the much longer one of 243 years. Here are some examples of that of 8 years; the instants are rounded to the nearest integer hour of Universal Time:

Venus in greatest western elongation			*Venus in conjunction (in right ascension) with Regulus*		
1972 Aug. 27	2^h	45°53′	1972 Oct. 4	23^h	Venus 18′ S
1980 Aug. 24	19	45°52′	1980 Oct. 4	16	Venus 16′ S
1988 Aug. 22	12	45°51′	1988 Oct. 4	8	Venus 14′ S
1996 Aug. 20	3	45°50′	1996 Oct. 4	0	Venus 12′ S

In March and April 1996 Venus was a bright evening star, well placed for observers in the northern hemisphere. Consequently, this will also be the case in March and April of the years 2004, 2012, etc., as it was in March and April 1988, 1980, 1972, etc. Here, however, we may not extrapolate too far into the past or into the future by reason of the 'error' of 2 or 3 days in the periodicity, which

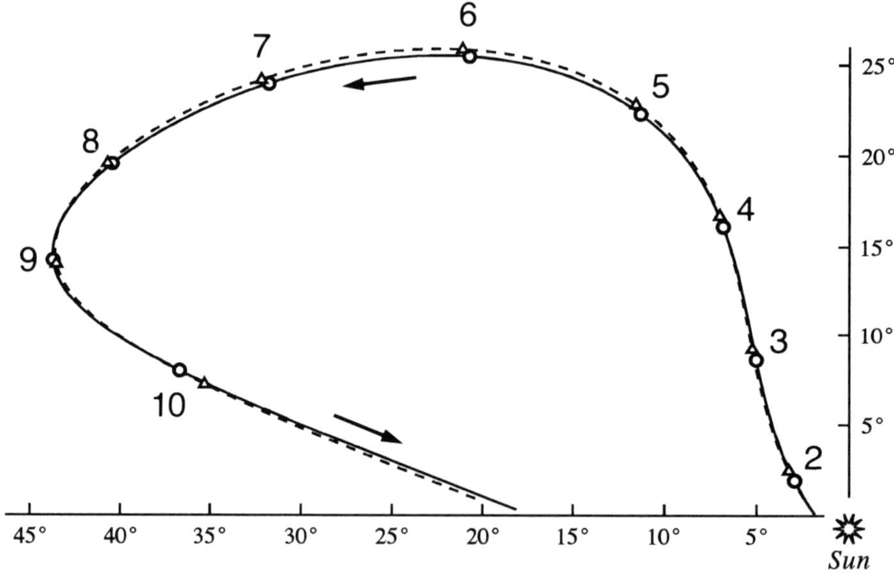

Fig. 38.b : The evening appearance of Venus at Boston, Mass., in 1994 (solid line) and in 2002 (dashed). The graph shows the position of the planet with respect to the setting Sun. Horizontally : the difference between the azimuths of Venus and the Sun. Vertically : the altitude of Venus. The points labelled 2, 3, etc., are the positions of Venus on February 1, March 1, etc. The small circles refer to the year 1994, the small triangles to 2002. Look how similar the two curves are.

sums up each time we add 8 years: 13 sidereal revolutions of Venus are 0.94 day shorter than 8 of the Earth, making 5 *synodic* revolutions of Venus 2.4 days shorter than 8 Julian years.

Was Venus visible on the morning of 1492 October 12 when Columbus discovered America? Here we should use the longer period of 243 years. We find 1492 + 243 + 243 = 1978. In an almanac for 1978 we read that in October Venus was not visible in the morning sky, so this also has been the case in 1492. If you don't have an almanac of the year 1978, you can use one for 8 or 16 years later. As you see, almanacs of previous years retain their value for some researches, so don't throw them away!

The sidereal revolution period of *Mars* is 686.979 852 days, so the ratio of the periods Earth / Mars is 0.531 684 24, and this yields the following fractions:

$$\frac{1}{2} \qquad \frac{8}{15} \qquad \frac{17}{32} \qquad \frac{25}{47} \qquad \frac{42}{79} \qquad \frac{151}{284} \qquad \text{etc.}$$

Besides the very rough periodicity of 2 years, corresponding to one synodic revolution of Mars, we have the ever better periods of 15, 32, 47, 79, and 284 years. Note that 47 = 15 + 32, while the period of 79 years is the sum of those of 32 and 47 years. The following opposition dates of Mars illustrate these periodicities. In the last column, note the jump of 10 extra days between the events of 1412 and 1696, due to the Gregorian calendar reform in A.D. 1582, when it was decided to suppress ten days.

15 years	*32 years*	*47 years*	*79 years*	*284 years*
1935 Apr. 6	1884 Feb. 1	1839 Mar. 12	1743 Feb. 16	1128 Feb. 8
1950 Mar. 23	1916 Feb. 10	1886 Mar. 6	1822 Feb. 19	1412 Feb. 9
1965 Mar. 9	1948 Feb. 17	1933 Mar. 1	1901 Feb. 22	1696 Feb. 20
1980 Feb. 25	1980 Feb. 25	1980 Feb. 25	1980 Feb. 25	1980 Feb. 25
1995 Feb. 12	2012 Mar. 3	2027 Feb. 19	2059 Feb. 27	2264 Feb. 28

Of course, the perihelic oppositions of Mars too repeat with the above-mentioned periodicities. For instance, that of August 2003 (least distance to Earth 0.373 AU) will occur 15 years after that of September 1988 (0.393 AU), 32 years after that of August 1971 (0.376 AU), 47 years after that of September 1956 (0.378 AU), and 79 years after that of August 1924 (0.373 AU).

On 1984 May 11, a transit of the Earth over the solar disk was visible from Mars, 79 years after a similar event on 1905 May 8. However, there will be none in May 2063, because even after such an accurate cycle of 79 years the displacement with respect to the nodes of the Martian orbit is rather large.

Figure 38.*c* is another illustration of the periodicity of 79 years. Here we see the loops described by Mars in the vicinity of the star Regulus in 1994-1995 and in 2073-2074. See how similar the two loops are. That of 2073-2074 occurs about 2 degrees to the east of that of 1994-1995.

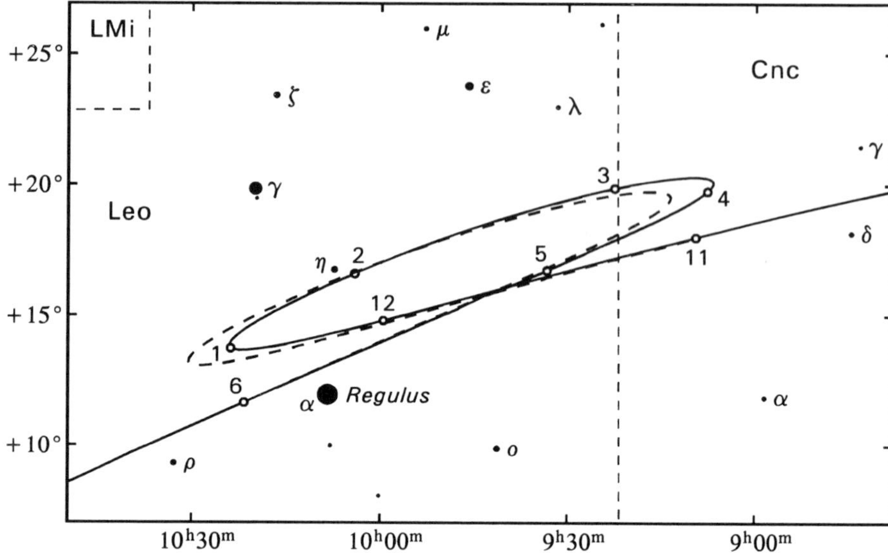

Fig. 38.c : *The back-and-forth motion of Mars in the vicinity of Regulus in 1994-1995 (solid line) and in 2073-2074 (dashed). The position labelled '11' is that of 1994 November 1, '12' = 1994 December 1, etc. till '6' = 1995 June 1, each time at 0h Universal Time.*

The sidereal revolution period of *Jupiter* is 11.86 years, so for this planet we have periodicities of 12 and of 83 years. Here are some examples:

Jupiter in opposition *with the Sun*	*Jupiter in conjunction* *with the Sun*
1968 Feb. 20	1806 June 25
1980 Feb. 24	1889 June 24
1992 Feb. 29	1972 June 24
2004 Mar. 4	2055 June 24

Was Jupiter visible on the morning of 1492 October 12? If to 1492 we add six times 83, we arrive at 1990. In an astronomical almanac of that year we read that in October Jupiter was visible in the morning, so that was also the case in 1492.

The revolution period of *Saturn* is 29.46 years. Hence, two periods of this planet yield the approximate periodicity of 59 years. For example, on 1960 July 7 Saturn was in opposition with the Sun; 59 years later, in A.D. 2019, the planet will be in opposition on July 9.

For *Uranus* there is the periodicity of 84 years, or one revolution period of the planet, which we already mentioned on page 161.

Approximate periodicities can be found for minor planets too. Oppositions of *Ceres* repeat at nearly the same date of the year after a period of 23 years, for instance on 1950 June 3, 1973 June 1, 1996 May 29, and 2019 May 28. (These are the dates of opposition in celestial longitude, not in right ascension).

For *Pallas* there also is a periodicity of 23 years. *Juno* and *Vesta* have periodicities of 48 and 29 years, respectively.

The shift of the opposition date after one such cycle is larger for perihelic oppositions than for oppositions taking place near the aphelion of the asteroid's orbit. Here are examples for the 29-year periodicity of Vesta:

perihelic oppositions	*aphelic oppositions*
1949 June 11	1954 Dec. 16
1978 June 5	1983 Dec. 13
2007 May 30	2012 Dec. 9
2036 May 23	2041 Dec. 6
(shift ≈ 6 days)	(shift ≈ 3 days)

For 433 *Eros*, there is a periodicity of 81 years. For instance, the very favorable perihelic opposition of January 1975 (least distance to Earth 0.151 AU) will repeat in January 2056 (distance 0.150 AU).

The case of minor planet 16 *Psyche* is remarkable. The sidereal revolution period of this body is almost exactly 5 years, so the oppositions repeat after five years on almost the same calendar dates. For instance, here are the dates of the oppositions of Psyche in celestial longitude, from 1987 to 2005:

1987 Mar. 1	1988 May 7	1989 Aug. 4	1990 Dec. 5
1992 Mar. 1	1993 May 8	1994 Aug. 3	1995 Dec. 6
1997 Mar. 1	1998 May 8	1999 Aug. 4	2000 Dec. 6
2002 Mar. 2	2003 May 9	2004 Aug. 4	2005 Dec. 7

Conjunctions of Psyche with a given star too repeat under almost identical circumstances every five years. For example, here are some conjunctions (in right ascension) of Psyche with α Tauri:

1980 Dec. 30	Psyche 1°06′ north of Aldebaran
1985 Dec. 25	Psyche 1°00′ north of Aldebaran
1990 Dec. 24	Psyche 0°58′ north of Aldebaran
1995 Dec. 24	Psyche 0°56′ north of Aldebaran
2000 Dec. 29	Psyche 1°05′ north of Aldebaran

Approximations of a decimal number by fractions

How can one obtain fractions, with integer numbers as numerator and denominator, and which approximate a given decimal number α ever more closely? We will not explain the mathematical theory here, but just illustrate the method by means of an example.

So, let us consider the number $\alpha = 4.152091$, which we came across in the case of Mercury earlier in this chapter. What we wish to obtain are the numerators $p_1, p_2, p_3, \ldots$ and the denominators q_1, q_2, q_3, etc., of the successive fractions. Let n be the serial number of a fraction, with numerator p_n and denominator q_n.

Write the given number α as a fraction with integer numerator A and integer denominator B, the denominator being a power of 10. This gives

$$\alpha = 4.152091 = \frac{4\,152\,091}{1\,000\,000}$$

so $A = 4\,152\,091$, $B = 1\,000\,000$. Now construct Table 38.A as follows. We begin with the calculation of the four first lines a, b, c, and r.

In column $n = 1$, put the values $a = A = 4\,152\,091$ and $b = B = 1\,000\,000$. Calculate the integer quotient c and the rest r of the division of a by b, namely $c = \text{INT}\,(a/b) = 4$, and $r = a - bc = 152\,091$.

In the next column ($n = 2$), take for a the value of b from the previous column, and for b the value of r of that same column. Then we calculate, as before, $c = \text{INT}\,(a/b) = 6$, and $r = a - bc = 87454$.

And so we continue for the successive columns, each time taking $a_n = b_{n-1}$, $b_n = r_{n-1}$, $c_n = \text{INT}\,(a_n/b_n)$, and $r_n = a_n - b_n c_n$. The operation stops when we reach a point where the rest r is zero.

Now we calculate the values p and q. In the first column ($n = 1$), we take $p_1 = c_1 = 4$, while q_1 is always 1. In the next column, we put $p_2 = c_1 c_2 + 1 = (4 \times 6) + 1 = 25$, and $q_2 = c_2 = 6$.

From then on, $p_n = c_n p_{n-1} + p_{n-2}$ and $q_n = c_n q_{n-1} + q_{n-2}$. For instance, in column $n = 5$, we have $p_5 = (2 \times 54) + 29 = 137$, $q_5 = (2 \times 13) + 7 = 33$.

TABLE 38.A

n	1	2	3	4	5
a	4 152 091	1 000 000	152 091	87 454	64 637
b	1 000 000	152 091	87 454	64 637	22 817
$c = \text{INT}\,(a/b)$	4	6	1	1	2
$r = a - bc$	152 091	87 454	64 637	22 817	19 003
p	4	25	29	54	137
q	1	6	7	13	33

The values p and q are the numerators and denominators of the required fractions. So in this manner we obtain the following fractions:

$$\frac{4}{1} \quad \frac{25}{6} \quad \frac{29}{7} \quad \frac{54}{13} \quad \frac{137}{33} \quad \frac{191}{46} \quad \frac{901}{217} \quad \frac{1092}{263} \quad \frac{60961}{14682} \quad \text{and so on.}$$

The remarkable fact is that these fractions are alternately less and greater than the given number α, but each fraction is closer to this value than the previous one. So we have

$$
\begin{array}{llll}
4/1 & = 4.000000 & \text{this is } 0.152091 & \text{too small} \\
25/6 & = 4.166667 & 0.014576 & \text{too large} \\
29/7 & = 4.142857 & 0.009234 & \text{too small} \\
54/13 & = 4.153846 & 0.001755 & \text{too large} \\
137/33 & = 4.151515 & 0.000576 & \text{too small} \\
191/46 & = 4.152174 & 0.000083 & \text{too large} \\
901/217 & = 4.152074 & 0.000017 & \text{too small}
\end{array}
$$

The next value yields $1092/263 = 4.152091$, to six decimals, and this is precisely the value of the given number α, so in practice we should stop here. The fraction $60961/14682$ and the next ones have no physical meaning, as the given number α has initially been rounded to six decimal places: its true value is *not* equal to 4.152091000000.

The method described above is based on the theory of the so-called *continued* or 'chain' fractions, and the obtained ordinary fractions are called the *convergents*. A simple continued fraction has the form

$$a + \cfrac{1}{b + \cfrac{1}{c + \cfrac{1}{d + \dots}}}$$

Any number can be written as a *simple* chain fraction — 'simple' because the numerators are unity. When the number is rational, that is, expressible as the quotient of two integers, the chain terminates; when irrational, such as $\sqrt{2}$, the chain repeats its terms periodically.

However, the method does *not* provide *all* successive fractions which converge to the given value α ever more closely. For instance, the fraction $710/171$ is a better approximation to $\alpha = 4.152091$ than $191/46$, but it is less accurate than $901/217$. Similarly, if we start from the approximate value 3.1415926536 for the number π, the method gives the successive fractions

$$\frac{3}{1} \quad \frac{22}{7} \quad \frac{333}{106} \quad \frac{355}{113} \quad \dots$$

but in this series many fractions are missing. The complete list of gradually better fractions starts as

$$\frac{3}{1} \quad \frac{13}{4} \quad \frac{16}{5} \quad \frac{19}{6} \quad \frac{22}{7} \quad \frac{179}{57} \quad \frac{201}{64} \quad \frac{223}{71} \quad \frac{245}{78} \quad \frac{267}{85} \quad \frac{289}{92} \quad \frac{311}{99} \quad \frac{333}{106} \quad \frac{355}{113} \quad \ldots$$

For instance, 179/57 gives a more accurate value for π than 22/7, but a less accurate one than 333/106.

If we really want *all* successive fractions which approach the given number α more and more closely, then we have to use the 'brute-force' method, which consists of trying all successive values 1, 2, 3, ... as the denominator, each time finding the best matching numerator, and then checking whether the obtained fraction represents an improvement over the previously retained result.

Which of the two methods should be preferred? The answer depends on one's aim. If really *all* successively better fractions are required, then the brute-force method should be used. This is the case, for instance, for the following problem: what fraction with integer numerator and integer denominator, and *whose denominator is not larger than 100*, most closely approaches the number π? The answer is 311/99, a fraction which is not obtained by the method based on continued fractions.

The brute-force method requires a much longer calculation time. Besides, in practice one seldom needs *all* successive fractions arranged in order of increasing accuracy. Indeed, in such a series there appear too many approximations which are nearly equivalent. The method based on continuous fractions is a procedure which selects *some* fractions, with the guarantee that the absolute value of their error decreases considerably faster to zero than the series obtained by the brute-force method.

In other words, the brute-force method yields a large number of fractions which generally are not interesting. In the case of Mercury, the error of the fraction 710/171 is more than twice the error of 901/217, so the cycle of 217 years, though a little longer than that of 171 years, really is much better than the latter. In the case of π, the fraction 179/57 is barely a better approximation than 22/7.

In this respect, we note that the above-mentioned fractions 179/57 to 333/106, which are successively better approximations for π, all can be deduced from the first one: the next fraction is obtained by increasing the numerator by 22, and the denominator by 7. The continued-fraction method gives only the last fraction of this group, namely 333/106, which is the most accurate one.

The advantage of the continued-fraction method is even better seen by starting, from instance, from $\alpha = 0.002478$, a value which is comprised between 1/403 and 1/404. The brute-force method gives as first approximation the fraction 0/1, and then *all* the successive fractions 1/202, 1/203, 1/204, etc. till 1/404, and then 2/807. Mathematically, these crazy results are completely justified (1/204 is a better approximation to α than 1/203, and 1/205 is a still better approximation, and so on), but they just are not interesting in practice.

39. *Opposition loops*

During the weeks around its opposition with the Sun, an exterior planet has an apparent retrograde (westward) motion among the stars. This is caused by the fact that the Earth overtakes and passes the planet, which thus seems to move 'backward'. But since the planetary orbits are not situated exactly in the plane of that of the Earth — the ecliptic — the motions are not just a simple back-and-forth shift on one and the same line. Instead, they occur along characteristic curves.

These elegant loops are most striking in the case of oppositions of Mars. At these times, this planet is relatively close to the Earth, and as seen from nearby the deviations in latitude are amplified — see Figure 39.*a*. In the case of Jupiter, Saturn, Uranus and Neptune, the loops are flat because these planets are situated at a much greater distance than Mars, and their orbital inclinations are small.

The inner planets Mercury and Venus, too, describe nice loops on the celestial sphere. However, these loops always occur in the vicinity of the Sun, near

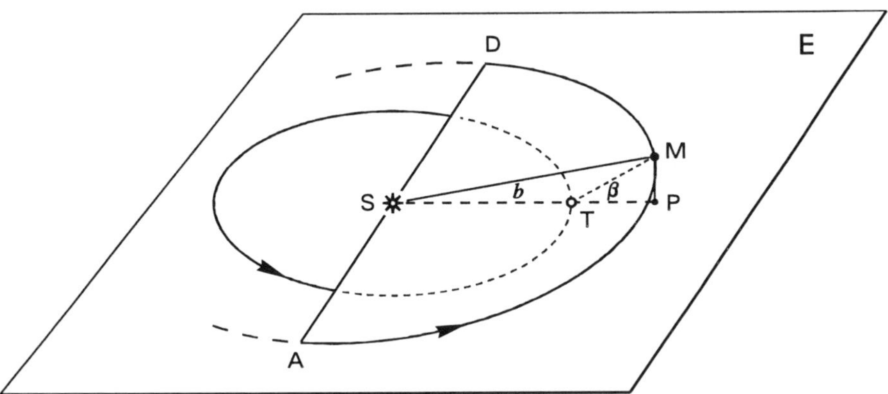

Fig. 39.a : A perspective view of the orbits of Earth and Mars. In reality, these orbits are almost circular. In the course of its revolution around the Sun S, the planet Mars (M) is alternately north ('above') and south ('below') the plane E of the Earth's orbit, the ecliptic. T = Earth. P = projection of M on the plane of the Earth's orbit. The intersection AD of the two planes is called the line of nodes.

On 1995 February 12 Mars was in opposition with the Sun, and this took place almost exactly midway between the ascending node A and the descending node D. The line SM was almost exactly perpendicular to AD, and the heliocentric latitude b of Mars was close to its greatest possible value, 1°51', which is the inclination of the orbit. The geocentric latitude β, however, was much larger. On 1995 February 12, the distances were SM = 1.662 and TM = 0.676 astronomical units. Therefore β was then approximately equal to 1°51' × (1.662/0.676) = 4°33'.

the planet's inferior conjunction, so they cannot be easily observed, or even not at all. On the other hand, opposition loops of Mars are interesting because the planet describes them in a dark sky, so they can be easily followed.

Figure 39.*b* shows the different shapes these opposition loops can have. The first case, No. 1 or *Na*, occurs when at the time of its opposition the planet is near the *ascending node* of its orbit, as it was the case for Mars in November 1958; it will occur again in November 2005. Indeed, at first the planet is south of the ecliptic; the next months, during the back-and-forth shift in longitude, the latitude of the planet is continuously increasing, so that the apparent path among the stars takes the shape of a 'Z', a zig-zag without a point of intersection.

Case No. 9 (*Nd*) occurs under similar circumstances, but at the planet's *descending* node, so the apparent path takes the shape of an 'S'.

When at the time of the opposition the exterior planet is at the point of its orbit that is farthest north from the ecliptic, in its greatest northern *heliocentric* latitude, approximately midway between the ascending node and the descending node, as in Figure 39.*a*, the loop is of type *M+*, No. 5 in Figure 39.*b*. In that case, the *geocentric* latitude of the planet also reaches its maximum value at the time of the opposition. Indeed, before and after the instant of the opposition the planet is more distant from the Earth *and* its heliocentric latitude is smaller, and these two factors act together to make the geocentric latitude smaller than at the instant of the opposition.

The counterpart of this case, No. 13 or *M−* in Figure 39.*b*, occurs under similar circumstances, but now at the planet's greatest *southern* heliocentric latitude.

Approximately halfway between the cases *Na*, *M+*, *Nd* and *M−* we find the shapes *S1*, *S2*, *S3* and *S4* (Nos. 3, 7, 11, and 15 in Figure 39.*b*). These types are the limiting cases between loops without and loops with a point of intersection. They have the shape of an S (or a Z) in which one of the two 'turnings' degenerates into a *cusp*. In this latter point the planet is simultaneously stationary in both right ascension and declination (and in both longitude and latitude), so there we have a truly stationary point.

In the bends of a loop there is a point where the planet's right ascension (or its longitude, according to the definition of the station) reaches a maximum or a minimum, so that during a moment the right ascension (or the longitude) doesn't change. At such a point the motion of the planet is said to be stationary. In most cases, however, it is *not* a truly stationary point: the 'stationary' planet does indeed not move in right ascension (or in longitude), but it is still moving northward or southward in declination (or in latitude).

To illustrate this, we give in Table 39.A the apparent right ascension α and declination δ of Mars near its stationary points in October 1958 (five weeks before the opposition of 1958 November 16) and in January 1944 (five weeks after the opposition of 1943 December 5). 'Apparent' means that the planet's position is referred to the equinox of the date, and that the effects of light-time, aberration and nutation have been taken into account. The values are for 0^h Universal Time. Besides, the daily variations $\Delta\alpha$ and $\Delta\delta$ are given.

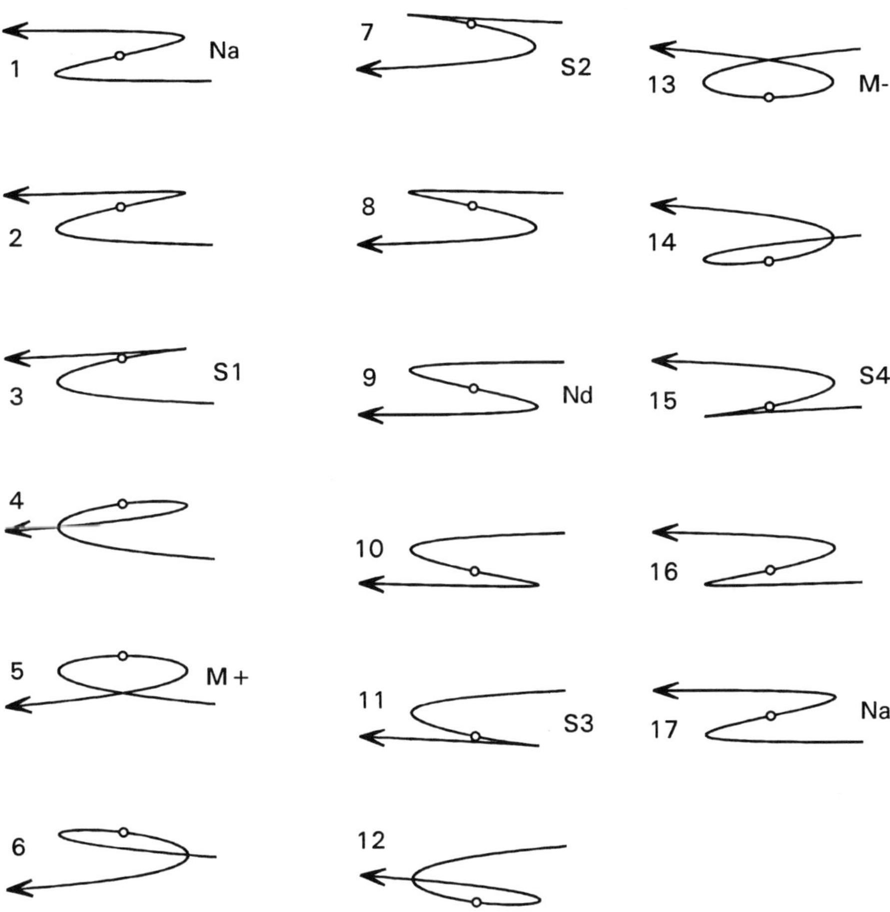

Fig. 39.b *: The different possible shapes of an opposition loop. See the explanation in the main text. The position of the planet at the instant of the opposition with the Sun is indicated by a small circle. The shape of the loop depends on the position of the planet with respect to the line of nodes of its orbit at the instant of the opposition. The first case occurs when at the time of opposition the planet passes at the ascending node. The cases 2, 3, 4, etc., take place at ever increasing celestial longitudes, and type No. 17 is merely a repetition of type No. 1.*

For the planets Jupiter to Neptune, the opposition loops are flatter than those of Mars. In the case of a minor planet with high orbital inclination and eccentricity, the opposition loop can be very distorted in comparison with the sketches shown here.

TABLE 39.A

Positions of Mars near its stationary points
in October 1958 and in January 1944

Date	α	$\Delta\alpha$	δ	$\Delta\delta$
	h m s	s	° ′ ″	″
1958 Oct. 2	4 01 07	+29	+18 56 35	+189
4	4 01 57	+21	+19 02 39	+175
6	4 02 33	+14	+19 08 14	+161
8	4 02 54	+ 7	+19 13 22	+146
10	4 03 01	− 0	+19 18 00	+132
12	4 02 53	− 8	+19 22 08	+117
14	4 02 30	−15	+19 25 47	+102
Date	α	$\Delta\alpha$	δ	$\Delta\delta$
	h m s	s	° ′ ″	″
1944 Jan. 8	4 09 37	− 7	+23 46 53	−13
9	4 09 31	− 4	+23 46 43	− 7
10	4 09 29	− 1	+23 46 39	− 0
11	4 09 30	+ 3	+23 46 42	+ 6
12	4 09 34	+ 6	+23 46 51	+12
13	4 09 42	+ 9	+23 47 07	+18

In the first case, Mars was stationary in right ascension on October 10 about 0^h Universal Time. But at that moment its declination was still increasing by about two arcminutes per day, as is seen in the column $\Delta\delta$. Hence, Mars described a 'turning', namely the turning at bottom left of shape No. 1 in Figure 39.*b*. So, on 1958 October 10 Mars moved due north on the celestial sphere.

On the other hand, in the second case we have a truly stationary point: on 1944 January 10 about 5^h UT Mars stood completely motionless on the starry sky, as his motions in right ascension and in declination were zero simultaneously — or practically so.

Finally, Figure 39.*b* shows the shapes No. 2, 4, 6, . . . , 16 which occur between the cases *Na*, *S1*, etc.

Table 39.B lists all the oppositions of Mars from 1941 to 2035. Next to the date of opposition, the type of the loop is given as defined in Figure 39.*b*.

Figures 39.*d* to 39.*g* illustrate the loops described by Pluto. Although this is a distant planet, its loops are wide open by reason of the high orbital inclination (17°). In these four maps, calculated with *Guide* software, Pluto's general motion is eastward (from the right to the left). In each case, the path of Pluto is plotted over a period of 1000 days, and tics are marked every 10 days. The planet's annual loops are a reflection of the annual motion of the Earth around the Sun. The height of each of the four maps corresponds to 4 degrees. North is up.

TABLE 39.B

Mars — Opposition dates and type of the loop, 1941 to 2035

1941	Oct.	10	15	1973	Oct.	25	16	2005 Nov. 7 — 1
1943	Dec.	5	3	1975	Dec.	15	4	2007 Dec. 24 — 4

1941 Oct. 10	15	1973 Oct. 25	16	2005 Nov. 7	1
1943 Dec. 5	3	1975 Dec. 15	4	2007 Dec. 24	4
1946 Jan. 14	4	1978 Jan. 22	4	2010 Jan. 29	5
1948 Feb. 17	5	1980 Feb. 25	6	2012 Mar. 3	6
1950 Mar. 23	6	1982 Mar. 31	6	2014 Apr. 8	6
1952 May 1	8	1984 May 11	9	2016 May 22	10
1954 June 24	12	1986 July 10	12	2018 July 27	12
1956 Sep. 10	14	1988 Sep. 28	14	2020 Oct. 13	14
1958 Nov. 16	1	1990 Nov. 27	2	2022 Dec. 8	3
1960 Dec. 30	4	1993 Jan. 7	4	2025 Jan. 16	4
1963 Feb. 4	5	1995 Feb. 12	5	2027 Feb. 19	5
1965 Mar. 9	6	1997 Mar. 17	6	2029 Mar. 25	6
1967 Apr. 15	7	1999 Apr. 24	8	2031 May 4	8
1969 May 31	10	2001 June 13	11	2033 June 28	12
1971 Aug. 10	13	2003 Aug. 28	14	2035 Sep. 15	14

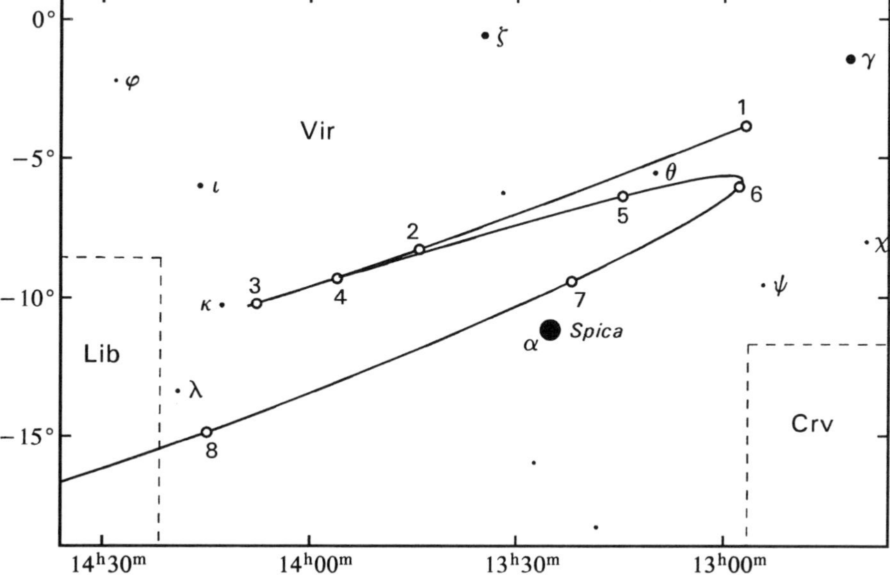

Fig. 39.c : *The path of Mars among the stars from January to August 1967. It has the shape No. 7 of Figure 39.b. On 1967 March 8 the planet was stationary in both right ascension and declination, so its apparent path presented a cusp there. The figure also illustrates a triple conjunction between Mars and the star Spica. The positions labelled 1 through 8 are those of Mars on 1967 January 1, February 1, etc. to August 1, at 0h Universal Time. The right ascensions and declinations, mentioned along the borders of the map, are 2000.0 coordinates.*

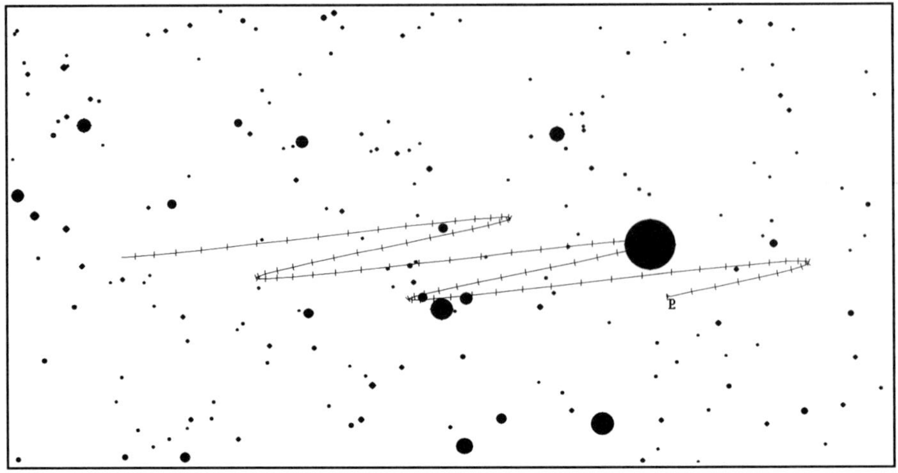

Fig. 39.d : *The path of Pluto during 1000 days, starting (at right) from 1929 January 1. As Pluto was near the ascending node of its orbit, the loops take the shape of a 'Z'. Pluto was discovered on plates taken in January 1930, when it was 3/4 degree east of δ Geminorum, the brightest star in this map.*

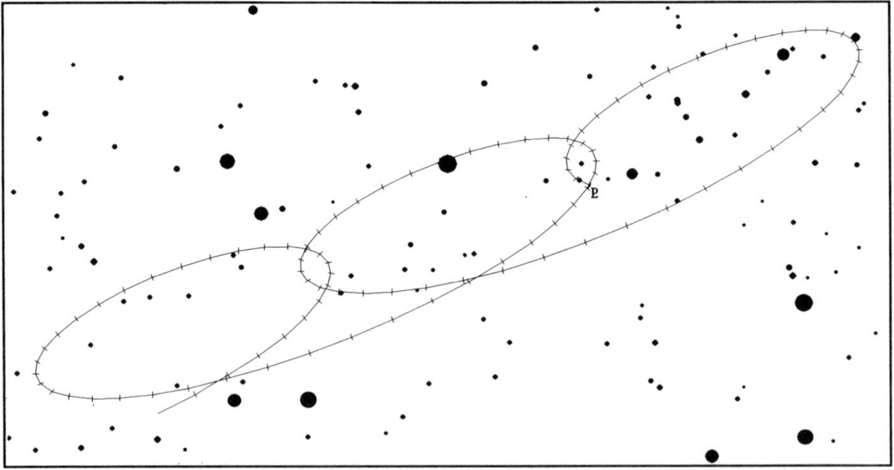

Fig. 39.e : *The path of Pluto during 1000 days, starting (at right) from 1979 January 1. The planet was near the border between Virgo and Bootes. As Pluto was near its greatest northern heliocentric latitude, its loops were shaped like No. 5 in Figure 39.b.*

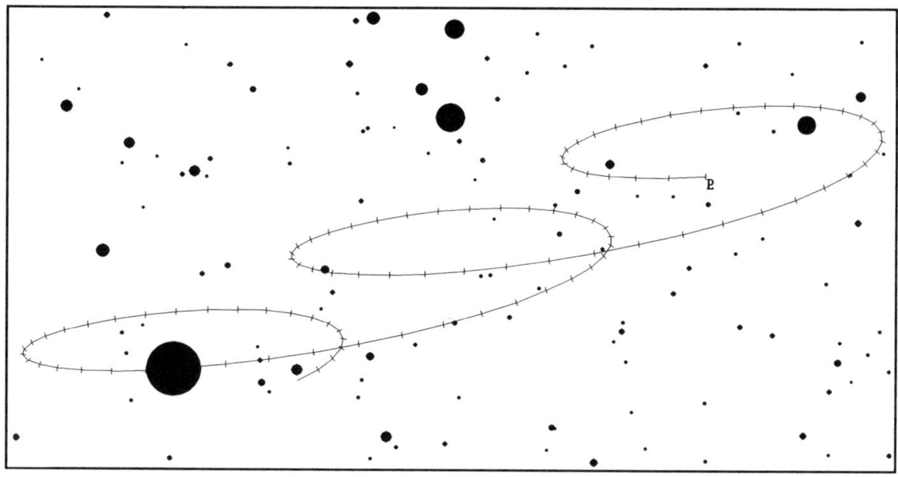

Fig. 39.f: *The path of Pluto during 1000 days, starting (at right) from 1997 January 1. On 1999 January 3, Pluto passes only 13" south of ζ Ophiuchi, a star of visual magnitude 2.7, the brightest star in the region shown here. The second brightest star on this map, above Pluto's second loop, is υ Ophiuchi. Here the loops are of type No. 6.*

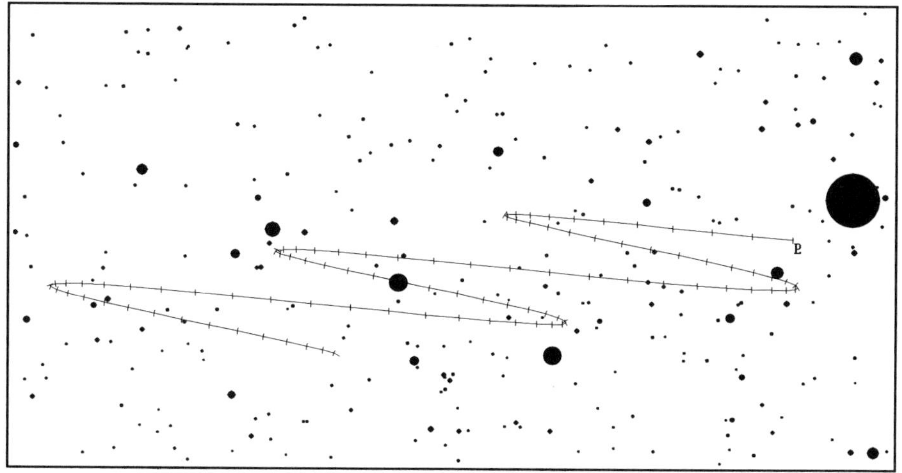

Fig. 39.g: *The path of Pluto during 1000 days, starting (at right) from 2017 January 1. As Pluto is now near the descending node of its orbit, the path has the shape of an 'S'. The brightest star on this map is π Sagittarii, visual magnitude 3.0. On 2018 July 3, Pluto will pass only 20 arcseconds north of 50 Sagittarii, a star of magnitude 5.6*

40. *Opposition places*

The text which follows was first published in the Belgian journal Heelal *of June 1983.*

Where in the sky is an exterior planet situated when it is in opposition with the Sun? The orbital inclinations of the planets Mars to Neptune are rather small: Mars 1°51′, Jupiter 1°18′, Saturn 2°29′, Uranus 0°46′, and Neptune 1°46′. For this reason, we always see these planets in the vicinity of the ecliptic including the moment of their opposition.

If we wish a more accurate reply, some calculation should be performed using the orbital elements of the planet concerned. Specifically, we need the following data: the semimajor axis a of the orbit expressed in astronomical units, the orbital eccentricity e, the orbit's inclination i, the argument of the perihelion ω, and the longitude of the ascending node Ω. *Attention!* ω is the *argument* of the perihelion, that is, the angular distance from the ascending node to the perihelion. This is *not* the same as the *longitude* of the perihelion, which is equal to the sum $\Omega + \omega$.

These orbital elements slowly vary over time, but for our purpose we may consider them to be constant, as we are interested in the situation near A.D. 2000. Moreover, we shall refer the elements i, ω and Ω to the mean equinox of 2000.0.

In this chapter, we will consider Mars and four asteroids. Their elements are given in the table below. The angles i, ω and Ω are in degrees and decimals. For the asteroids, the elements are taken from the Russian yearbook *Ephemerides of Minor Planets* for 1996; strictly speaking, they are the osculating elements for the Epoch 1996 November 13.0.

Planet or minor planet	a	e	i	ω	Ω
Mars	1.523 679	0.093 401	1.8497	286.5021	49.5581
Ceres	2.769 957	0.076 083	10.5954	72.4164	80.6562
Pallas	2.771 108	0.233 558	34.8068	309.7243	173.2914
Juno	2.668 742	0.257 479	12.9664	247.9748	170.1870
Eros	1.458 399	0.222 922	10.8311	178.5815	304.4368

Now we can start with the calculation. We wish to obtain the *opposition place* for any arbitrarily given ecliptical (celestial) longitude λ. In other words: suppose that the Earth and the planet have the same heliocentric longitude λ; then the planet is in opposition (in longitude) with the Sun, and we wish to find its right ascension α and declination δ at that instant. This can be performed as follows. The formulae will be given without proof.

Radius-vector of the Earth (this is the distance of the Earth to the Sun), in astronomical units:

$$R = \frac{0.9997218}{1 + 0.016709 \cos(\lambda - 102°937)}$$

Find the planet's argument of latitude by means of the formula

$$\tan u = \frac{\tan(\lambda - \Omega)}{\cos i}$$

The angle u lies in the same quadrant as $(\lambda - \Omega)$. If u is obtained by means of the ordinary ATN (arctangent) function, which gives the angle in the first or in the fourth quadrant only, then the following test should be performed: if the cosine of $(\lambda - \Omega)$ is negative, increase the result by 180 degrees or π radians.

Radius-vector of the planet:

$$r = \frac{a(1 - e^2)}{1 + e \cos(u - \omega)}$$

Then calculate the planet's heliocentric latitude b, distance to the Earth Δ, and geocentric latitude β:

$$\sin b = \sin u \sin i$$

$$\Delta = \sqrt{R^2 + r^2 - 2rR \cos b} \qquad (\cos b > 0)$$

$$\sin \beta = \frac{r}{\Delta} \sin b$$

Finally, α and δ are found by means of the classical transformation formulae

$$\tan \alpha = \frac{\sin \lambda \cos \varepsilon - \tan \beta \sin \varepsilon}{\cos \lambda}$$

$$\sin \delta = \sin \beta \cos \varepsilon + \cos \beta \sin \varepsilon \sin \lambda$$

where ε is the obliquity of the ecliptic. For the year 2000 we can take $\varepsilon = 23°4393$, whence $\sin \varepsilon = 0.397777$ and $\cos \varepsilon = 0.917482$.

If we perform the calculation for several values of λ, we obtain as many opposition places of that planet. Then we can plot these positions in a graph. We did it for Mars and the four minor planets Ceres, Pallas, Juno and Eros — see the graphs on the next two pages.

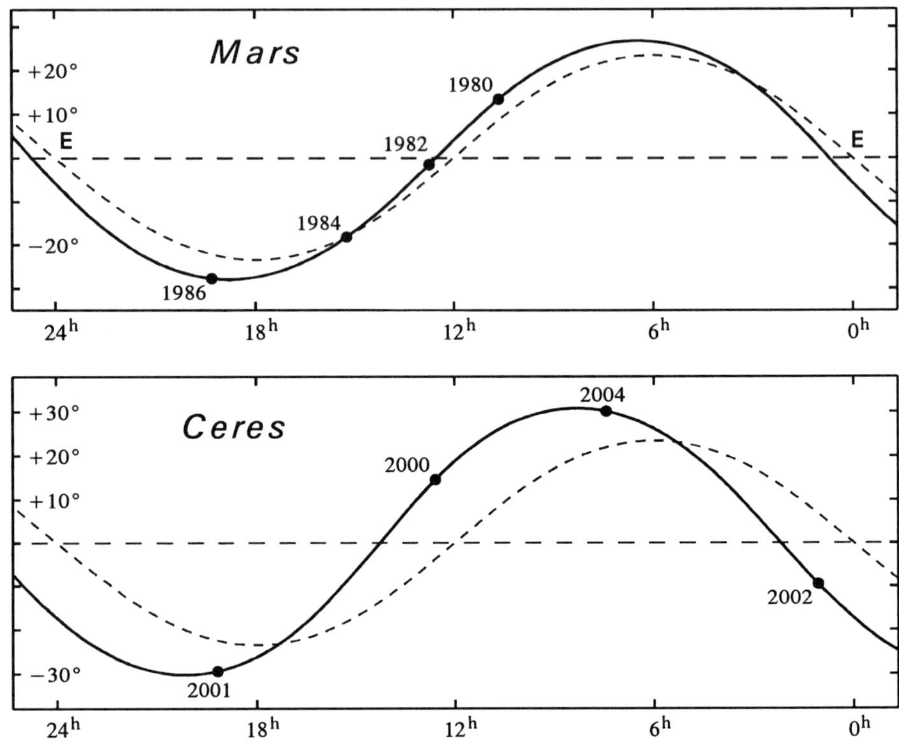

In each of these graphs, the horizontal dashed line represents the celestial equator. The vernal equinox E is situated near the right border *and* near the left one, as on a geographic map in Mercator projection: the width (from right to left) of each drawing corresponds to the whole circumference of the sky along the equator, plus two small overlapping regions at extreme left and right. The horizontal scale indicates the right ascensions α (in hours), and the vertical scale the declinations δ in degrees. The undulating dashed line is the ecliptic, which passes through the vernal equinox E, reaches its greatest northern declination at $\alpha = 6^h = 90°$, crosses the equator again at $\alpha = 12^h = 180°$, etc.

The thicker, solid line is the locus of all the opposition places of the planet. *Very important: this line is **not** the apparent path of the planet on the celestial sphere!!* But: if the planet or the minor planet is in opposition with the Sun, then it is situated *somewhere* on that line. To elucidate this, we plotted in the drawing for Juno the apparent path of this asteroid from January to November 1995 — the short, dotted curve. At the time of its opposition, on 1995 June 18, Juno was situated at the place indicated by the dot.

In the graphs for Mars, Ceres, Pallas and Eros some opposition times are indicated (dots with corresponding years). For Eros, some bright stars are added.

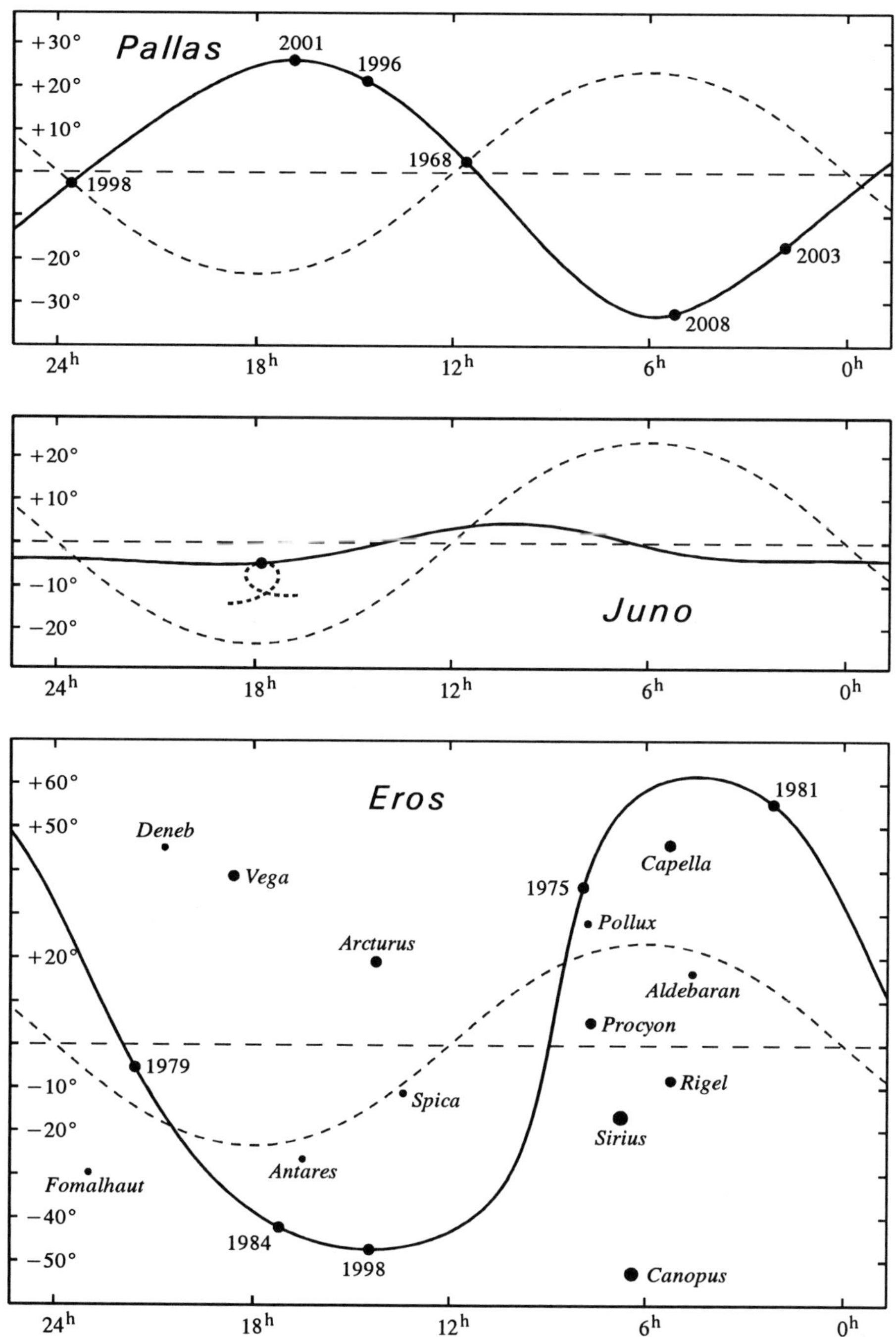
Pallas
+30°
+20°
+10°
−20°
−30°
2001
1996
1998
1968
2003
2008
24ʰ
18ʰ
12ʰ
6ʰ
0ʰ
+20°
+10°
−10°
−20°
Juno
24ʰ
18ʰ
12ʰ
6ʰ
0ʰ
+60°
+50°
+20°
−10°
−20°
−40°
−50°
Eros
1981
Deneb
Vega
Capella
1975
Pollux
Arcturus
Aldebaran
Procyon
1979
Spica
Rigel
Sirius
Antares
Fomalhaut
1984
1998
Canopus
24ʰ
18ʰ
12ʰ
6ʰ
0ʰ

The graph for *Mars* is not very sensational. As earlier explained, Mars always remains in the vicinity of the ecliptic. Its orbital inclination is $1°51'$. But as seen from the *Earth*, the planet can reach larger celestial latitudes than $1°51'$ — see Figure 39.*a*. At the instant of the opposition, Mars can be up to $4°34'$ north and up to $6°52'$ south of the ecliptic. As can be seen on the graph on page 242, Mars can depart farther south than north from the ecliptic; this is because an opposition near longitude 320°, about halfway between the descending and the ascending node, occurs near the planet's *perihelion*. The closer a planet is to the Earth, the larger the ratio between its geocentric and its heliocentric latitudes. For a distant planet such as Neptune, the difference between the two ($\beta - b$ in Figure 39.*a*) is very small.

In the drawing for Mars, the planet's position at the oppositions in 1980, 1982, 1984 and 1986 is indicated. At the oppositions of 1980 and 1982, the planet was north of the ecliptic, in the constellations Leo and Virgo, respectively. The opposition of 1984 occurred almost exactly at the descending node of the planet's orbit. This resulted in a rare event: a *transit of the Earth* over the Sun's disk was visible from Mars. The opposition of 1986 took place south of the ecliptic, and moreover near its southernmost point, so this resulted in the very southern declination of $-28°$.

In the realm of the minor planets we find more interesting cases, as many of these bodies have large orbital inclinations.

The graph for *Ceres* doesn't differ much from that of Mars. By reason of its larger orbital inclination ($10°36'$), Ceres can depart somewhat farther from the ecliptic than Mars does. The extreme values of the declination of Ceres, at the time of its opposition with the Sun, are $+31°$ and $-30°$. At the opposition of 1992 July 25, Ceres was situated at declination $-30°$, while at that of 1995 February 3 its declination was $+30°$. See the list on page 50 of my *Astronomical Tables of the Sun, Moon and Planets* (Willmann-Bell, Inc.; 2nd ed., 1995).

With *Pallas* we have a completely different situation. By reason of its large orbital inclination ($34°48'$), this asteroid can depart much more from the ecliptic than Mars or Ceres: up to $48°43'$ to the north, and up to $56°37'$ to the south — as before, at the instant of its opposition with the Sun. But, moreover, the longitude of the ascending node is 173°, and hence the *descending* node is close to the vernal equinox. The consequence is the Pallas opposition-curve as shown in the drawing. We could call it an 'anti-ecliptic'. At an opposition in December or in January, a normal planet lies in northern declinations, near the constellations Taurus or Gemini. Not so for Pallas, which then stays in southern celestial regions, far from the ecliptic. At the opposition of 1985 December 22, for instance, Pallas was in the constellation Columba, at declination $-33°$. At that of 2008 December 4, Pallas will again be in Columba. On the other hand, the asteroid can reach uncommon (for a planet) northern constellations such as Bootes (opposition of April 1996) and Hercules (May 2001).

As shown in the drawing, at the opposition of 1968 March 13 Pallas was very close to the ecliptic, at the ascending node of its orbit. As seen from Pallas, a *transit of the Earth* over the solar disk could be seen. One may assume that this must be a very rare event for an asteroid with such a high orbital inclination. But the fact is that only thirty years later, at the opposition of 1998 September 16, another transit of the Earth over the Sun is observable from Pallas, this time at the descending node of its orbit!

Juno presents us again another different picture. The descending node of this minor planet also lies close to the vernal equinox, but the orbital inclination is smaller than that of Pallas. The consequence is seen in the drawing: for all longitudes, the 'opposition curve' of Juno is close to the celestial equator! At the time of the opposition with the Sun, the declination of Juno is always between $+4°$ and $-5°$. However, this does not mean that Juno cannot reach larger declinations, but then it is not in opposition. See the short dotted curve, which is the path of Juno from January to November 1995.

The last graph concerns minor planet 433 *Eros*. The orbital inclination is not particularly large, but Eros is much closer to the Earth's orbit than minor planets such as Ceres or Juno. At a perihelic opposition such as those of January 1975 or January 2056, Eros approaches us to 0.15 AU. For this reason, as seen from the Earth this minor planet sometimes is situated far from the ecliptic. Its extreme declinations-at-opposition are $+62°$ and $-47°$.

The interested reader can calculate curves for some other minor planets. What would be the results for an object of the Apollo type, or for the periodic comet Halley? In these cases a part of the orbit is situated inside that of the Earth, so for a certain range of longitudes the body cannot be in opposition with the Sun. Some interesting orbits are listed in the following table; as before, the elements i, ω and Ω are expressed in degrees and are referred to the equinox of 2000.0. *Ganymed* is minor planet 1036, *not* the third satellite of Jupiter!

Body	a	e	i	ω	Ω
Hidalgo	5.76375	0.65811	42.527	56.413	21.627
Ganymed	2.65883	0.53845	26.634	132.270	215.796
Icarus	1.07788	0.82689	22.881	31.225	88.149
Betulia	2.19429	0.49012	52.125	159.324	62.377
Apollo	1.47103	0.56002	6.356	285.639	35.923
Tantalus	1.29016	0.29866	64.012	61.606	94.403
Phaethon	1.27131	0.89008	22.104	321.830	265.580
Damocles	11.85935	0.86627	61.710	191.270	314.196
P/Halley	17.94163	0.96730	162.242	111.866	58.860

41. *Triple conjunctions*

The outer (exterior) planets usually move eastward against the stars. But for a few months around their opposition with the Sun they appear to change direction and move backward, or 'retrograde', because our moving Earth is overtaking them.

This is illustrated in Figures 38.*c* and 39.*c*. In the first case, we see that Mars is three times in conjunction with the star Regulus, while in the second case it is three times in conjunction with Spica. These are examples of so-called *triple conjunctions* (TCs).

Two planets also can give rise to a TC. For instance, in 1980–1981 there was a TC between Jupiter and Saturn. The first conjunction (in celestial longitude) took place on 1980 December 31. Because Jupiter's apparent motion is faster, it overtook Saturn again while retrograding, on 1981 March 4. Then, on 1981 July 24, a third and final conjunction happened after the two planets had resumed their eastward motion. More distant planets have shorter arcs of retrogradation.

Notice that *a triple conjunction is **not** a grouping of three planets*. A TC occurs when in the course of a little less than a year two planets, or a planet and a star, share three times the same ecliptical longitude (or the same right ascension, if the conjunctions are defined that way). In the case of a TC of two bright outer planets, these celestial bodies remain near each other in the night sky during several months and their slow relative motion can be quite spectacular. So when a TC of two bright planets or a TC of a bright planet and a bright star takes place, people become interested in future and past occurrences.

TCs can also occur among an interior planet (Mercury, Venus) and an outer planet or a star. But often this takes place in the vicinity of the Sun. For instance, in 1991 there was a TC

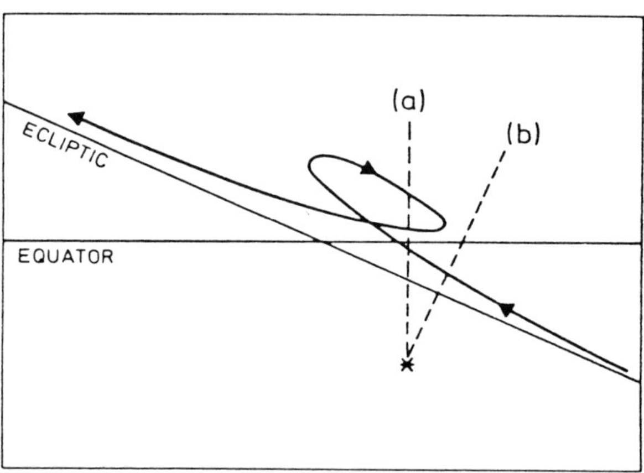

Fig. 41.a : In some cases a triple conjunction takes place in right ascension, but not in longitude. Of course, the opposite can take place too. This drawing shows a retrograde loop described by a planet in the vicinity of a star. There is a triple conjunction in right ascension (a), but only a single conjunction in longitude (b). The line (a) is perpendicular to the celestial equator and is directed toward the northern celestial pole. Line (b) is perpendicular to the ecliptic.

between Venus and Jupiter; that same year there was also a TC Venus–Regulus. In this chapter, TCs of the outer planets mutually as well as TCs of the outer planets with first-magnitude zodiacal stars are discussed.

The definition of a triple conjunction — and of a conjunction generally — depends on the choice of the coordinates. A TC may, for instance, take place in *right ascension* or in *longitude*. In most cases, both will happen, but there are exceptions. In 1982, for instance, a TC of Saturn and Spica took place in right ascension (on January 8, February 25 and September 21), but only one conjunction in longitude (on October 5). Such a situation is illustrated schematically in Figure 41.*a*. So one should make a distinction between conjunctions in *right ascension* and conjunctions in *celestial longitude*. The instant of a conjunction between two planets, or between a planet and a star, as given by modern astronomical tables and almanacs, generally is the time when the two celestial bodies have the same right ascension. This choice of definition results from the fact that modern astronomers generally work with tables which give the planets' positions in the equatorial system (right ascension and declination), not in the ecliptical one (which is based on the Earth's orbital plane). Under these conditions, it is much easier, indeed, to calculate conjunctions in right ascension.

However, the planets' orbits are in very nearly the same plane as the Earth's, so their motions as seen from the Earth are approximately parallel to the ecliptic. Hence, it might be more logical to consider the conjunctions in *longitude*. At the instant of such a conjunction in longitude, the shortest line joining the two bodies is perpendicular to the ecliptic. Moreover, it is generally near the time of the conjunction in longitude, not around that of the conjunction in right ascension, that the angular separation of the two objects is at a minimum and they appear closest together on the sky. For this reason, in this chapter *we shall refer only to conjunctions in longitude*.

A TC between two outer planets is possible only if these bodies are in opposition with the Sun at nearly the same time. Table 41.A summarizes this parameter for all couples of outer planets. Here, Δt represents the extreme difference between the opposition times of the planets, where a TC remains possible. The values have been calculated for *circular* orbits.

TABLE 41.A

Greatest possible difference Δt between the times of opposition where a triple conjunction is possible (circular orbits!)

Planets	Δt	*Planets*	Δt
Mars–Jupiter	3.9 days	Jupiter–Uranus	3.1 days
Mars–Saturn	5.4	Jupiter–Neptune	3.7
Mars–Uranus	6.5	Saturn–Uranus	1.4
Mars–Neptune	7.0	Saturn–Neptune	2.1
Jupiter–Saturn	1.7	Uranus–Neptune	0.63

Jupiter – Saturn

Let us first consider the Jupiter–Saturn conjunctions, which are the rarest ones among bright planets by reason of the slow motions of these bodies.

As seen from the Earth, Jupiter and Saturn move eastward along their orbits at average rates of 30.363 and 12.235 degrees per year, respectively (with respect to the slowly moving vernal equinox). Thus, Jupiter pulls ahead of Saturn at a rate of 18.13 degrees each year, so it will gain a full 360° in 19.86 years. This means that Jupiter and Saturn appear to line up in heliocentric conjunction every 19.86 years. (There are actually deviations with respect to this mean value, due primarily to the slight eccentricities of the planets' orbits).

In 20 years Jupiter runs through about five-thirds of its orbit, and Saturn through two-thirds. Hence, there is a displacement of about 120 degrees (more exactly, 117°) towards the west, or clockwise as seen from above the Sun's north pole, between one heliocentric conjunction and the next one. After 60 years, or three heliocentric conjunctions, both planets come again together in the same region of the sky, with a mean eastward displacement of 9 degrees with respect to the vernal point, or 8 degrees with respect to the background stars.

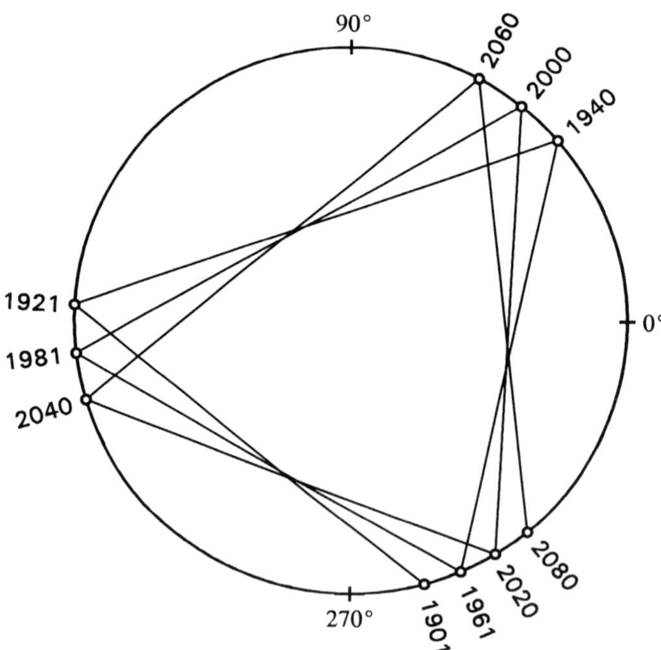

Fig. 41.b : *Positions of the Jupiter–Saturn heliocentric conjunctions from 1901 to 2080. The vernal equinox (longitude zero) is at right.*

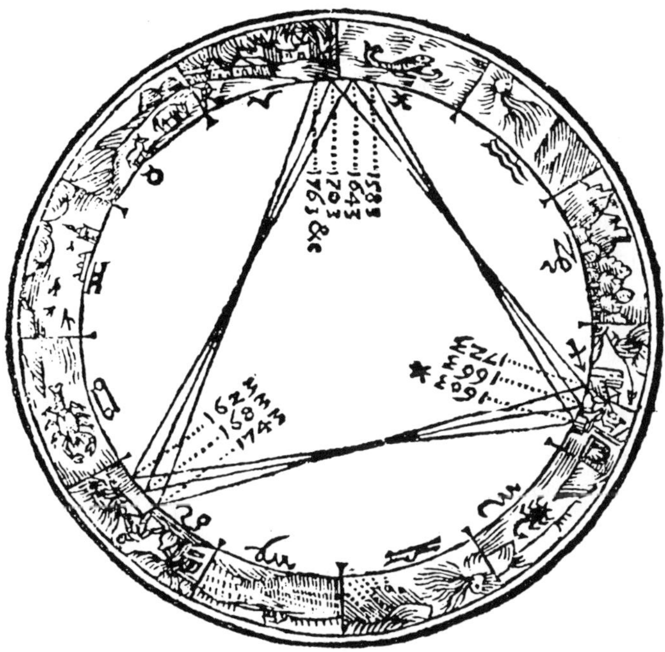

Fig. 41.c : *Drawing taken from Kepler's* De Stella Nova *(Prague, 1606).*

As an example, the left part of Table 41.B lists the heliocentric conjunctions of the two planets from A.D. 1861 to 2080, together with their common heliocentric longitude at these instants, measured from the mean vernal equinox of the date. Compare, for instance, the longitudes at the conjunctions of 1881, 1940, 2000, and 2060. The heliocentric longitudes corresponding to the conjunctions of 1901 to 2080 have been represented in Figure 41.*b*. Note that these longitudes are situated at the vertices of a triangle which slowly rotates nine degrees per 60 years. Figure 41.*c*, due to Kepler, is a similar diagram for older conjunctions.

For an observer on the planet Earth, the situation is somewhat more complicated. Since the Earth's orbit is small with respect to the orbits of Jupiter and Saturn, the geocentric conjunctions always occur close to the times of the heliocentric conjunctions. However, they can precede or follow them a few weeks or even two or three months, depending on the Earth's position at the time of the heliocentric Jupiter–Saturn conjunction.

Since Jupiter and Saturn are superior planets, they move along their orbits more slowly than the Earth. Thus, near the time of a heliocentric Jupiter–Saturn conjunction they are close to each other for several months.

TABLE 41.B

The conjunctions (in longitude) between Jupiter and Saturn, 1861 to 2080

Heliocentric conjunction		Geocentric conjunction		
Date	*Longitude*	*Date*	$\Delta\beta$	*Constel.*
	° ′		° ′	
1861 Dec. 28	166 50	1861 Oct. 21	−0 48	Leo
1881 Apr. 13	31 45	1881 Apr. 18	+1 13	Aries
1901 Sep. 28	285 38	1901 Nov. 28	−0 26	Sagittarius
1921 Aug. 23	177 00	1921 Sep. 10	−0 57	Virgo
1940 Nov. 15	41 43	⎡ 1940 Aug. 8	+1 11	Aries
		1940 Oct. 20	+1 14	Aries
		⎣ 1941 Feb. 15	+1 17	Aries
1961 Apr. 16	293 41	1961 Feb. 19	−0 14	Sagittarius
1981 Apr. 16	187 08	⎡ 1980 Dec. 31	−1 03	Virgo
		1981 Mar. 4	−1 03	Virgo
		⎣ 1981 July 24	−1 06	Virgo
2000 June 22	52 01	2000 May 28	+1 09	Taurus
2020 Nov. 2	301 50	2020 Dec. 21	−0 06	Capricornus
2040 Dec. 7	197 05	2040 Oct. 31	−1 08	Virgo
2060 Feb. 2	62 35	2060 Apr. 7	+1 07	Taurus
2080 May 21	310 01	2080 Mar. 15	+0 06	Capricornus

TABLE 41.C

Triple conjunctions (in longitude) Jupiter–Saturn
from 1000 B.C. to A.D. 3000

−979	−6	1008 *p*	2239 *p*
−860 *f*	333 *p*	1306 *p*	2279
−819 *p*	411 *f*	1425	2656 *p*
−562 *f*	452	1683 *p*	2794 *f*
−521 *p*	710 *p*	1940 *f*	2913 *f*
−145 *f*	967 *f*	1981 *p*	

Since the Earth itself moves a significant distance during those weeks or months, a heliocentric conjunction can give rise to either a *single* geocentric conjunction (this is the most frequent case), or to a *triple* conjunction. (A 'double' conjunction is impossible, of course!).

The right part of Table 41.B lists all Jupiter–Saturn conjunctions in celestial longitude as seen from the Earth from A.D. 1861 to 2080. Triple conjunctions are indicated by a bracket. Column $\Delta\beta$ indicates at what angular distance Jupiter passes to the north (+) or to the south (−) of Saturn. The last column mentions in what constellation the conjunction takes place.

The conjunctions of 1881 and 1921 occurred close to the Sun in the sky and hence were not easily observable. On 2020 December 21, Jupiter will pass only 6 minutes of arc from Saturn; that will be the closest approach of the two planets on the sky since the conjunction of 1623 July 16, when the separation was 5′.

It is important to bear in mind that a triple conjunction is not necessarily a close one. The conjunctions of 2020 and 2080 are close ones, but they are single. The triple conjunctions of 1940–1941 and 1980–1981 were not close, because the planets remained more than one degree from each other.

Table 41.C gives the list of all geocentric triple conjunctions of Jupiter and Saturn during a time period of 40 centuries, from −1000 to the year +3000.

- Notice that in this book the 'B.C.' years are counted astronomically. That is, the year before the year +1 is the year zero, and the year preceding the latter is the year −1. The year −6 is called 7 B.C. by the historians.

- In many cases, a triple conjunction extends over two successive calendar years. This will be indicatd by the letter *p* or *f* after the year number, meaning that one of the three events occurs in the *preceding* or in the *following* year, respectively. For instance, 1306 *p* means that the first conjunction occurred in A.D. 1305 and the next two in 1306, while 1940 *f* means that the first two conjunctions took place in 1940 and the third one in 1941. When no letter is given, as for 1425 and 2279, this indicates that all three conjunctions occur during that year.

There was *almost* a TC in 1384-1385, when the two planets failed to have the same longitude by only *one minute of arc* in October 1384.

We see from Table 41.C that the TC of 1980-1981 occurred only 40 years after another one, but that the next one will not take place before 2238-2239. Forty years later there will be another TC between Jupiter and Saturn, but after that mankind will have to wait for nearly four centuries for the next one!

In 1265-1266 and in 1821 there was a TC of Jupiter and Saturn in *right ascension*, but not in celestial longitude.

There is no rigorous periodicity for the TCs. In the list we find six times an interval of 40 or 41 years, and seven times an interval of 257 or 258 years. An exact cycle does not exist. Indeed, a rigorous periodicity is impossible for several reasons, including the orbital eccentricities and the slow, so-called 'secular' variations in the orbits of the two planets.

Mars – Jupiter

Let us first consider the successive conjunctions of these planets. We note that 2 synodic revolutions of Jupiter, or 797.77 days, are only a little longer than one synodic revolution of Mars (779.94 days), so that during several years the Mars–Jupiter conjunctions repeat under similar circumstances *with respect to the Sun*; see Figure 41.*d*. For instance, the Mars–Jupiter conjunctions of 1935 and the next years were all visible in the evening sky, but gradually closer to the Sun:

1935 Aug. 26	74° east from the Sun	
1937 Oct. 30	74° — —
1940 Jan. 6	77° — —
1942 Apr. 3	63° — —
1944 July 5	43° — —
1946 Sep. 24	29° — —
1948 Dec. 1	24° — —
1951 Feb. 7	25° — —
1953 Apr. 27	21° — —

(Due to the eccentricities of the orbits, the gradual decrease of the elongation from the Sun is not regular). The next two conjunctions took place close to the Sun on the sky and hence were not observable:

1955 July 25	Jupiter in conj. with the Sun on August 4
1957 Oct. 16	Jupiter in conj. with the Sun on October 5

Thereafter, the Mars–Jupiter conjunctions were visible before sunrise in the eastern sky:

1959 Dec. 28	18° west from the Sun	
1962 Mar. 6	20° — —
1964 May 19	20° — —
1966 Aug. 12	28° — —
1968 Nov. 6	46° — —

These conjunctions occurred at gradually greater elongations from the Sun. Finally, on 1980 February 24-25 the two planets came in opposition with a time difference of only twelve hours, so in 1979-1980 they had a triple conjunction.

The complete cycle of these variations has a duration of 47.8 years, so TCs of Mars and Jupiter are *possible* only at intervals of 48 years. Will there be a TC 48 years after that of 1980? In 2027 Mars will be in opposition with the Sun on February 19, eight days *after* Jupiter; and the next opposition of Mars, on 2029 March 25, will occur 18 days *before* the Jupiter opposition. In both cases the time difference is too large and no TC will occur. There will be the same bad luck 48 years later. Not before A.D. 2123 will there again be a TC of Mars and Jupiter.

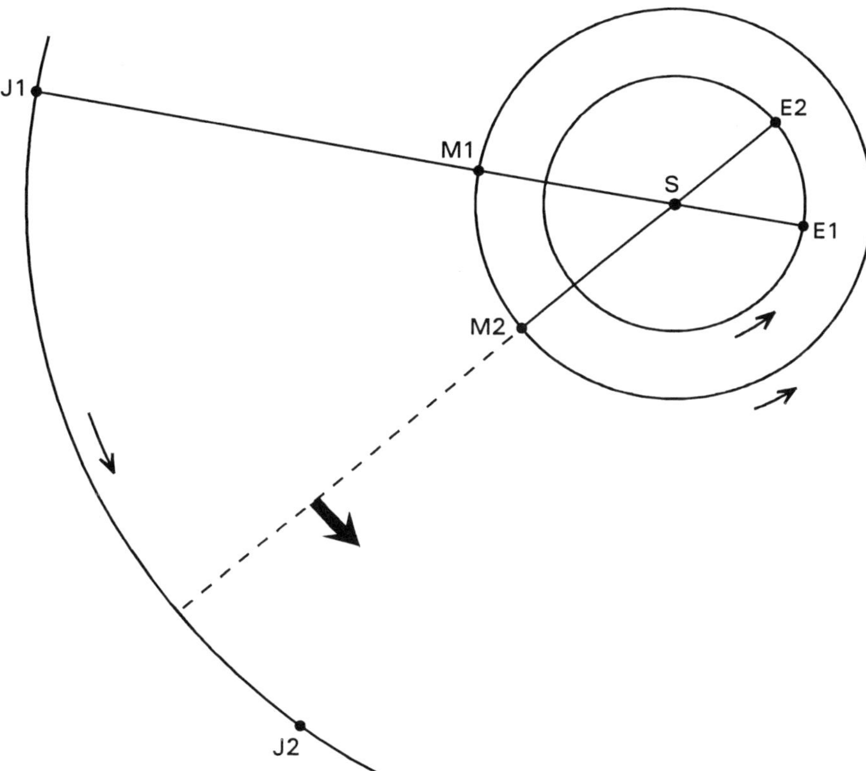

Fig. 41.d : *This drawing represents the orbits of Earth and Mars, and a part of Jupiter's. Suppose that Mars and Jupiter are simultaneously in conjunction with the Sun. Then the Earth (E1), the Sun (S), Mars (M1) and Jupiter (J1) are on a straight line. One synodic revolution of Mars has a duration of 780 days, or 2 years and 50 days. After this interval Mars is again in conjunction with the Sun: the Earth has described its orbit twice + the distance E1–E2; and Mars has described its orbit once + the distance M1–M2. The points E2, S and M2 are on a straight line again. Because two synodic revolutions of Jupiter are 18 days longer than one synodic revolution of Mars, Jupiter is not yet again in conjunction with the Sun. That is, the line Earth–Sun extended, which moves in the direction of the thick arrow, has not yet reached Jupiter (J2). So the line Earth–Mars has not yet reached this planet: Mars did not yet return in conjunction with Jupiter. This occurs only a few weeks later, after the Mars–Sun conjunction. Therefore the Mars–Jupiter conjunction occurs when the two planets are again visible in the morning sky, west from the Sun.*

Until now, we have considered only uniform, circular motions. The influence of the orbital eccentricities on TCs has two important effects. Firstly, the quantity Δt, which we defined earlier, will be a function of the opposition date. In the case of Mars–Jupiter, we found $\Delta t = 3.9$ days in the assumption of circular orbits — see Table 41.A. In fact, G. P. Können and Jean Meeus (*Journal of the British Astronomical Association*, Vo. 93, No. 1, page 21; December 1982) found that the extreme values are 1.5 and 5.1 days, Δt reaching its minimum value near the perihelion of Mars. Hence, a TC is relatively easily realised near the aphelion of the orbit of Mars.

Secondly, the time between two successive oppositions varies with the opposition date. This time is smaller when the opposition takes place near Mars' aphelion. Oppositions of Mars repeat after a little more than two years. The precise time interval may vary between 764 and 811 days, depending on the planet's initial position with respect to its perihelion. For Jupiter the time between three oppositions is about 2 years and 2 months, more exactly between 791 and 806 days. This means that under favorable conditions *two* TCs may be produced with a time interval of only two years. In this case, even a third TC is possible after another two years.

If in a given year Mars and Jupiter are, for instance, in opposition with the Sun on May 31 and June 1, respectively, then two years later the opposition dates will be August 10 and August 10, and again two years later October 25 and October 24, respectively, so that *three* successive TCs of the two planets will have taken place! Obviously, such 'twins' or 'triplets' will happen only if the oppositions take place not far from the perihelion of Mars, and also if the perihelia of Mars and Jupiter differ by only 38° approximately, as is the present case.

Table 41.D lists all the TCs of Mars–Jupiter from the year -1000 to A.D. 3000. It appears that during these 40 centuries there are four twins, but no triplet.

TABLE 41.D

Triple conjunctions of Mars –Jupiter from -1000 to $+3000$

$-984\ p$	$22\ p$	1027	$2170\ p$
twin { -933	$165\ p$	$1170\ f$	2313
twin { -931	$308\ p$	$1313\ f$	2456
twin { -888	$355\ p$	$1456\ f$	2599
twin { -886	$498\ p$	$1504\ p$	$2699\ f$
-788	641	$1647\ p$	twin { 2742
-645	784	$1790\ p$	twin { 2744
$-502\ f$	884	$1837\ p$	2791
$-359\ f$	twin { 927	$1980\ p$	$2842\ f$
$-312\ f$	twin { 929	2123	$2986\ p$
$-169\ f$	976		

Mars – Saturn

While two synodic revolutions of Jupiter are about 18 days *longer* than one synodic revolution of Mars, in the case of Saturn 2 synodic revolutions are 24 days *shorter*. Therefore, successive conjunctions of Mars–Saturn move the other way with respect to the Sun: when they occur in the evening sky, they shift gradually toward the opposition, not toward the Sun as is the case for Jupiter. For instance:

2002 May 4	elongation = 30° East	
2004 May 25	37° East	
2006 June 18	42° East	
2008 July 10	47° East	

In the case of Mars–Saturn, the cycle of these variations has a length of 34.0 years. So, every 34 years a TC *can* take place. And, just as for Mars–Jupiter, two subsequent TCs of Mars and Saturn are possible within a time separation of one synodic period. If we consider uniform, circular motions we find $\Delta t = 5.4$ days for the TCs of Mars–Saturn (Table 41.A). Due to the eccentricities of the orbits, the actual value of Δt varies between the extremes 2.8 and 6.8 days.

Table 41.E lists all the TCs Mars–Saturn from the year −1000 to A.D. 3000. During these forty centuries there are eight twins.

In 1503-1504 there was a TC of Mars with Jupiter as well as with Saturn, but no TC of Jupiter with Saturn. This is illustrated in Figure 41.*e*. Curiously, in the whole 21st century no TC between bright outer planets will take place: 'we' have to wait until A.D. 2123 for the next one!

TABLE 41.E

Triple conjunctions of Mars –Saturn from −1000 to +3000

−907 *p*	twin { 177 *f* / 180 *p*	twin { 1061 *f* / 1064 *p*	1946 *p*
−741	216	1100	2148 *f*
−705 *f*	380 *f*	1198	twin { 2185 / 2187
−668	419 *p*	1264 *f*	2221
−602	519	1303 *p*	2319
−465 *p*	twin { 619 *f* / 622 *p*	twin { 1503 *f* / 1506 *p*	2388 *p*
−365	658	1542	twin { 2627 *p* / 2629
twin { −265 *f* / −262 *p*	756	1640	2663
−226	822 *f*	1706 *f*	2761
− 62 *f*	861 *p*	twin { 1743 *p* / 1745 *p*	2830 *p*
− 23 *p*		1779	2866
77			

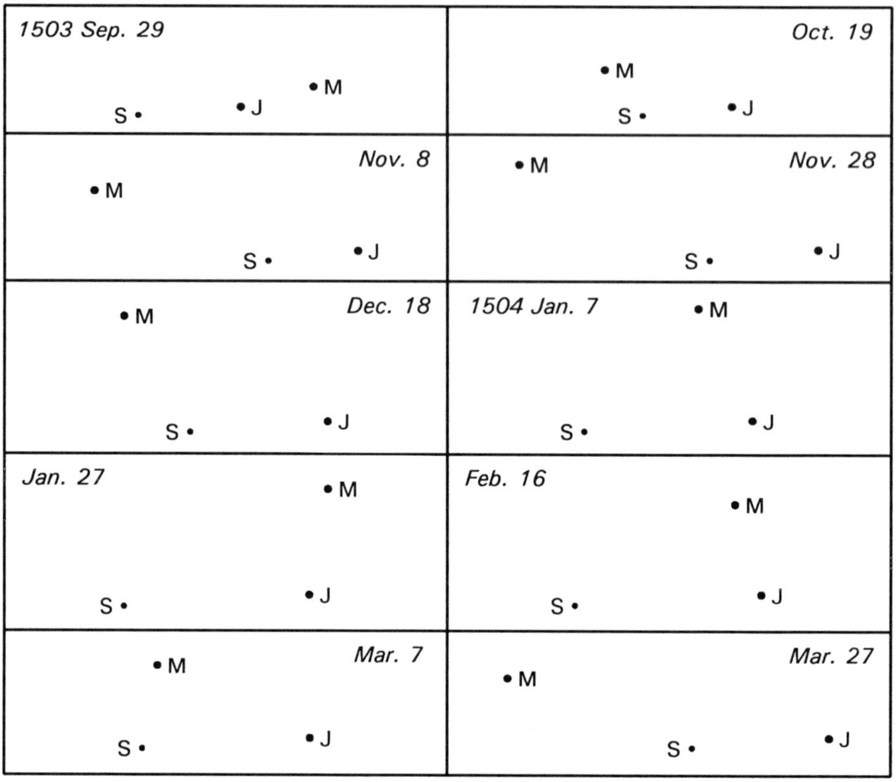

Fig. 41.e : *Positions of Mars (M), Jupiter (J) and Saturn (S) at intervals of 20 days in 1503-1504. Ecliptic north is up. Each box is 15 degrees wide. The opposition dates of the planets were 1503 Dec. 26, Dec. 22 and Dec. 27, respectively.*

Triple conjunctions between a bright planet and a star

Triple conjunctions are by no means rare. In any century, many are produced between a bright planet and a bright star. Table 41.F gives all TCs of Mars, Jupiter and Saturn with the first-magnitude stars near the ecliptic from A.D. 1900 to 2100. In total there are 60 cases, that is one about every three years.

In 1979-1980 there was a TC between Mars and Jupiter, *and* a TC between Mars and Regulus, but *no* TC Jupiter–Regulus.

Notice that there are many more conjunctions betweeen a bright outer planet and a first-magnitude zodiacal star than mentioned in Table 41.F, but they are *single*, not triple conjunctions.

TABLE 41.F

Triple conjunctions in longitude between a bright outer planet
and a first-magnitude zodiacal star, 1900 to 2100

Mars – Aldebaran	:	1911 *f*, 1943 *f*, 1990 *f*, 2022 *f*, 2069 *f*
Mars – Regulus	:	1901 *p*, 1916 *p*, 1948 *p*, 1980 *p*, 1995 *p*, 2027 *p*, 2059 *p*, 2074 *p*
Mars – Spica	:	1920, 1935, 1967, 2014, 2046, 2093
Mars – Antares	:	1969, 2048, 2095
Jupiter – Aldebaran	:	1917 *f*, 1929 *f*, 2000 *f*, 2012 *f*, 2083 *f*, 2095 *f*
Jupiter – Regulus	:	1956 *p*, 1968 *p*, 2039 *p*, 2051 *p*, 2063 *p*
Jupiter – Spica	:	1934, 1946 *p*, 1958 *p*, 2029 *p*, 2041 *p*
Jupiter – Antares	:	1900, 1912, 1983, 1995, 2066, 2078, 2090
Saturn – Aldebaran	:	1942 *f*, 2001 *f*, 2061 *p*, 2089 *f*
Saturn – Regulus	:	1978 *p*, 2037 *p*, 2096 *p*
Saturn – Spica	:	1953 *p*, 2012 *p*, 2041, 2100 *p*
Saturn – Antares	:	1957 *p*, 1986, 2016 *p*, 2045

From the table, periodicities can be found, namely 79 years for Mars and 83 years for Jupiter. (These periods were already mentioned in Chapter 38). For Jupiter and Saturn, two TCs may also be separated by only one sidereal period of the planet, 11.9 and 29.5 years, respectively. As far as Mars is concerned, the minimum time between two TCs is 15 years (for example, with Spica in 1920 and 1935).

Of course, not every opportunity of this kind results in a TC. The probability for this is $P = R/\Delta\lambda$, where R is the length of the retrograde path and $\Delta\lambda$ the difference of the longitudes of the planet at two successive favorable oppositions. Können found that, for Mars, $P = 0.13$ at a perihelic opposition and $P = 0.58$ for an aphelic one. The mean frequency of a TC with Regulus (requiring an opposition of Mars near its aphelion) will be one in about 26 years. The variation of P also explains why TCs of Mars–Antares are less frequent than Mars–Spica.

Incidentally, the difference in longitude between Spica and Antares is almost exactly equal to the difference in longitude of Mars after two successive oppositions in this part of the ecliptic. Consequently a TC of Mars–Antares is *always* preceded by a TC of Mars–Spica two years earlier. See Table 41.F. However, because of the difference in P for these stars, the reverse does not hold.

As to the other outer planets, it can be proved that P increases with the distance to the Sun. For Jupiter, $P = 0.3$ (indicating that only 30 % of its heliocentric conjunctions with a star give rise to a triple conjunction, as seen from the Earth). For Saturn, $P = 0.5$.

The TCs between *Jupiter* and a given star can be arranged as a big 'panorama'. Two examples are given in Table 41.G. In each case, the year is given when Jupiter is in opposition with the Sun. (For instance, for Jupiter–Regulus, the event of A.D. 1968 is actually 1968 p in our notation we explained earlier). Each of these two panoramas contains all TCs from the year zero to the middle of the 21st century. In fact, each panorama extends upwards infinitely far into the past, and downwards infinitely far into the future.

TABLE 41.G

Panoramas of triple conjunctions Jupiter–star

Jupiter – Regulus								Jupiter – Aldebaran						
70	82							31	43					
153	165	177						114	126					
	248	260						197	209					
	331	343							292	304				
	414	426							375	387				
	497	509	521						458	470				
	580	592	604						541	553				
		675	687						624	636	648			
		758	770							719	731			
		841	853							802	814			
		924	936	948						885	897			
		1007	1019	1031						968	980			
			1102	1114							1063	1075		
			1185	1197							1146	1158		
			1268	1280	1292						1229	1241		
				1363	1375						1312	1324		
				1446	1458							1407	1419	
				1529	1541							1490	1502	
				1612	1624							1573	1585	
				1695	1707	1719						1656	1668	
				1778	1790	1802						1739	1751	
					1873	1885							1834	1846
					1956	1968							1917	1929
					2039	2051	2063						2000	2012

In these panoramas, a shift of one place toward the right means 12 years later, while a shift of one place downwards gives a conjunction 83 years later.

In the case of Jupiter–Regulus, after 2, 3 or 4 'duets' of TCs (such as 1612-1624) we have the start of a new vertical series (for instance 1719), where one or two 'triplets' of TCs take place (1695-1707-1719 and 1778-1790-1802). Each vertical, 83-year series contains 10 or 11 TCs and lasts for about eight centuries.

In the case of the Jupiter–Aldebaran, the vertical series contain only nine TCs, and for this reason the triplets are much rarer than for Regulus. Between the years 0 and 2100 there is only one triplet of TCs Jupiter–Aldebaran (in 624-636-648), while there are nine for Regulus. This difference is due to the eccentricity of the orbit of Jupiter. In the constellation Taurus, Jupiter is closer to its perihelion, in Leo it is closer to the aphelion.

Triple conjunctions of *Mars* and a given star too can be grouped in a panorama, in such a way that they are arranged vertically in 79-year series, and horizontally in 363-year series. One gets an interesting view when the panorama is drawn as in Figure 41.*f*, where every thick dot represents a TC. Only for the points situated between the upper and lower solid lines is there a TC. In A.D. 1006, for instance, represented by the small dot just above '1085', no TC occurred between Mars and Antares. See also Figure 41.*g*.

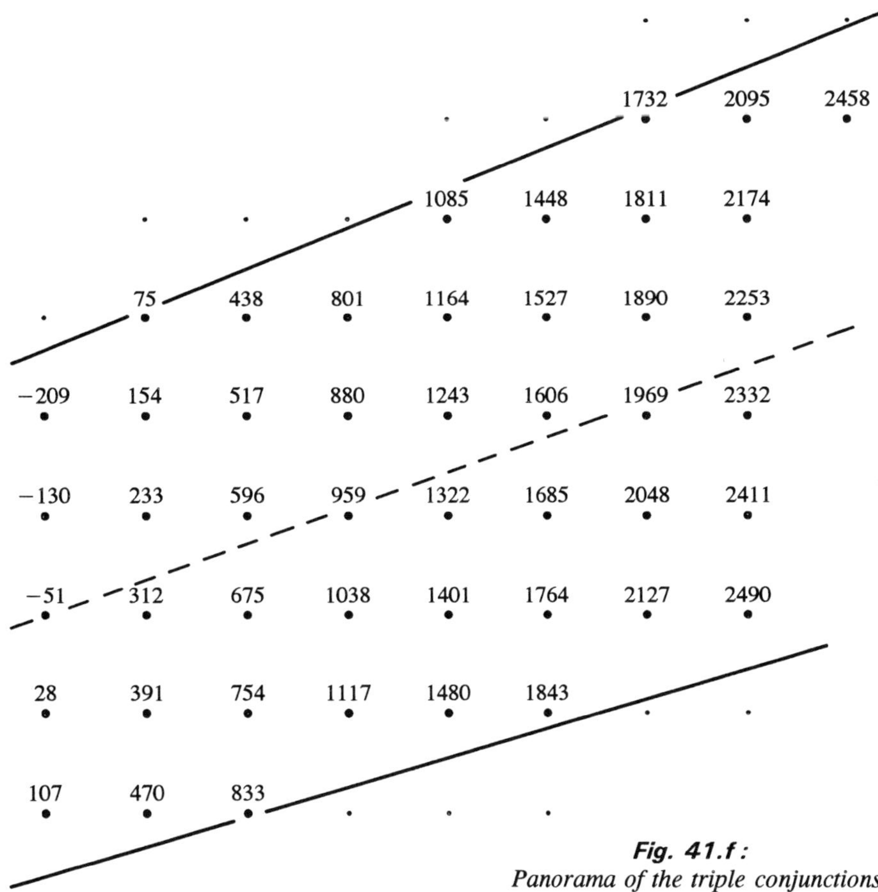

Fig. 41.f :

*Panorama of the triple conjunctions
of Mars and Antares from the second
century B.C. to the middle of the third millennium.*

Midway between the two limits a 'central line' is drawn. For a year number that would be exactly on this line, Mars and Antares are simultaneously in opposition with the Sun. This was almost the case in 1606, when on May 25 the planet and the star came in opposition with the Sun with a time difference of only 8 hours. For all points situated above the central line, for instance 1811, the opposition of Mars precedes that of Antares. The opposite holds for the points situated below the central line.

We further note that the two limits are not exactly parallel: in the future they are more distant from each other than in the past. The reason is easily found: the perihelion of Mars' orbit gradually moves away from the star. Hence, Mars' retrogradation loop in the vicinity of Antares is lengthening with time, so that the probability of a TC increases.

Year	Longitude of Mars' perihelion	Longitude of Antares	Difference
0	299°	222°	77°
1000	318°	236°	82°
2000	336°	250°	86°

These longitudes are measured along the ecliptic, from the vernal equinox of the date. The increase of Antares' longitude is due to the precession.

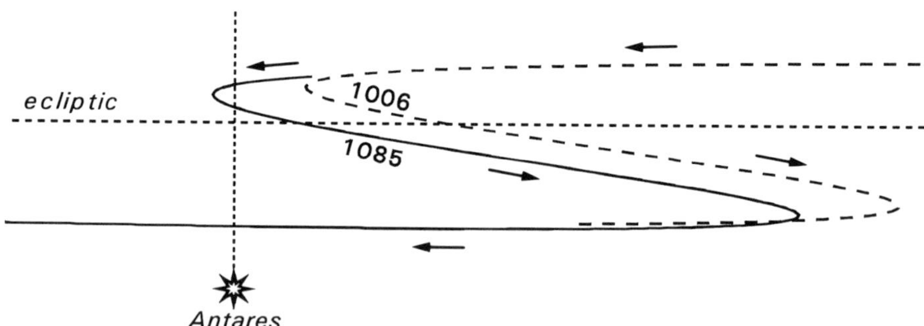

Fig. 41.g : *The beginning of a 79-year series. This figure shows the retrogradation loops of Mars in the years 1006 and 1085. In 1006 there was only a single conjunction between the planet and the star. But 79 years later the loop is shifted some degrees towards the east, so that in 1085 a triple conjunction took place.*

TABLE 41.H

Triple conjunctions in longitude involving Uranus or Neptune, 1900 to 2100

Mars – Uranus	:	1907, 1943 *f*, 1965 *p*, 2042 *p*, 2063
Mars – Neptune	:	1933 *p*, 2072 *p*
Jupiter – Uranus	:	1927 *f*, 1955 *p*, 1969 *p*, 1983, 2010 *f*, 2038 *p*, 2052 *p*, 2066, 2093
Jupiter – Neptune	:	1920 *p*, 1971, 2009, 2047 *f*, 2086 *p*
Saturn – Uranus	:	1988, 2079
Saturn – Neptune	:	1953 *p*, 1989
Uranus – Neptune	:	1993
Uranus – Aldebaran	:	None
Uranus – Regulus	:	1962 *p*, 2046 *p*
Uranus – Spica	:	None
Uranus – Antares	:	1984 *p*, 2068 *p*
Neptune – Aldebaran	:	2056 *f*
Neptune – Regulus	:	1927–1929 (*), 2093 *p*
Neptune – Spica	:	1953 *p*
Neptune – Antares	:	1974–1975 (*)

(*) Quintuple conjunction

Uranus and Neptune

Table 41.H lists all TCs involving the distant planets Uranus and Neptune in the period A.D. 1900-2100.

If we don't take Pluto into consideration, the rarest planet-planet conjunctions are those of Uranus–Neptune. The mean interval between successive heliocentric conjunctions of these planets is 171.4 years. Before the TC of 1993 there were TCs in 1478-1479, 1650, and 1821. In the future, there will be a *single* conjunction in January 2165, and TCs in 2336-2337 and 2508-2509. (In 2164-2165, Uranus and Neptune will have a TC in *right ascension*, but not in celestial longitude).

There is something interesting about the stations of Uranus which does not apply to Jupiter and Saturn. The interval in longitude (or in right ascension) between the eastern station preceding an opposition and the western station following the *next* opposition is small. Moreover, that interval is almost zero when the planet is near the aphelion of its orbit, as in 1925. The above-mentioned formula $P = R/\Delta\lambda$ yields a mean value of $P = 0.93$ for Uranus. Therefore, a star is almost always passed by Uranus with a triple conjunction.

Finally, for Neptune, $P = 1.3$. This indicates not only that a heliocentric conjunction will always give rise to a TC, but also that there is a probability of 30% for two additional conjunctions the next year. This is called a *quintuple* conjunction. For Regulus, this was the case in 1927-1929.

Indeed, in the case of Neptune the opposition loop is larger than the planet's annual displacement. That this is so can be easily checked. During a time span of 1½ years, Neptune moves over 3.28 degrees in its orbit. This corresponds to a distance in space of 1.72 astronomical units. What can you deduce from this? — Repeat the calculation for Uranus.

42. *Planetary groupings*

It may happen that three or more planets are near each other on the sky. Such a grouping is sometimes erroneously called a 'conjunction'. *Two* celestial bodies are said to be in conjunction if their celestial longitudes or their right ascensions are equal. But *three* or more planets never can be simultaneously in conjunction. Although this is theoretically possible, in practice the probability of the event is zero.

Trios

Let us first consider the groupings of *three* planets, A, B, and C. For any instant, it is possible to calculate the angular diameter Δ of the smallest circle containing these three bodies. For this calculation, use can be made of the method described in Chapter 19 of my *Astronomical Algorithms* (Willmann-Bell, ed.; 1991). Two cases can occur (Figure 42.*a*):
— the smallest circle has as diameter the longest side of the triangle A, B, C, and one point lies inside of the circle;
— the smallest circle is the circle passing through the points A, B, C.

Of course, as the planets move, the diameter Δ of the circle varies with time, and at a certain instant it reaches its minimum value. It is this smallest circle we are interested in.

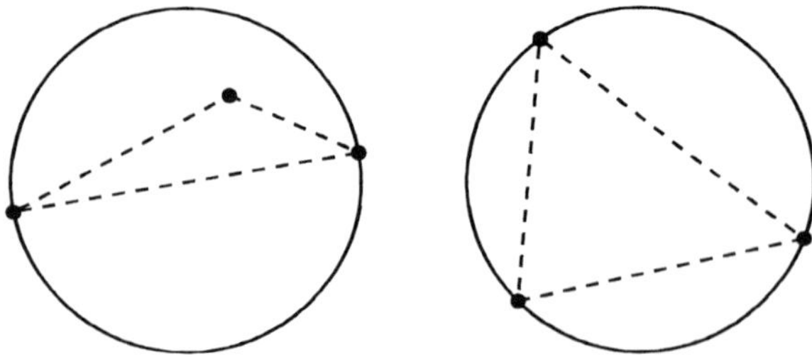

Fig. 42.a : *The two possible types of the smallest circle containing three points.*

We have calculated all planetary trios fitting in a circle with a minimum diameter smaller than 5 degrees, from A.D. 1980 to 2050. That limit of 5° has been chosen more or less arbitrarily, but we have to make a choice. The data for the years 1980-2005 had been published earlier in the journal *L'Astronomie* (Société Astronomique de France) of December 1977, pages 488-489. The *search* for the trios from 2006 to 2050 has been enormously facilitated by using a file provided by Mr. J. V. Uptain (see further in this chapter).

Our results are given in Table 42.A. The first and the second column give the time of the smallest circle, rounded to the nearest integer hour (Universal Time). The next column mentions the abbreviated names of the three planets (ME = Mercury, etc.). The angular diameter of the smallest circle containing these planets is mentioned next. The last column gives, for that instant, the angular distance from the Sun; more precisely, it is the elongation of that planet of the trio which is closest to the Sun: E = East from the Sun = visible in the evening in the West; W = West from the Sun = visible in the morning in the East. A dash in the last column means that some of the planets are to the East of the Sun, and some to the West. For reason of completeness, the list mentions all trios, even those which occur close to the Sun in the sky and hence are not observable.

TABLE 42.A

The planetary trios, years 1980 to 2050

Date	UT	Planets	Diameter	Elong.
	h		° ′	°
1980 Nov. 3	19	VE–JU–SA	3 53	36 W
1981 Aug. 25	16	VE–JU–SA	2 44	36 E
1981 Sep. 10	13	ME–JU–SA	4 14	22 E
1985 Dec. 4	1	ME–VE–SA	1 52	10 W
1986 Feb. 9	21	ME–VE–JU	1 42	5 E
1987 Aug. 21	10	ME–VE–MA	1 49	—
1991 June 18	9	VE–MA–JU	1 48	45 E
1992 Feb. 29	1	VE–MA–SA	4 28	27 W
1994 Jan. 1	4	ME–VE–MA	2 35	2 W
1995 Nov. 19	2	VE–MA–JU	2 03	23 E
1996 Mar. 23	6	ME–MA–SA	1 23	4 W
2000 Apr. 15	6	MA–JU–SA	4 54	17 E
2000 May 9	10	ME–JU–SA	2 27	—
2000 May 18	13	VE–JU–SA	1 35	6 W
2002 May 7	1	VE–MA–SA	2 49	27 E

Table continues on next page

TABLE 42.A (Cont.)

Date	UT	Planets	Diameter	Elong.
	h		° ′	°
2004 Sep. 29	7	ME–MA–JU	1 04	5 W
2005 June 26	11	ME–VE–SA	1 22	23 E
2006 Dec. 10	13	ME–MA–JU	0 59	15 W
2008 Aug. 15	20	ME–VE–SA	2 23	16 E
2008 Sep. 7	15	ME–VE–MA	3 35	24 E
2008 Sep. 12	17	ME–VE–MA	3 34	25 E
2009 Feb. 24	6	ME–MA–JU	3 40	21 W
2010 Aug. 8	7	VE–MA–SA	4 49	46 E
2011 May 11	20	ME–VE–JU	2 03	26 W
2011 May 21	8	ME–VE–MA	2 08	22 W
2013 May 27	9	ME–VE–JU	2 26	16 E
2015 Oct. 26	3	VE–MA–JU	3 35	43 W
2021 Jan. 10	19	ME–JU–SA	2 23	12 E
2021 Feb. 13	11	ME–VE–JU	4 36	10 W
2026 Apr. 20	18	ME–MA–SA	1 40	22 W
2028 June 16	1	ME–VE–MA	4 51	18 W
2032 June 2	11	ME–MA–SA	2 51	11 E
2036 July 22	11	ME–MA–SA	1 12	20 E
2038 Aug. 26	18	ME–VE–JU	3 38	14 W
2040 Sep. 6	10	VE–MA–SA	3 56	26 E
2040 Sep. 10	17	ME–JU–SA	4 54	18 E
2040 Oct. 31	7	ME–JU–SA	4 07	17 W
2041 Nov. 3	12	ME–VE–SA	2 15	12 W
2044 Dec. 19	4	ME–VE–SA	1 16	22 W
2048 May 27	14	ME–VE–JU	1 46	—

Quadruplets

Groupings of *four* planets are much rarer than groupings of three, of course. We have searched for the occurrence of such *quadruplets* during a period of 35 centuries, from the year zero to +3500. Here too, we limited ourselves to the naked-eye planets Mercury to Saturn, and to groupings inside of a circle with a diameter of at most five degrees.

Our results are given in Table 42.B. This list has been first published in *L'Astronomie* (Société Astronomique de France), Vol. 104, No. 4, pages 183-186; April 1990.

TABLE 42.B

The planetary quadruplets, years 0 to 3500

Date	UT	Planets	Diameter	Elong.
	h		° ′	°
1 Nov. 4	23	ME–VE–MA–JU	2 32	16 W
243 Mar. 19	14	ME–VE–MA–JU	4 25	18 W
273 Sep. 30	23	ME–VE–JU–SA	4 41	6 W
302 Sep. 5	15	ME–VE–MA–SA	2 02	8 E
491 Feb. 5	18	ME–VE–JU–SA	4 54	14 W
590 June 8	12	ME–VE–JU–SA	4 51	3 E
710 June 24	24	ME–VE–MA–SA	4 08	20 E
714 Dec. 13	17	ME–VE–MA–JU	4 27	7 W
842 Dec. 30	15	ME–VE–MA–SA	3 21	2 E
1365 Oct. 17	7	ME–VE–JU–SA	4 44	3 E
1524 Feb. 10	2	VE–MA–JU–SA	3 12	10 E
1568 Nov. 19	5	ME–VE–MA–JU	3 24	11 E
1586 Apr. 22	4	ME–VE–MA–SA	2 09	7 W
1624 Sep. 5	1	ME–VE–MA–JU	4 50	1 W
1680 June 13	2	ME–VE–MA–JU	4 38	22 W
1725 Mar. 17	8	ME–VE–MA–JU	1 21	24 W
1821 May 4	14	ME–MA–JU–SA	4 43	24 W
1861 Sep. 5	5	ME–MA–JU–SA	4 08	—
1910 Oct. 28	12	ME–VE–MA–JU	3 03	7 W
2100 Nov. 6	13	ME–MA–JU–SA	4 47	8 W

Table continues on next page

Again, the table gives the diameter of the smallest circle containing the four planets, and the corresponding date and instant (Universal Time). The last column gives, for that instant, the angular distance to the Sun; more precisely, it is the elongation of that planet of the quadruplet which is closest to the Sun in the sky.

We note that in most cases the planets Mercury and Venus are present. In only two cases (1524 and 2378) is Mercury absent from the quadruplet. In the case of the quadruplets of A.D. 2576 and 2623, the Sun lies *inside* of the smallest circle containing the four planets.

The grouping of January 3431 is remarkable. On January 25 at 7^h UT, the four planets will lie inside of a geocentric sector of only 0°23′ in *longitude* (but in latitude the separation will be 2°21′).

The most compact quadruplet of the whole period 0–3500 is that of 1725 March 17, the smallest diameter of the circle being 1°21′ — see Figure 42.*b*.

TABLE 42.B (Cont.)

Date	UT	Planets	Diameter	Elong.
	h		° ′	°
2297 July 16	7	ME–VE–MA–SA	3 58	20 W
2378 Feb. 2	13	VE–MA–JU–SA	2 37	32 E
2470 May 15	12	ME–VE–MA–SA	4 54	10 W
2576 Dec. 19	11	ME–MA–JU–SA	4 57	—
2581 Jan. 9	17	ME–VE–MA–SA	4 40	19 E
2623 Dec. 5	4	ME–VE–MA–JU	2 06	—
2713 July 30	8	ME–VE–MA–SA	4 10	10 E
2876 Mar. 17	4	ME–VE–MA–SA	4 10	27 W
2978 Sep. 6	20	ME–VE–MA–SA	3 15	16 W
3006 Feb. 14	7	ME–VE–MA–JU	4 06	7 E
3013 Nov. 14	2	ME–VE–JU–SA	4 09	10 W
3033 May 31	21	ME–VE–JU–SA	3 24	18 E
3191 Sep. 29	23	ME–VE–MA–JU	4 20	22 E
3247 July 8	9	ME–VE–MA–JU	3 09	6 E
3292 Apr. 11	17	ME–VE–MA–JU	4 42	6 E
3324 Apr. 26	19	ME–VE–MA–SA	3 22	9 E
3424 Oct. 25	1	ME–VE–MA–SA	4 23	4 W
3430 Dec. 11	4	ME–MA–JU–SA	4 40	15 E
3431 Jan. 26	5	ME–VE–JU–SA	2 14	21 W
3477 Nov. 22	7	ME–VE–MA–JU	4 48	18 E

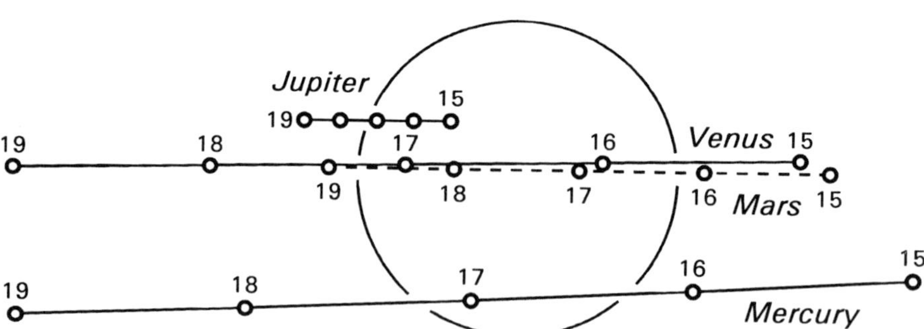

Fig. 42.b : *The exceptional planetary grouping of March 1725. The positions of Mercury, Venus, Mars and Jupiter are indicated from March 15 to March 19, at 0^h Universal Time. Ecliptic North is up. The circle has a diameter of two degrees.*

Quintuplets

Some years ago, Mr. Jerald V. Uptain, of Aberdeen, Mississippi, calculated all groupings of two, three, four and five planets taking place from 4000 B.C. to A.D. 2750. He adopted the following criteria as limits for the four types of groupings: a smallest distance of 3° for the planetary pairs, a smallest circle with a diameter of 9° for the trios, of 16° for the quadruplets, and of 25° for the quintuplets. For the period 2000 B.C. to A.D. 2750, Mr. Uptain found 86 quintuplets (diameter of circle < 25°), nine of which fit in a circle with a diameter of 10° or less. We recalculated these nine groupings and could confirm the values of Mr. Uptain. Our results are given in Table 42.C.

TABLE 42.C

The planetary quintuplets, years −2000 to +2750

Date	UT	Diameter	Elong.
	h	° ′	°
−1952 Feb. 26	21	4 22	27 W
−1058 May 28	17	6 27	21 E
− 184 Mar. 25	5	6 44	27 W
− 144 July 26	24	9 57	—
− 46 Nov. 28	15	9 23	15 W
332 Oct. 4	11	8 43	10 W
710 June 26	3	5 55	19 E
1186 Sep. 17	10	8 52	1 E
2040 Sep. 8	0	9 18	21 E

On July 26 of the year −144 (145 B.C.), the bright star Regulus (α Leonis) was situated inside of the same area as the planets, but — alas! — the Sun too, so that remarkable grouping could not be observed.

On 332 October 4, the star Spica (α Virginis) was inside of the same area as the planets, and the lunar crescent was in the vicinity.

On 1186 September 17, Spica was near the planetary grouping, but so was the Sun, so the planetary gathering could not be seen.

Mr. Uptain wrote: "Planetary clustering has fascinated people throughout history. Those of the more spectacular variety have always evoked comment, a sense of wonder and, not infrequently, outright fear. For example, records from A.D. 1186 show that the close grouping of five planets that year caused near panic among the citizens of Europe after an 'authority' predicted that world-wide disasters would result. In modern times, planetary close approaches continue to attract viewers, including the general public, which must depend on the news media for most of its information on the subject. Amateur astronomers, many of whom specialize in planetary study, find a particular fascination with the subject, drawing upon their own knowledge and information gleaned from astronomy publications for news of impending clusterings. Interests include those of the historian, who may be able to assign a date to the occurrence of an historic event associated with known or suspected close approaches."

The interested reader can find more details about groupings of five planets in the article "Quintuple planetary groupings" by S. De Meis and J. Meeus in the *Journal* of the British Astronomical Association, Vol. 104, No. 6, pages 293-297 (December 1994).

Other groupings

If we include the four first-magnitude zodiacal stars, then many more groupings can be found. We leave this subject to the interested calculator, and just mention the following cases.

On 1978 July 11, at 16^h UT, Venus, Saturn and Regulus were inside of a circle with a diameter of 1°33'.

In 1980, from February 29 to May 20, Mars, Jupiter and Regulus were inside of a circle with a diameter of six degrees. The smallest circle containing the two planets and the star had a minimum diameter of 1°34', on May 3 at 4^h UT.

On 1987 August 21, Mercury, Venus, Mars, Regulus and the (center of the) Sun were inside of a circle with a diameter of three degrees.

On 1989 August 4, Mercury, Mars and Regulus were in a circle of 1°18'.

The most compact grouping in modern times which we know is that of Mercury, Jupiter and Regulus in September 1991. The smallest circle containing these three bodies had a diameter of only 0°20', on September 10 at 13^h UT.

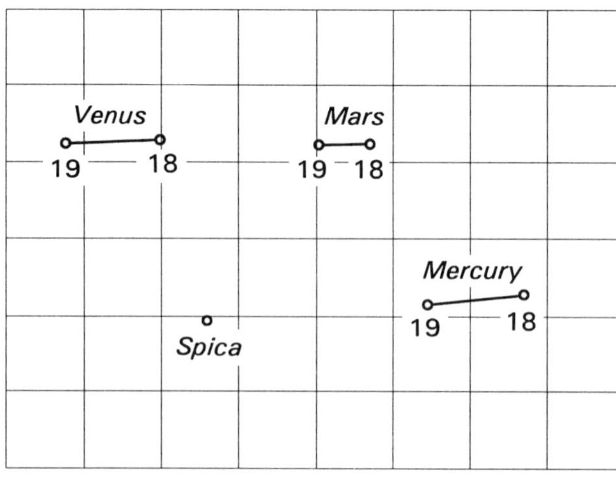

Fig. 42.c : On 2040 Sept. 18, the planets Mercury, Venus and Mars, and the star Spica will be in a circle with a diameter of 5°07'. In the drawing, ecliptic North is up, and the grid lines are drawn at intervals of 1°. The positions of the planets are shown for 0^h UT on September 18 and 19.

43. Periodicities in the phenomena of the satellites of Jupiter

The original version of the following text was published in the Belgian journal *Ciel et Terre*, Vol. 90, No. 2, pages 123-135 (March-April 1974).

The four great satellites of Jupiter describe their orbits in relatively short times, and their ever changing configurations suggest at first sight that their motions are chaotic. In fact, order and regularity reign in the small world of Jupiter.

Table 43.A contains some values concerning these satellites. The sidereal period is the revolution period with respect to the stars. The synodic period is measured with respect to the Sun, and its value is a little larger than the sidereal period by reason of the motion of Jupiter in its orbit. Because, as seen from Jupiter, the Earth is always in the vicinity of the Sun, the numbers in the last column of Table 43.A are also the mean synodic revolution periods with respect to the Earth. In this chapter we are interested in the synodic periods, not the sidereal ones. So, in what follows the synodic periods will simply be called 'periods'.

TABLE 43.A

Satellite		Mean distance to center of Jupiter in equatorial radii of Jupiter	Mean sidereal period (days)	Mean synodic period (days)
I	Io	5.9073	1.769 138	1.769 860
II	Europa	9.3991	3.551 181	3.554 094
III	Ganymede	14.9924	7.154 553	7.166 387
IV	Callisto	26.3699	16.689 018	16.753 552

Approximate commensurabilities

Two sidereal revolution periods of Saturn have *nearly* the same duration as five revolution periods of Jupiter. There are other similar approximations. For instance, 8 periods of the Earth $\approx$ 13 periods of Venus (see Chapter 38), and 2 periods of Uranus $\approx$ 1 period of Neptune.

There are similar near-periodicities for the first three satellites of Jupiter. Two periods of satellite I have *nearly* the same duration as one period of satellite II, and two periods of II have *nearly* the same duration as one period of III:

$$2 \times 1.769\,860 \;=\; 3.539\,720 \quad \text{instead of} \quad 3.554\,094$$
$$2 \times 3.554\,094 \;=\; 7.108\,188 \quad \text{instead of} \quad 7.166\,387$$

From this it results that some phenomena repeat at intervals of 3.55 or 7.1 days, so for instance:

— in 1996, satellites I and II were simultaneously eclipsed in Jupiter's shadow cone on May 28 and 31; June 4, 7, 11, 14, 18, 22, 25, and 29; July 2, 6, and 9;
— in 1996, there were simultaneous shadow transits of II and III across the disk of Jupiter on June 6, 13, 20, and 27, and on July 4, so this latter series contained only five events.

Important remark. — When we speak of 'simultaneous' eclipses or transits, this does *not* mean that the two eclipses or the two transits begin and end at the same instants. We mean that *during some time* the two phenomena occur simultaneously. For instance, on 1996 June 13 the transit of the shadow of satellite III began at $2^{h}55^{m}$ UT and ended at $6^{h}00^{m}$; the transit of the shadow of II began at $4^{h}26^{m}$ and ended at $7^{h}12^{m}$. Therefore, the shadows of the two satellites fell simultaneously on Jupiter from $4^{h}26^{m}$ to $6^{h}00^{m}$.

Exact commensurability

Besides the above-mentioned commensurabilities, which are only approximate, there exists an *exact* commensurability between the motions of the satellites I, II and III. In what follows, we will suppose to be on the Sun, so all the phenomena we will describe will concern an heliocentric observer. At the very end of this chapter, we will return to geocentrism.

At a certain instant, one of the satellites is exactly in front of Jupiter, in inferior conjunction. We will call this the position $u = 0°$. After a quarter of a revolution (*), the satellite reaches its greatest western elongation ($u = 90°$). Then we have superior conjunction ($u = 180°$), the greatest eastern elongation ($u = 270°$) and finally, after one complete synodic revolution, inferior conjunction again ($u = 360° = 0°$).

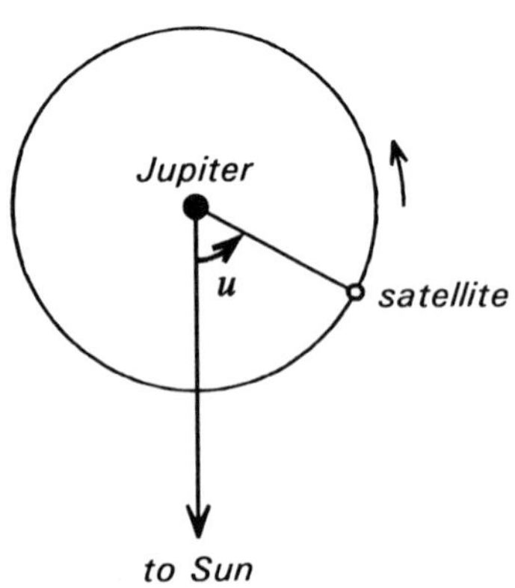

Fig. 43.a

(*) Many authors incorrectly use the word 'orbit' instead of *revolution*, for instance in the sentence "A typical comet loses about 0.1 % of its mass on each orbit around the Sun." This should be *on each revolution*. The word *orbit* refers to the path of the body, for instance when we say that the orbit of Mars has an inclination of $1°51'$. But *revolution* refers to the *motion* along that path: in 1 century, Mars performs 53 revolutions (not 53 orbits!) around the Sun. Mars has only one orbit.

So, we will call u the arc described by the satellite since its last inferior conjunction. In other words, u is the angle Sun–Jupiter–satellite (Figure 43.a). The angle u of satellite I will be designed by u_1, that of satellite II by u_2, etc. The orbital eccentricities and inclinations, which are rather small, will be neglected here. So we will work with uniform, circular motions. Then we have

$$
\begin{aligned}
u_1 &= 163.8067 + 203.405\,8643\,d \\
u_2 &= 358.4108 + 101.291\,6334\,d \\
u_3 &= 5.7129 + 50.234\,5179\,d \\
u_4 &= 224.8151 + 21.487\,9801\,d
\end{aligned}
\qquad \text{(A)}
$$

where the angles u are expressed in degrees, and d is the time measured in days from 2000 January 1 at 12^h Dynamical Time. (In modern times, the difference between Dynamical Time and Universal Time is small; see pages 7 and 8). Note that for instants earlier than 2000 January 1.5 the quantity d is negative.

From expressions (A) it is easy to deduce that

$$
u_1 - 3\,u_2 + 2\,u_3 = 180^\circ \qquad \text{(B)}
$$

This is not an approximate relation, but a *rigorous commensurability*, and hence a permanent characteristic of the system of the three inner satellites. The fourth satellite escapes from this law, and for this reason in what follows we will limit ourselves to the satellites I, II and III only.

Instead of (B), we may also write

$$
u_1 - u_2 + 2\,(u_3 - u_2) = 180^\circ \qquad \text{(C)}
$$
$$
3\,(u_1 - u_2) + 2\,(u_3 - u_1) = 180^\circ \qquad \text{(D)}
$$

from which one deduces several remarkable consequences:

1. It is not possible to have simultaneously $u_1 - u_2 = 0$ and $u_2 - u_3 = 0$. Consequently, *as seen from Jupiter*, the first three satellites never can be simultaneously in conjunction. Hence, they never can be simultaneously in front of the planet, nor be eclipsed simultaneously in the planet's umbral cone.

2. If $u_1 - u_2$ is equal to $0°$ or to $360°$, then from (C) it results that $u_3 - u_2$ is equal to either $+90°$ or $-90°$. Hence, if the first two satellites are in conjunction (as seen from Jupiter), their common radius vector is perpendicular to the radius vector of satellite III. In other words, satellite III is then 90 degrees ahead or behind I and II (see the left drawing of Figure 43.b).

3. If $u_3 - u_2 = 0$, then from (C) it results that $u_1 - u_2 = 180°$. That is, if the satellites II and III are in conjunction, then satellite I is in opposition with them, at the other side of Jupiter. See the middle drawing of Figure 43.b.

For instance, on 1996 June 13 the satellites II and III were simultaneously in transit while satellite I was behind the planet. Consequently, during more than two hours IV was the only visible satellite.

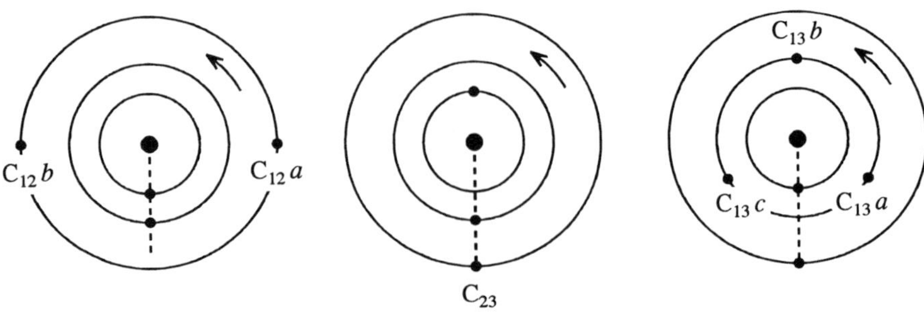

Fig. 43.b

4. If $u_3 - u_1$ is equal to $0°$, to $360°$ or to $-360°$, then according to (D) the difference $u_2 - u_1$ is equal to either $+60°$, $+180°$ or $-60°$. Hence, when I and III are in conjunction, satellite II can be at only one of three well-determined positions: either it is in opposition with them, or it precedes or follows them by 60 degrees. See the right drawing of Figure 43.*b*.

So in total there are six remarkable configurations, which we will designate by means of the following symbols (see Figure 43.*b*):

C_{23} Conjunction of the satellites II and III. Satellite I is then in opposition with them. (The index of C_{23} should be read as 'two-three', not as twenty-three).

$C_{12}a$ Satellites I and II are in conjunction, and III precedes them by $90°$.

$C_{12}b$ Satellites I and II are in conjunction, and III follows them by $90°$

$C_{13}a$ Satellites I and III are in conjunction; II precedes them by $60°$.

$C_{13}b$ Satellites I and III are in conjunction; II is in opposition with them.

$C_{13}c$ Satellites I and III are in conjunction; II follows them by $60°$.

Let us now examine how these various configurations occur successively.

The period of 7.0509 days

Let us first search for the frequency of the conjunctions between the satellites I and II. In other words: suppose that I and II are in conjunction (that is, $u_1 = u_2$); after what time will they be in conjunction again (as seen from Jupiter)?

According to formulae (A), satellite I moves at a rate of 203.405 8643 degrees per day in its orbit, while satellite II has a daily motion of 101.291 6334 degrees. The difference, or 102.114 2309, is the number of degrees satellite I moves more than II every day. This value is given in Table 43.B: the daily variation of Δu, that is, of the difference $u_1 - u_2$ of the longitudes of I and II. Consequently, the next conjunction of the two satellites will take place after $360 / 102.114 2309 =$ 3.525 463 560 days. This value is given in the last column of Table 43.B.

TABLE 43.B

Satellites	Variation of Δu in degrees per day	Time needed for a variation of 360 degrees
I — II	102.1142309	3.525463560 days $= P_{12}$
I — III	153.1713464	2.350309039 days $= P_{13}$
II — III	51.0571155	7.050927113 days $= P_{23}$

In the same way, the values are calculated for the couples I–III and II–III.

We note that P_{12}, the time difference between two successive conjunctions of the satellites I and II, is *exactly* equal to the half of P_{23}, and that P_{13} is *exactly* equal to one third of P_{23}. This is a consequence of the exact commensurability (B).

Hence, P_{23} can be considered as a fundamental period. During this time period of 7.0509 days there occurs one conjunction between the satellites II and III, two conjunctions between I and II, and three conjunctions between I and III.

Let us designate by p the twelfth part of P_{23}, that is, 0.587577259 day, or 14 hours 06 minutes. During a time interval p, the satellites I, II and III move over 119.51666, 59.51666, and 29.51666 degrees, respectively, in their orbits. Consequently, during the period p,

I travels 60 degrees more then II,
I travels 90 degrees more than III,
II travels 30 degrees more than III.

Now, suppose that II and III are in conjunction (C_{23}), and that they are exactly in front of Jupiter, in their inferior conjunction. Then, as we have seen, satellite I is exactly in superior conjunction. That is, we suppose that $u_1 = 180°$ and $u_2 = u_3 = 0°$. This disposition is represented in drawing '0' of Figure 43.*c*. Let us call t_0 the instant at which this particular disposition of the three satellites takes place.

After one period p, that is, at time $t_0 + p$, the satellites are placed as indicated in drawing '1' of Figure 43.*c*. The radius vector of I is now perpendicular to that of III, and we have $u_1 = 180° + 119°52 = 299°52$; $u_2 = 59°52$; and $u_3 = 29°52$. See the line $p = 1$ in Table 43.C.

After another period p, that is, at time $t_0 + 2p$, satellite I has caught up satellite III ($u_1 = u_3 = 59°03$), while II is 60° ahead of them ($u_2 = 119°03$). We now have configuration $C_{13}a$ of Figure 43.*b*.

At time $t_0 + 3p$ (drawing '3'), I has caught up to II, and these two satellites are 60° ahead of III. We now have configuration $C_{12}b$.

The reader now can follow the continuation of this celestial roundabout by means of Figure 43.*c* and Table 43.C. The configurations $C_{13}b$, $C_{12}a$ and $C_{13}c$ occur 6, 9 and 10 periods p, respectively, after the starting time t_0.

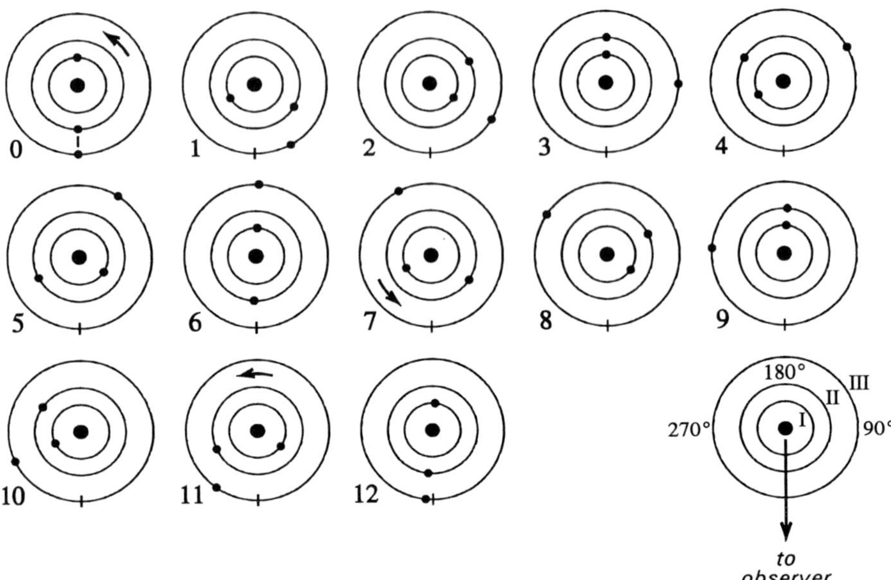

Fig. 43.c : *Successive positions of the three inner Galilean satellites of Jupiter 1, 2, 3, etc., periods 'p' after the starting time 'zero'. The observer is supposed to be at the bottom, in the direction of the big arrow.*

TABLE 43.C

Positions of the three inner satellites 1, 2, ... periods 'p' after time t_0

p	days after time t_0	u_1	u_2	u_3	$u_1 - u_2$	$u_1 - u_3$	$u_2 - u_3$	Config.
0	0.0000	180.00	0.00	0.00	180	180	0	C_{23}
1	0.5876	299.52	59.52	29.52	240	270	30	
2	1.1752	59.03	119.03	59.03	300	0	60	$C_{13}a$
3	1.7627	178.55	178.55	88.55	0	90	90	$C_{12}b$
4	2.3503	298.07	238.07	118.07	60	180	120	
5	2.9379	57.58	297.58	147.58	120	270	150	
6	3.5255	177.10	357.10	177.10	180	0	180	$C_{13}b$
7	4.1130	296.62	56.62	206.62	240	90	210	
8	4.7006	56.13	116.13	236.13	300	180	240	
9	5.2882	175.65	175.65	265.65	0	270	270	$C_{12}a$
10	5.8758	295.17	235.17	295.17	60	0	300	$C_{13}c$
11	6.4633	54.68	294.68	324.68	120	90	330	
12	7.0509	174.20	354.20	354.20	180	180	0	C_{23}

Finally, after 12 periods p, or 7.0509 days after the starting time t_0, we find the initial disposition C_{23} again. However, the longitudes are now $u_1 = 174°20$ instead of 180°, and $u_2 = u_3 = 354°20$ instead of 360°. In the course of one period P_{23} of 7.0509 days, each of the six configurations (conjunctions) illustrated in Figure 43.*b* occurs once; see the last column of Table 43.C.

The period of 437.64 days

From what precedes, it results that after one 'fundamental' period P_{23} of 7.0509 days, the satellites I, II and III return in the same relative positions, but that their longitudes u (with respect to the Sun) have decreased by 5.80 degrees. In other words, after one period P_{23}, satellite III has performed one complete revolution *minus* 5°80; satellite II has made two revolutions minus 5°80; and satellite I has made four revolutions minus 5°80.

One obtains a more accurate value by multiplying 50.2345179 degrees (the daily motion of III) by 7.050927113, and subtracting the result from 360°. One then finds 5.800076 degrees.

After another period P_{23}, the shifting will be twice 5.80 degrees with respect to the initial disposition, and so on. After what time will the position of the conjunction C_{23} have made a complete turn (as seen from the Sun)? Dividing 360° by 5°800076 one finds 62.06816 periods P_{23}, which is equal to 437.64 days, or one year and 2½ months.

Let us now have again a look at Figure 43.*c* and at Table 43.C. We started from the instant t_0, when satellites II and III were simultaneously in inferior conjunction, and satellite I in superior conjunction: I, Jupiter, II and III were exactly on a straight line, and this line was exactly directed to the Sun. After 7.0509 days, that is, at time $t_0 + 12p$, or $t_0 + P_{23}$, the conjunction line I–Jupiter–II–III has rotated over 5.8 degrees, so satellites II and III no longer reach their inferior conjunctions simultaneously. However, the difference is still sufficiently small to allow another 'simultaneous' transit over the planet's disk. And, for the same reason, there was a simultaneous transit 7.0509 days *before* t_0.

But that's not all. At time $t_0 + 6p$ (drawing 6 in Figure 43.*c*), I and III pass simultaneously behind Jupiter, while II is in transit. And at the instants $t_0 + 3p$ and $t_0 + 9p$, I and II are simultaneously behind the planet.

Consequently, during several weeks before and after the time t_0 there are:

(a) a series of simultaneous transits of II and III over the disk of Jupiter. These phenomena occur at intervals of 7.05 days. Example: there were simultaneous shadow transits of II and III on 1996 June 6, 13, 20, 27, and July 4;

(b) halfway between the phenomena *(a)*, that is, at intervals of 7.05 days too, there are simultaneous eclipses of I and III. Example: on 1996 June 16 and 23;

(c) halfway between the phenomena *(a)* and *(b)*, hence at intervals of 3.53 days, there are simultaneous eclipses of I and II; then satellite III is at its greatest elongation, successively east and west from Jupiter.

This situation repeats at intervals of 437.64 days. For instance, there are simultaneous shadow transits of satellites II and III on the following dates:

in 1997: August 17, 24, 31; September 7 and 14;
in 1998: October 28; November 4, 11 and 18;
in 2000: January 8, 15 and 22;
in 2001: March 13, 20, 27; and April 4;
and so on.

We note that the series of January 2000 contains only three cases, while that of August-September 1997 has five. This difference is due mainly to the inclination of the orbits of the satellites on the orbital plane of Jupiter. In 1997, the Sun lies almost exactly in the plane of the satellite orbits, so the eclipses and the transits are nearly central and have a long duration. On the contrary, in January 2000, approximately a quarter of a revolution of Jupiter later, the phenomena occur at high latitude and hence have a shorter duration, principally those of satellite III. In August and September 1997, the shadow transits of III have a duration of $3\,{}^2/_3$ hours, but in January 2000 the duration is only 2 hours, whence a smaller probability for a simultaneous transit with satellite II.

Table 43.D mentions the times t_0 during the years 1960–2040. So these are the epochs when, as seen from the Sun, satellites II and III are simultaneously in inferior conjunction with Jupiter.

TABLE 43.D

The instants t_0 from 1960 to 2040

1960 July 3	1980 Nov. 15	2001 Mar. 29	2021 Aug. 11
1961 Sep. 13	1982 Jan. 26	2002 June 10	2022 Oct. 23
1962 Nov. 25	1983 Apr. 9	2003 Aug. 22	2024 Jan. 4
1964 Feb. 6	1984 June 20	2004 Nov. 1	2025 Mar. 16
1965 Apr. 18	1985 Aug. 31	2006 Jan. 13	2026 May 28
1966 June 30	1986 Nov. 12	2007 Mar. 27	2027 Aug. 8
1967 Sep. 11	1988 Jan. 23	2008 June 6	2028 Oct. 19
1968 Nov. 21	1989 Apr. 5	2009 Aug. 18	2029 Dec. 31
1970 Feb. 2	1990 June 17	2010 Oct. 30	2031 Mar. 13
1971 Apr. 16	1991 Aug. 28	2012 Jan. 10	2032 May 24
1972 June 26	1992 Nov. 8	2013 Mar. 23	2033 Aug. 5
1973 Sep. 7	1994 Jan. 20	2014 June 3	2034 Oct. 16
1974 Nov. 18	1995 Apr. 2	2015 Aug. 15	2035 Dec. 28
1976 Jan. 30	1996 June 13	2016 Oct. 26	2037 Mar. 10
1977 Apr. 12	1997 Aug. 25	2018 Jan. 6	2038 May 21
1978 June 23	1998 Nov. 5	2019 Mar. 20	2039 Aug. 2
1979 Sep. 4	2000 Jan. 17	2020 May 31	2040 Oct. 13

It is *important* to note that these dates are those of an 'ideal' and hence *imaginary* inferior conjunction! In 1989, for instance, the satellites II and III were in inferior conjunction on April 1 and on April 8, and *not* at the date April 5 which is mentioned in Table 43.D. But *if* the date had been April 5, then the two inferior conjunctions would have been exactly simultaneous ($u_2 = u_3 = 0°$).

Moreover, in the calculation of the dates the eccentricity of the orbit of Jupiter has not been taken into account. For this reason, the error can amount to one week, which is not important in the present study. For instance, the middle of the 1997 series August 17 – 24 – 31 and September 7 – 14 differs by six days from the date 1997 August 25 mentioned in the table.

During some weeks around the dates given in Table 43.D, we thus have the three above-mentioned series *(a)*, *(b)* and *(c)*. But this does not mean that there are no simultaneous phenomena at intermediate epochs.

Let us again look at Figure 43.c and at Table 43.C. At time $t_0 + 2p$, or 1.18 day after time t_0, satellites I and III are in conjunction but this occurs at $u = 59°.03$. But 7.05 days later this conjunctions takes place at $u = 59.03 - 5.80 = 53.23$, and so on. At time $t_0 + 2p + 10\,P_{23}$, or 71.68 days after t_0, the I–III conjunction occurs at $u = 1°.03$, practically at inferior conjunction. Near this epoch, then, there is another series of simultaneous phenomena: this time, it concerns shadow transits of I and III. On the average, the middle of the series occurs 72.94 days after the instant t_0. Note that $72.94 = 437.64/6$. One can reason similarly for the other conjunctions. The results are shown in Table 43.E.

TABLE 43.E

Near the instant	Middle of the series of simultaneous phenomena	Time interval between the phenomena of the series
t_0 (given in Table 43.D)	*(a)* transits of II and III	7.05 days
	(b) eclipses of I and III	7.05 days
	(c) eclipses of I and II	3.53 days
$t_0 + 73$ days	*(d)* transits of I and III	7.05 days
$t_0 + 146$ days	*(e)* eclipses of I and III	7.05 days
$t_0 + 219$ days	*(f)* eclipses of II and III	7.05 days
	(g) transits of I and III	7.05 days
	(h) transits of I and II	3.53 days
$t_0 + 292$ days	*(i)* eclipses of I and III	7.05 days
$t_0 + 365$ days	*(j)* transits of I and III	7.05 days

The series *(f)*, *(g)* and *(h)* occur halfway between two successive instants t_0. For instance, in 1994 there were simultaneous shadow transits of I and II from August 11 to September 19.

Moreover, we see that the following series occur only once in the course of the period of 437.64 days: simultaneous transits of I and II *(h)*, simultaneous eclipses of I and II *(c)*, simultaneous transits of II and III *(a)*, and simultaneous eclipses of II and III *(f)*.

On the other hand, there are *three* series of simultaneous transits and of simultaneous eclipses of I and III per period of 437.64 days, namely the series *(d)*, *(g)*, *(j)*, and *(b)*, *(e)*, *(i)*. Hence, these series occur at intervals of 146 days. For instance, there were simultaneous eclipses of I and III on the following dates:

1971 April 15 and 22 (t_0),
1971 September 12 and 19 (t_0 + 146 days),
1972 February 3 and 10 (t_0 + 292 days),
1972 June 25, July 2 and 9 (t_0),
1972 November 22-23 and 30 (t_0 + 146 days),
1973 April 15, 22 and 29 (t_0 + 292 days),
1973 September 5, 12 and 19 (t_0),
and so on.

Example. — When will there be simultaneous shadow transits in A.D. 2000? In Table 43.D we find t_0 = 2000 January 17, and consequently there will be, according to Table 43.E:
— simultaneous transits of II and III near time t_0, hence near 2000 January 17;
— simultaneous transits of I and III near t_0 + 73 days, hence near 2000 March 30;
— simultaneous transits of I and II, and of I and III, near t_0 + 219 days, that is, about 2000 August 23.

Geocentric phenomena

Until now, we supposed the observer to be on the Sun. For such an heliocentric observer the transits of the satellites coincide with the transits of their shadows, and their eclipses with the occultations behind Jupiter.

For a terrestrial observer there is no longer coincidence, except on the very day of the opposition with the Sun. But as seen from Jupiter the elongation of the Earth from the Sun is at most 12°. Hence, the differences between the heliocentric and the geocentric phenomena never are very large. For instance, near the epoch t_0 + 73 days an observer on the Earth will witness simultaneous transits of the shadows of I and III as well as simultaneous transits of these satellites themselves. At most, the series of the simultaneous transits of the satellites will be a little displaced with respect to that of their shadows. So, for example, there were
— simultaneous transits of the shadows of satellites I and III on 1972 September 8, 15 and 22;
— simultaneous transits of these satellites on 1972 September 22 and 29.

44. *Jupiter and triple shadow phenomena*

As we have seen in the preceding chapter, simultaneous transits of the shadows of *two* satellites of Jupiter are reasonably frequent. Simultaneous shadow transits of *three* satellites, however, are much rarer events.

Since satellites I, II and III can never be in conjunction simultaneously, satellite IV is always involved in such *triple shadow phenomena.*

In 1979, the Belgian amateur astronomer and calculator Christian Steyaert calculated all triple shadow phenomena taking place from 1900 to 2100. His results were published in the *Journal* of the British Astronomical Association, Vol. 90, No. 1, pages 42-44 (December 1979). [In his list, the date 1932 Sept. 6 is a transcription error and should be 1938 June 1].

The list we present here (Table 44.A) has been calculated from scratch, and our calculations are based on the E-2 theory of the satellites of Jupiter by J. H. Lieske, improved by the corrections by the same author as published in *Astronomy and Astrophysics*, Vol. 176, pages 146-158 (1987).

The second column gives the mid-time of the event, in *Dynamical Time*. To convert to Universal Time, subtract 1 minute near A.D. 2000, and about 4 minutes near A.D. 2100 (see pages 7 and 8). The third column contains the duration of the event, in minutes. These values refer to the *center* of the shadows.

The next column mentions which satellites are involved in the triple shadow event. For convenience, the satellites have been designated by Arabic numbers instead of Roman ones.

There exists a period of 17507.46 days, or 48 Julian years minus 24½ days, after which the four satellites have again nearly exactly the same positions with respect to the Sun and Jupiter. This period corresponds to 9892 synodic revolutions of satellite I, 4926 revolutions of II, 2443 of III, and 1045 of IV. Phenomena which are separated by this period are indicated by the same letter in the fifth column. Example: series *d*, containing the events of 1949 February, 1997 January, 2045 January, and 2092 December.

Besides this periodicity, we find in the list several times the interval of 234.6 days (66 synodic revolutions of satellite II, or 14 of IV) and of 284.9 days (161 synodic revolutions of I, or 17 of IV).

The last column of the table gives the elongation of Jupiter from the Sun. E = elongation East = visible in the evening sky; W = elongation West = visible in the morning sky. When the elongation is small, for instance at the events of 1985 January 23 and 1997 January 30, the event is not observable by reason of the vicinity of the Sun. However, we included these invisible cases for the sake of completeness.

Photographs by H. E. Dall of the triple shadow transit of 1956 April 21 have been published in the *Journal* of the British Astronomical Association, Vol. 88, No. 4, page 361 (June 1978).

TABLE 44.A

Triple shadow phenomena of the satellites of Jupiter, 1900 to 2100

Date	Dynamical Time	Duration	Satellites	Series	Elong.
	h m	m			°
1901 Dec. 30	6 12	77	1 3 4	a	13 E
1908 May 16	11 02	134	1 3 4	b	72 E
1909 July 25	20 29	77	1 2 4		42 E
1915 July 29	10 23	113	1 3 4		127 W
1919 Mar. 14	8 40	36	1 2 4	c	103 E
1938 June 1	14 06	12	1 2 4		99 W
1949 Feb. 23	19 38	44	1 2 4	d	43 W
1949 Dec. 5	17 37	127	1 3 4	a	48 E
1956 Apr. 21	22 00	114	1 3 4	b	110 E
1966 June 28	7 16	73	2 3 4	e	5 E
1967 Feb. 17	19 39	32	1 2 4	c	148 E
1974 May 1	18 10	6	1 2 4		60 W
1985 Jan. 23	23 03	39	1 2 4	f	7 W
1997 Jan. 30	6 16	99	1 2 4	d	8 W
1997 Nov. 11	4 42	137	1 3 4	a	85 E
2004 Mar. 28	8 10	18	1 3 4	b	153 E
2013 Oct. 12	5 05	64	1 2 4	g	90 W
2014 June 3	18 57	95	2 3 4	e	38 E
2015 Jan. 24	6 41	25	1 2 4	c	165 W
2032 Mar. 20	12 00	135	1 3 4	h	64 W
2032 Dec. 30	10 29	84	1 2 4	f	27 E
2038 Aug. 5	17 46	12	1 3 4		2 W
2045 Jan. 5	17 29	81	1 2 4	d	26 E
2045 Oct. 17	15 57	115	1 3 4	a	128 E
2061 Sep. 17	16 32	108	1 2 4	g	53 W
2062 May 10	5 53	62	2 3 4	e	72 E
2062 Dec. 30	17 42	20	1 2 4	c	119 W
2073 Sep. 23	23 45	27	1 2 4		54 W
2080 Feb. 24	23 14	123	1 3 4	h	28 W
2080 Dec. 5	21 49	119	1 2 4	f	63 E
2091 Aug. 31	2 29	17	1 2 4		125 E
2092 Dec. 12	5 09	7	1 2 4	d	62 E

45. *Jupiter without satellites*

The original version of the following text was published in the Journal *of the British Astronomical Association, Vol. 98, No. 1, pages 35–37 (December 1987). For this chapter we used improved elements for the motions of the Galilean satellites, so our results differ slightly from the data given in our original article.*

The planet Jupiter sometimes appears to be without moons, when the four Galilean satellites are simultaneously invisible because they are passing across the planet's disk (transit), are behind the planet (occultation), or are in its shadow (eclipse). In the first years of the 20th century, a list of such events between A.D. 1800 and 2000 was calculated by the Italian amateur astronomer Enzo Mora; his article (in French), dated 1909 October 20, was published in the *Astronomische Nachrichten* in 1910.

Enzo Mora

Enzo Mora is short for Gian Vincenzo Mora. He was born in Sequals, a small village near Udine, in 1870 into a noble family. Soon interested in art and mathematics, he studied drawing and architecture at the Fine Arts School of Venice. At the same time he studied astronomy and languages. (He had full command of Latin, Greek, English, French, German and Spanish). From 1910 to 1915 he worked as a draughtsman in Padua and then, until 1943 (aged 73 years!), in Sesto S. Giovanni, near Milan, living a simple, honest life.

He kept an astronomical observation diary (comets, variable stars, planets) but mainly devoted his time to calculations: *Tables of the Sun from 1600 to 2199*; *Tables of the lunar phases* (both unpublished); *Tables of the satellites of Jupiter* (for which he received a golden medal from the Mexican Astronomical Society), and many papers in *Astronomische Nachrichten*, *Coelum*, *L'Astronomie* and in English and American journals.

He corresponded with Flammarion, Poincaré, Danjon, Cerulli, Horn d'Arturo, and Antoniadi. Camille Flammarion wrote to him: "You are a great astronomer; it is a pity that you are relegated in that small Italian village". Gabrielle Flammarion (Camille's wife), in a letter from Juvisy, wrote that Enzo Mora had "the most beautiful handwriting I have ever seen".

The studies of Mora were serious and meticulous, requiring a large amount of time, without modern computing aids.

He died in Sequals in 1953.

The events, 1900 to 2100

In his article, Enzo Mora listed 36 occurrences of Jupiter without satellites (or JWS for short) taking place between the years 1800 and 2000, and he wrote that "la recherche des dates de disparition a été assez pénible."

We have made a similar investigation for the period 1900 to 2100. The search was made by means of a computer, thus avoiding a 'pénible recherche', although many hours were spent writing the computer program.

Our calculations are based on the E-2 theory of the Galilean satellites by the American astronomer Jay H. Lieske, improved by the corrections by the same author as published in *Astronomy and Astrophysics*, Vol. 176, pages 146-158 (1987).

Our results are given in Table 45.A. The number of each event (first column) is the same as in Mora's article, but it has been extended beyond No. 36 after A.D. 2000. There is an additional No. 31a, as we shall see.

The instants are expressed in the uniform time-scale known as Dynamical Time. To convert them to Universal Time, subtract a quantity which is zero near A.D. 1900, increasing to 1 minute in 1994, and to approximately 4 minutes in the year 2100 (see pages 7 and 8). The times given in the table refer to the centers of the disks of the satellites, and for the calculation of the eclipses the Sun has been supposed to be a point.

The situation of the satellites is indicated by the following letters: O = the satellite is occulted behind Jupiter; T = the satellite is in transit; E = the satellite is eclipsed in the planet's shadow. If a satellite is eclipsed *and* occulted, only O is indicated. OE means that the invisible satellite is first occulted, then eclipsed; etc.

Finally, the last column gives the angular distance of Jupiter from the Sun. If the elongation is east (E), the planet is east of the Sun and hence visible in the evening sky; in western elongation (W), the planet is in the morning sky. Of course, the planet is visible all night if the elongation is close to 180°, as for event No. 42.

Events Nos. 27 and 32 occurred close to the Sun and hence could not be observed. Events 29, 30, 31a, 42 and 50 are of short duration (less than ten minutes). In our list, the event with the longest duration is No. 45, which will last for 2 hours and 16 minutes. The longest *possible* duration for a JWS is 2 hours and 55 minutes and, when this happens, satellite II is in transit, and I is occulted and eclipsed. Indeed, when during the JWS satellite II is in transit, then satellite I is necessarily behind Jupiter, occulted or eclipsed. The reason is that, if I and II were in conjunction (as seen from Jupiter), then because of the exact commensurability between the motions of the first three satellites, III would be in quadrature (see Chapter 43); in that case, no JWS could possibly take place as seen from Earth.

One may see from the list that, during a JWS, either satellite I or satellite II is in transit but never both. On the other hand, during a JWS, the satellites I and III can both be in transit (as on 1914 May 11), or both satellites II and III (as on 1913 October 22).

In the Table, many events are separated by 48 years minus 24 days 13 hours. See, for instance, the events Nos. 20, 31 and 39. As we mentioned in the previous chapter, this interval of 17507.46 days corresponds to 9892 synodic revolutions of satellite I, 4926 of II, 2443 of III, and 1045 of IV. However, this period cannot be used for predicting future events. For instance, in 2028, forty-eight years after the JWS of 1980, no other JWS will take place.

Comparison with Mora's results

In his article, Enzo Mora rounded the instants to the nearest tenth of an hour. For the period 1900–2000, we have compared these times with ours.

For the satellites I and II, the agreement is excellent. For the third satellite, this is generally the case too, the difference reaching 0.1 hour in only two cases. But larger errors occur for satellite IV, Mora's error being sometimes as large as a quarter of an hour.

Event No. 27 is remarkable in this respect. This JWS began and ended with the occultation of satellite IV. According to Mora, the duration of this occultation was 1.2 hours, or approximately 72 minutes, while we obtain a duration of only 46 minutes. It should be noted, however, that this occultation was the first one of a series of occultations of satellite IV, so it took place in the polar regions of Jupiter, where a small error in the calculated latitude of the satellite leads to an appreciable error in the calculated duration. The *Nautical Almanac* did not provide information for that date, Jupiter being too close to the Sun.

For the event No. 30, on 1949 September 21, Mora gives a duration of only 0.1 hour; in his text, he writes that the duration of this JWS is 9 minutes, the eclipse of IV ending this lapse of time after the beginning of the occultation of I. According to our calculations, the eclipse of IV ended less than one minute after the beginning of the occultation of I, the instants, rounded to the nearest integer minute of Dynamical Time, being 9^h53^m and 9^h53^m, respectively. The *Nautical Almanac* gives 9^h53^m UT for both phenomena.

Event No. 31a is a JWS that was missed by Mora. It was of very short duration, however, and hence the Italian astronomer may be excused. We find that on 1962 April 16 the eclipse of IV began one minute before the transit of I ended. The *Nautical Almanac* gives 17^h53^m UT for both events.

We also made calculations for the years 1800 to 1900, although these JWSs are not given in Table 45.A. We found that Mora's event No. 15 (1879 August 3) did not happen, the transit of satellite III ending at 10^h06^m TD, two minutes before the beginning of the transit of I. Mora said that these events coincided. Again, he can be excused for such a small difference. The *Nautical Almanac* gave 10^h07^m UT for both events.

More serious is the fact that Mora missed two events of longer duration. One of them is the JWS of 1859 October 27, which lasted for 19 minutes, from the beginning of the occultation of satellite III (10^h20^m Dynamical Time) to the end of

TABLE 45.A — *Jupiter without satellites, 1900 to 2100*

No.	Limits of the disappearance (Dynamical Time)			Satellite that		Situation of the satellites				Elongation from the Sun
	Date	Begin	End	disappears last	reappears first	I	II	III	IV	
		h m	h m							°
19	1907 Oct. 3	19 49	19 59	II	III	O	T	E	O	61 W
20	1913 Oct. 22	5 00	5 29	I	III	O	T	T	E	73 E
21	1914 May 11	5 41	6 58	IV	I	T	O	T	E	89 W
22	1919 Mar. 5	9 58	10 59	IV	III	T	O	O	O	112 E
23	1931 Feb. 14	20 51	23 05	I	I	T	OE	E	T	136 E
24	1932 May 4	3 25	4 20	III	I	T	OE	E	E	90 E
25	1932 Nov. 21	2 13	2 51	IV	III	O	T	O	E	69 W
26	1939 July 17	4 11	5 02	I	IV	T	E	E	E	105 W
27	1942 July 10	13 45	14 31	IV	IV	O	T	EO	O	11 W
28	1943 Sep. 27	20 13	20 59	II	I	O	T	E	T	45 W
29	1949 Mar. 4	17 11	17 20	III	I	T	O	E	O	50 W
30	1949 Sep. 21	9 53	9 53	I	IV	O	T	E	E	114 E
31	1961 Sep. 27	15 59	17 30	IV	III	O	T	T	E	113 E
31a	1962 Apr. 16	17 52	17 53	IV	I	T	O	T	E	52 W
32	1966 June 28	6 10	6 41	III	IV	O	T	T	T	5 E
33	1980 Apr. 9	13 12	14 16	I	III	T	O	O	E	131 E
—	1980 Apr. 9	14 36	15 28	III	I	T	OE	E	E	131 E
34	1990 June 15	22 48	24 22	I	IV	O	T	O	E	21 E
35	1991 Jan. 2	20 44	21 54	IV	III	T	O	O	E	150 W
36	1997 Aug. 27	21 38	21 55	I	IV	O	T	O	E	160 E
37	2001 Nov. 8	16 28	16 43	III	I	T	O	E	T	121 W
38	2008 May 22	3 51	4 10	I	III	E	T	O	E	129 W
39	2009 Sep. 3	4 44	6 30	III	I	OE	T	T	E	159 E

TABLE 45.A (Cont.)

No.	Limits of the disappearance (Dynamical Time)			Satellite that		Situation of the satellites				Elon-gation from the Sun
	Date	Begin	End	disappears last	reappears first	I	II	III	IV	
		h m	h m							°
40	2019 Nov. 9	12 17	12 56	I	IV	T	O	T	E	38 E
41	2020 May 28	11 18	13 12	IV	I	O	T	T	E	131 W
42	2021 Aug. 15	15 40	15 48	I	IV	E	T	T	T	175 W
43	2033 July 28	3 08	5 01	I	III	EO	T	T	O	150 W
44	2038 May 22	9 10	10 49	I	III	O	T	O	E	55 E
45	2038 Dec. 9	8 20	10 36	I	I	T	O	O	E	106 W
46	2049 Oct. 15	3 47	4 01	III	I	T	EO	E	T	80 W
47	2050 May 28	17 23	18 34	III	I	OE	T	T	E	53 E
48	2056 Apr. 27	15 21	16 22	II	IV	EO	T	O	E	88 W
49	2057 July 15	22 55	24 06	IV	I	O	T	O	T	127 W
50	2061 Mar. 16	18 48	18 56	IV	I	E	T	O	T	90 E
51	2067 Oct. 15	22 45	24 34	I	III	T	O	T	E	74 E
52	2068 May 3	23 31	24 41	III	I	O	T	T	E	89 W
53	2073 Feb. 26	3 03	3 57	III	IV	O	T	O	O	113 E
54	2080 Nov. 26	23 12	24 12	I	III	T	OE	E	O	70 E
55	2082 Feb. 14	6 32	8 00	IV	I	T	OE	E	T	34 E
56	2085 Feb. 7	14 36	15 24	I	III	O	T	O	T	137 E
57	2086 Apr. 27	19 49	21 12	I	III	O	T	O	E	91 E
58	2086 Nov. 14	19 18	21 22	I	IV	T	O	O	E	68 W
59	2097 Jan. 20	2 29	2 56	I	III	O	T	E	E	162 E
60	2097 Feb. 13	22 21	23 42	IV	II	OE	T	T	T	135 E
61	2098 May 4	3 27	5 10	III	II	OE	T	T	E	89 E

the occultation of I (10^h39^m); satellite II was in transit, while IV was occulted. According to the *Nautical Almanac*, the JWS even had a duration of 27 minutes, from 10^h15^m to 10^h42^m Universal Time. This JWS took place 17507 days before that of 1907 October 3.

The other JWS omitted by Mora took place on 1895 March 16, for which we found a duration of 10 minutes, from the beginning of the transit of IV (16^h39^m) to the end of the transit of III (16^h49^m); satellite I was in transit, while II was occulted. It is rather strange that Mora did not mention this JWS because he could have deduced it from the phenomena listed in the *Nautical Almanac*, where the duration of the JWS is even given as being 25 minutes, from 16^h26^m to 16^h51^m!

A list of JWS events, for the years 1800 to 2001, has been calculated independently by Roger W. Sinnott; it was published in *Sky and Telescope* of September 1982, page 214. Sinnott, too, found the events of 1859, 1895 and 1962, which were overlooked by Mora.

Acknowledgement

Thanks are due to Dr. Ing. Salvatore De Meis, of Milan (Italy), for supplying information on Enzo Mora.

Bibliography

Enzo Mora, 'Disparitions simultanées des satellites de Jupiter', *Astronomische Nachrichten*, No. 4379, pages 167–170 (1910).

Enzo Mora, 'Tavole dei satelliti di Giove', in *Boletin de la Sociedad Astronomica de Mexico* (1912).

Enzo Mora, *Effemeridi del Sole dal 1600 al 2199* (unpublished).

Enzo Mora, *Tavole delle fasi lunari* (9 pages, unpublished; Sequals, 1943).

Giacomo Serafini, 'Alla Ricerca di Enzo Mora da Sequals', *Il Noncello*, No. 15 (1960).

46. *Heliacal risings and settings*

We will consider stars which are situated not too far from the ecliptic. Some time after its conjunction with the Sun, such a star reappears in the morning sky: the beginning of its new period of visibility.

First, we can ask on what date of the year a star and the Sun are rising simultaneously. This is called the *cosmic rising* of the star. In the case of Aldebaran (α Tauri) and at latitude $45°$N, the date is June 8, as we shall later see. Actually, the date is different for other geographical latitudes. Moreover, it slowly varies with time by reason of the phenomenon known as the precession of the equinoxes, which continuously changes the right ascensions and the declinations of the stars. In what follows, we shall consider the situation near A.D. 2000.

However, when the star rises at the same instant as the Sun, it cannot be seen. To become visible in the morning twilight, the star should be situated slightly *above* the horizon, for instance at an altitude of 3 degrees. Moreover, it must be sufficiently dark, so the Sun should be situated at some distance *below* the horizon; otherwise the star cannot be seen with the naked eye in the bright twilight. This depression of the Sun depends on several factors:

1. the brightness of the star;
2. the difference between the azimuth of the star and that of the Sun;
3. the color of the star;
4. the transparency of the air.

Hence, the calculation of the date of the first visibility of a star is a rather complicated problem. We have adopted a depression of $7°$ for the Sun. In the case of Aldebaran this is perhaps rather optimistic. But a celestial body such as Sirius, which is brighter than Aldebaran by more than two magnitudes and which appears above the horizon far from the place of sunrise, certainly is visible for a solar depression smaller than $7°$. On August 18, the date of the reappearance of Sirius in the morning sky (see Table 46.B), the declination of the Sun is $+13°$, while that of Sirius is $-17°$.

More realistic values for the altitudes may be: in the case of Aldebaran, $+3°$ for the star, $-8°$ for the Sun; for Sirius, $+2°$ for the star, and $-5°$ for the Sun. However, in what follows we have adopted in all cases, for reason of uniformity, an altitude of $+3°$ for the star, and of $-7°$ for the Sun.

Under these conditions, it is possible to calculate the so-called *heliacal rising* of a given star, that is, to find the date on which this star is again visible in the morning sky. For 20 bright stars, Table 46.A provides the following data: the star's name; its designation by Greek letter + official abbreviation of the constellation name; its visual magnitude according to the Revised Harvard Photometry; its mean right ascension α and declination δ for the epoch 1998.5; and finally its latitude β rounded to the nearest integer degree. So, the positions are for the middle of the year 1998. We have chosen a year lying halfway between two bissextile years.

TABLE 46.A

Data for 20 bright stars — Positions are for 1998.5

Star		Magn.	α	δ	β
			h m s	° ' "	°
Altair	α Aql	0.9	19 50 42.6	+ 8 51 51	+29
Fomalhaut	α PsA	1.3	22 57 34.1	−29 37 49	−21
Diphda	β Cet	2.2	0 43 30.9	−17 59 41	−21
Hamal	α Ari	2.2	2 07 05.3	+23 27 19	+10
Alcyone	η Tau	3.0	3 47 23.7	+24 06 02	+ 4
Aldebaran	α Tau	1.1	4 35 50.1	+16 30 23	− 5
Rigel	β Ori	0.3	5 14 27.9	− 8 12 12	−31
Nath	β Tau	1.8	5 26 11.8	+28 36 23	+ 5
Betelgeuse	α Ori	0.5	5 55 05.4	+ 7 24 25	−16
Sirius	α CMa	−1.4	6 45 04.9	−16 42 50	−40
Castor	α Gem	1.6	7 34 30.3	+31 53 31	+10
Procyon	α CMi	0.5	7 39 13.4	+ 5 13 44	−16
Pollux	β Gem	1.2	7 45 13.4	+28 01 48	+ 7
Regulus	α Leo	1.3	10 08 17.5	+11 58 28	+ 0
Denebola	β Leo	2.2	11 48 59.0	+14 34 50	+12
Spica	α Vir	1.2	13 25 06.8	−11 09 13	− 2
Arcturus	α Boo	0.2	14 15 35.6	+19 11 25	+31
Gemma	α CrB	2.3	15 34 37.5	+26 43 11	+44
Antares	α Sco	1.2	16 29 18.9	−26 25 44	− 5
Ras Alhague	α Oph	2.1	17 34 51.9	+12 33 40	+36

We first calculate the local sidereal time at which the star is 3° above the horizon. (The sidereal time is the hour angle of the vernal equinox). We will neglect the effect of the atmospheric refraction. The well-known formula is

$$\sin h = \sin \varphi \sin \delta + \cos \varphi \cos \delta \cos H \qquad (1)$$

where h is the star's altitude (3° in our case), φ is the observer's geographical latitude, δ the declination, and H the hour angle of the star. By means of this formula we can calculate H, and then the sidereal time $\theta = H + \alpha$, where α is the right ascension of the star. For Aldebaran we have (Table 46.A)

$$\alpha = 4^{\text{h}}35^{\text{m}}50^{\text{s}}.1 = 68.9587 \text{ degrees}, \qquad \delta = +16°30'23'' = +16°.5064.$$

We will perform the calculation for 45° N, the latitude of Minneapolis (USA), Ottawa (Canada) and Milan (Italy). For $h = 3°$ and $\varphi = 45°$, formula (1) gives $\cos H = -0.219139$, whence $H = -102°.6585 = -6^{\text{h}}50^{\text{m}}38^{\text{s}}$.

TABLE 46.B

Star	Conj.	Cr	Hr	Hs	Cs
Altair	Jan. 16	Dec. 15	Dec. 30	Jan. 28	Feb. 8
Fomalhaut	Mar. 3	Apr. 25	May 22	Jan. 29	Feb. 10
Diphda	Apr. 1	May 12	June 3	Mar. 3	Mar. 14
Hamal	Apr. 24	Apr. 1	Apr. 30	Apr. 21	May 4
Alcyone	May 20	May 12	June 4	May 11	May 24
Aldebaran	June 1	June 8	June 26	May 13	May 26
Rigel	June 10	July 11	July 24	May 2	May 14
Nath	June 13	June 5	June 23	June 4	June 20
Betelgeuse	June 20	July 6	July 20	May 21	June 4
Sirius	July 2	Aug. 6	Aug. 18	May 11	May 24
Castor	July 14	July 1	July 17	July 10	Aug. 4
Procyon	July 15	July 30	Aug. 11	June 10	June 25
Pollux	July 17	July 9	July 23	July 6	July 29
Regulus	Aug. 23	Aug. 22	Sep. 3	July 29	Aug. 24
Denebola	Sep. 20	Sep. 9	Sep. 20	Sep. 19	Oct. 16
Spica	Oct. 16	Oct. 18	Oct. 29	Sep. 14	Oct. 12
Arcturus	Oct. 29	Oct. 3	Oct. 15	Nov. 25	Dec. 12
Gemma	Nov. 18	Oct. 11	Oct. 22	Dec. 24	Jan. 7
Antares	Dec. 1	Dec. 6	Dec. 20	Nov. 1	Nov. 24
Ras Alhague	Dec. 16	Nov. 15	Nov. 27	Jan. 4	Jan. 17

We reckon the hour angles from -12 to $+12$ hours, that is, from $-180°$ to $+180°$, one hour corresponding to 15 degrees. The value found for $\cos H$ yields two solutions for H: a negative value which corresponds to the rising of the star, and a positive value corresponding to its setting. Indeed, on the star's diurnal arc there are *two* points which are $3°$ above the horizon: one in the East, the other in the West. In our example for the heliacal rise of Aldebaran, we should take the negative value of H, and the required local sidereal time is

$$\theta = H + \alpha = -6^h50^m38^s + 4^h35^m50^s = -2^h14^m48^s = 21^h45^m12^s.$$

The next question is: when is the Sun situated $7°$ below the horizon for this value of the sidereal time? On which date does this take place? At a given sidereal time θ and for an observer at geographical latitude φ, the ecliptic has a well-defined position with respect to the horizon. Therefore, it is possible to find on this ecliptic a point that is $7°$ below the horizon.

Formula (1) is valid for the Sun too. We can write it as follows:

$$\sin h_0 = \sin \varphi \sin \delta_0 + \cos \varphi \cos \delta_0 \cos (\theta - \alpha_0)$$

where δ_0, etc. now refer to the Sun, not to the star. This formula contains *two* unknowns, namely α_0 and δ_0. But we want but only one unknown: the longitude of the Sun on the ecliptic. To get clear of the equatorial system, we first replace, according to the well-known trigonometric formula, $\cos(\theta - \alpha_0)$ by $\cos\theta\cos\alpha_0 + \sin\theta\sin\alpha_0$. We thus obtain

$$\sin h_0 = \sin\varphi\sin\delta_0 + \cos\varphi\cos\delta_0\cos\theta\cos\alpha_0 + \cos\varphi\cos\delta_0\sin\theta\sin\alpha_0.$$

On the other hand, we have the following expressions for the Sun, which is situated on the ecliptic (see the books on spherical astronomy):

$$\sin\delta_0 = \sin\varepsilon\sin\lambda_0$$
$$\cos\delta_0\sin\alpha_0 = \cos\varepsilon\sin\lambda_0$$
$$\cos\delta_0\cos\alpha_0 = \cos\lambda_0$$

where ε is the obliquity of the ecliptic on the celestial equator (obliquity of the Earth's rotational axis), and λ_0 the Sun's longitude on the ecliptic. Placing these expressions in the last formula, we obtain

$$\sin h_0 = (\sin\varphi\sin\varepsilon + \cos\varphi\cos\varepsilon\sin\theta)\sin\lambda_0 + \cos\varphi\cos\theta\cos\lambda_0$$

This is the relation between the observer's latitude φ, the obliquity ε of the ecliptic, the local sidereal time θ, the altitude h_0 of the Sun, and its celestial longitude λ_0. Let us write, for short,

$$\begin{aligned} S &= \sin\varphi\sin\varepsilon + \cos\varphi\cos\varepsilon\sin\theta \\ C &= \cos\varphi\cos\theta \end{aligned} \tag{2}$$

All values in the right-hand sides are known, and hence S and C are known quantities. This gives

$$S\sin\lambda_0 + C\cos\lambda_0 = \sin h_0 \tag{3}$$

Here the unknown is the Sun's longitude λ_0. To solve this equation, we introduce an auxiliary angle A whose tangent is equal to C/S, so that

$$\cos A = \frac{S}{\sqrt{S^2 + C^2}}$$

Dividing both members of (3) by S and multiplying by $\cos A$ we get, after two steps of elementary trigonometry,

$$\sin(\lambda_0 + A) = \frac{\sin h_0}{\sqrt{S^2 + C^2}} \tag{4}$$

In our example for Aldebaran, we obtain $\theta = 326°300$, $\varepsilon = 23°4395$ (obliquity for 1998.5), $S = -0.07869$, and $C = +0.58828$. The auxiliary angle A is obtained from $\tan A = C/S = -7.4759$. This gives for A a solution in the second or in the fourth quadrant. To identify the appropriate quadrant, we note that the sign of $\cos A$ is the same as that of S. In our example, $S < 0$, so $\cos A$ is negative too and A is in the second quadrant: $A = 97°37'$. Then equation (4) becomes $\sin(\lambda_0 + A) = -0.20533$. This yields two values for the angle $(\lambda_0 + A)$, and hence also for λ_0, namely $94°14'$ and $-109°28'$. We have to select the right one! With some insight there is little danger that the wrong part of the ecliptic is calculated. Because the Sun–Aldebaran conjunction occurs on June 1 (see Table 46.B), we may expect that the star reappears in the morning sky at the end of June or in early July, hence when the Sun is near the summer point of the ecliptic, which is at longitude $+90°$. So the correct value for the Sun's longitude is $94°14'$. The Sun reaches this longitude on June 26. This is the required date. On June 26, Aldebaran is 3 degrees above the eastern horizon at the instant the Sun is 7 degrees below this horizon, for latitude $45°\,N$.

In the same way we can calculate the end of the star's evening visibility, that is, the date when $h = +3°$ and $h_0 = -7°$ with respect to the *western* horizon (heliacal setting). In this case the hour angles are between 0 and $+12$ hours, that is, between $0°$ and $+180°$.

We also can calculate, for a given star, the times of its *cosmic* rising or setting, when the star and the center of the solar disk are rising or setting simultaneously. If again no allowance is made for the refraction at the horizon, we have $h = h_0 = 0°$, and the formulae become much simpler.

We performed the calculation for the stars listed in Table 46.A and for latitude $45°\,N$. The results are given in Table 46.B. They are strictly valid for the year 1998, but for other years between, say, 1900 and 2100 the difference will be at most two days. The second column of Table 46.B gives the date (in 1998) of the conjunction in right ascension between the star and the Sun. The next columns give the dates of the star's cosmic rising (*Cr*), heliacal rising (*Hr*), heliacal setting (*Hs*) and cosmic setting (*Cs*).

From the table it appears that some stars, for instance Altair and Arcturus, become visible in the morning sky before they disappear in the evening twilight. Other stars, such as Fomalhaut and Sirius, remain invisible with the naked eye for three or four months, due to their southerly position.

For Regulus the dates *Cr* and *Cs* almost coincide with that of the conjunction with the Sun, as might have been expected: this star is close to the ecliptic.

For some stars, the dates *Hr* and *Hs* lie nearly symmetrically with respect to the conjunction with the Sun. Examples are Altair, Rigel and Ras Alhague. For other stars the difference can be quite large, and this may happen even when the star is close to the ecliptic, for instance Regulus. Here the obliquity of the ecliptic on the horizon plays an important role. In northern latitudes, when Regulus reappears in the morning sky, the ecliptic is steep on the horizon. Hence it is clear that the star rapidly reappears from the Sun's glare: its heliacal rising occurs only 11 days after the conjunction with the Sun. But when Regulus is near the western horizon, the ecliptic makes a small angle with the horizon, and the heliacal setting occurs as much as 25 days before the conjunction. For higher northern latitudes the difference is even larger: at latitude 52°N, the time intervals for Regulus are 11 and 36 days.

We invite the reader to observe by himself the beginning and the end of the visibility of some bright stars. Compare your results with the calculated times Hr and Hs. Observe only under good air transparency conditions. How many days can you 'win' by observing with a binocular instead of the naked eye?

It is evident that if the dates of Hr, etc., are required for an epoch far remote from A.D. 2000, for instance for antiquity, the star's coordinates given in Table 46.A cannot be used. Instead, one should calculate its right ascension and declination for that epoch, using the formulae for precession. But this is not a subject for this book.

Solution by iteration

It may be better to solve equation (3) by successive approximations. This equation can be written as

$$\text{either} \quad \sin \lambda_0 = (\sin h_0 - C \cos \lambda_0) / S$$
$$\text{or} \quad \cos \lambda_0 = (\sin h_0 - S \sin \lambda_0) / C \tag{5}$$

The first expression should be used when λ_0 is in the vicinity of $0°$ or $180°$, the second when λ_0 is near $90°$ or $270°$. Let us take again our example of Aldebaran. Here equation (3) is

$$-0.07869 \sin \lambda_0 + 0.58828 \cos \lambda_0 = -0.12187$$

This equation has two solutions in λ_0, but we know that the correct one is near $90°$. Therefore we take the second of the equations (5), which is numerically

$$\cos \lambda_0 = (-0.12187 + 0.07869 \sin \lambda_0) / 0.58828$$

Starting with the approximate value $\lambda_0 = 90°$, we put $\sin \lambda_0 = +1$ in the second member, and we find $\cos \lambda_0 = -0.07340$. This gives $\lambda_0 = 94°13'$, which is a better approximation for λ_0. In the second member of the equation we now put $\lambda_0 = 94°13'$, or $\sin \lambda_0 = +0.99729$, and we get $\cos \lambda_0 = -0.07376$, which gives $\lambda_0 = 94°14'$. A new iteration gives $94°14'$ again, so this is the correct solution.

We thought that this method of iteration (from the Latin *iterare* = to repeat) would be of interest to the reader. In our example the successive approximations are $90°$, $94°13'$, $94°14'$ and $94°14'$. Not always does one have such a rapid convergence. In some cases there even can be no convergence at all, or the convergence may be too slow for practical use.

The planets

Until now we have only considered stars. For the planets too, of course, one may calculate the times of their heliacal risings and settings.

The superior planets (Mars, Jupiter, Saturn) behave like the stars: after their conjunction with the Sun they reappear in the eastern morning sky, and several months after their opposition they disappear in the evening in the West. However, for the interior planets the situation is different.

Consider the case of Venus. During each synodic period (19 months), Venus becomes invisible twice: once when it passes between the Sun and the Earth (inferior conjunction), and once when it passes behind the Sun (superior conjunction). The durations of the periods of invisibility vary greatly, and depend on the planet's celestial latitude at the time of conjunction with the Sun, and on the geographic latitude of the observer.

In general, the period of invisibility near superior conjunction is much longer than that near inferior conjunction, due to the smaller angular speed of the planet with respect to the Sun. The daily motion of the Sun in celestial longitude has a mean value of 59'. For Venus, the daily motion is $+75'$ at superior conjunction, and $-38'$ at inferior conjunction. Hence, the daily variation of Venus' longitude with respect to the Sun is 97' at inferior conjunction, but only 16' at superior conjunction.

Some time after its superior conjunction with the Sun, Mercury or Venus becomes visible in the western evening sky. But this can hardly be called a heliacal 'rising', because the planet is setting! Hence, in the case of Mercury and Venus, it is better to use the expressions *evening first* for the first evening visibility after superior conjunction, *evening last* for the last evening visibility before inferior conjunction, *morning first* for the first morning visibility after inferior conjunction, and *morning last* for the last visibility before superior conjunction.

For the stars, we used the values $h = +3°$ and $h_0 = -7°$ in the calculation of the heliacal risings and settings. For Venus, which always is very bright, we may use the values $h = +2°$ and $h_0 = -4°$. For Jupiter, good values are $h = +2°$, $h_0 = -6°$. For Saturn, $h = +3°$, $h_0 = -7°$. For Mars, $h = +3°$, $h_0 = -8°$.

But certainly these are not the best values. For example, at its conjunctions with the Sun, Saturn can be as bright as visual magnitude $+0.2$, as in June 2002, or as faint as magnitude $+1.3$, as in March 1996. These values have been calculated from the classical formulae of G. Müller (1893). The variation is due to the variable distance of Saturn to the Sun (eccentricity of the orbit) and to the variable presentation of the planet's ring. Hence, the variable magnitude of Saturn certainly influences the choice of the values for h and h_0.

The case of Mercury is even more complicated. This planet is of visual magnitude $+2$ or $+3$ near its inferior conjunctions, but may almost reach -2 at superior conjunction. Moreover, near its inferior conjunction its brightness varies sensibly from day to day.

Because the planets are moving, their heliacal risings and settings cannot be calculated by means of the method we described for the stars. In the past, several authors constructed tables to perform this type of calculation, principally for antiquity. An example is *Tafeln zur Berechnung der jährlichen Auf- und Untergänge der Planeten* by P. V. Neugebauer, in *Astronomische Nachrichten*, No. 6331, pages 313-322 (1938).

Of course, an alternative method consists of calculating for several dates, near the expected time of the first or last visibility, the altitude of the planet at the instant the Sun has altitude h_0, and then to deduce the exact date by interpolation.

As an example, let us calculate the date of the heliacal rising of Saturn in A.D. 2002 at Washington, D.C. (longitude 77°04′ W, latitude 38°55′ N). The conjunction of Saturn with the Sun takes place on 2002 June 9, so we may expect that the planet's heliacal rising will occur at the end of June or in early July. Using the values $h = +3°$ and $h_0 = -7°$, we perform the calculation at intervals of five days. Of course, this needs a special calculation program. The instants in Universal Time are:

Date in 2002	Instant when the Sun is 7° under the horizon	Altitude (degrees) of Saturn at that instant
June 21	$9^h04^m26^s$	-1.33
June 26	$9^h05^m54^s$	$+1.86$
July 1	$9^h08^m07^s$	$+5.26$
July 6	$9^h11^m01^s$	$+8.87$

By interpolation, we find that $h \geq 3°$ occurs on June 28, so this is the calculated date of Saturn's first morning visibility at Washington, D.C., in 2002.

47. The positions of Uranus, Neptune, Pluto and Ceres at their discovery dates

The maps on pages 298 and 299 show the positions of the planets Uranus, Neptune, and Pluto, and of the minor planet Ceres among the stars at the time of their discovery. Each map is 8° wide and 4°30′ degrees high. North is up, and the right ascensions and declinations mentioned along the borders are referred to the standard equinox of 2000.0. Stars are plotted to visual magnitude 7.7.

Uranus was in opposition with the Sun on 1780 December 17, and stationary on 1781 March 2. The planet had just resumed direct motion with respect to the stars, when it was discovered by William Herschel (1738–1822) in England on the evening of 1781 March 13. Under the title 'Account of a comet', Herschel wrote in the *Philosophical Transactions*, LXXI (1781), page 492:

> "On Tuesday the 13th of March, between ten and eleven in the evening, while I was examining the small stars in the neighbourhoud of H [i.e. η] Geminorum, I perceived one that appeared visibly larger than the rest; being struck with its uncommon magnitude, I compared it to H Geminorum and the small star in the quartile between Auriga and Gemini, and finding it to be so much larger than either of them, suspected it to be a comet."

Using the VSOP87 planetary theory of P. Bretagnon, I calculated that on 1781 March 13, at 22^h30^m UT, Uranus was at position $\alpha = 5^h49^m08^s.2$, $\delta = +23°38'23''$ (2000.0 coordinates). The planet was 56' almost due south of the fifth-magnitude star 132 Tauri; see Figure 47.*a*. At the time of its discovery, Uranus was 18.98 astronomical units from the Sun, and 18.94 AU from the Earth (2839 and 2834 million kilometers, respectively). The elongation to the Sun was 91°. The conjunction with the Sun took place on 1781 June 19.

The story of the discovery of *Neptune* is well-known and will not be repeated here. On 1846 September 23, J. G. Galle (1812–1910), of the Berlin Observatory, received a letter from the French astronomer Leverrier, who requested him to search for the new planet he had discovered theoretically. That same evening, Galle found the object not far from the predicted position. It was situated at $\alpha = 22^h01^m31^s$, $\delta = -12°40'.3$ (2000.0 coordinates), 1°42' northwest from the star ι (Iota) Aquarii; see Figure 47.*b*. Neptune was 30.01 astronomical units from the Sun, and 29.18 AU from the Earth (4490 and 4365 million kilometers, respectively). The opposition with the Sun had occurred on 1846 August 20.

For *Pluto* the situation is more complicated. This planet was discovered by Clyde Tombaugh on 1930 February 18 on photographic plates taken at the Lowell Observatory on 1930 January 21, 23 and 29, and the discovery was announced on March 13. What is *the* date of the discovery of Pluto? We might choose either January 21 or February 18.

Using the method described in Chapter 36 of my *Astronomical Algorithms*, I calculated the positions of Pluto given in Table 47.A. The right ascensions α and declinations δ (2000.0 coordinates) are for 0^h Universal Time. The distances to the Sun (r) and to the Earth (Δ) are in astronomical units. Pluto was in opposition with the Sun on 1930 January 9, and at the time of its discovery the distant planet was close to the bright star δ Geminorum — see Figure 47.*c*. In fact, there was a close conjunction on March 8, when Pluto passed less than 1' south of the star. After Pluto had resumed its direct motion, it passed 2½ arcminutes north of δ Gem on 1930 April 22.

Pluto passed through the ascending node of its orbit in September 1930, so at the time of its discovery the planet was close to the ecliptic.

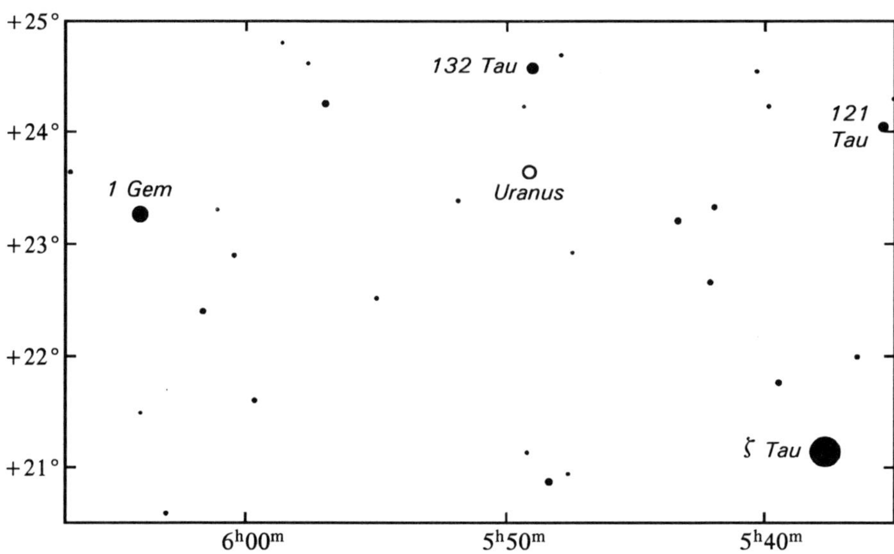

Fig. 47.a : *The position of Uranus on the evening of 1781 March 13 when the planet was discovered by William Herschel.*

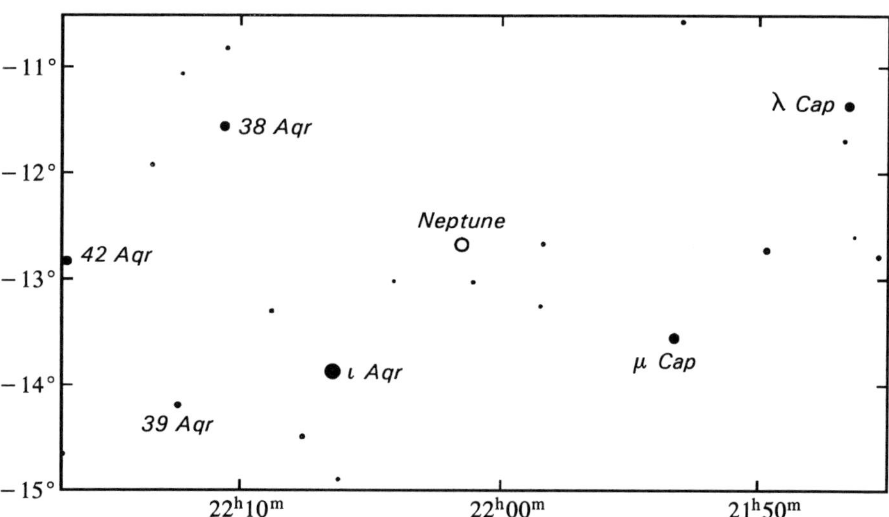

Fig. 47.b : *The position of Neptune on the evening of 1846 September 23 when the planet was seen by Galle at Berlin.*

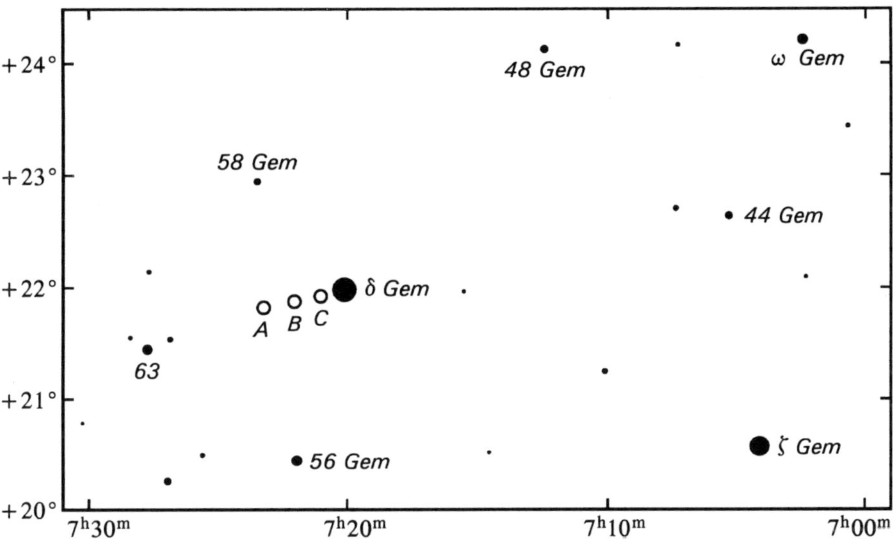

Fig. 47.c : *A, B and C are the positions of Pluto at 0h Universal Time on 1930 January 22, February 5, and February 19, respectively.*

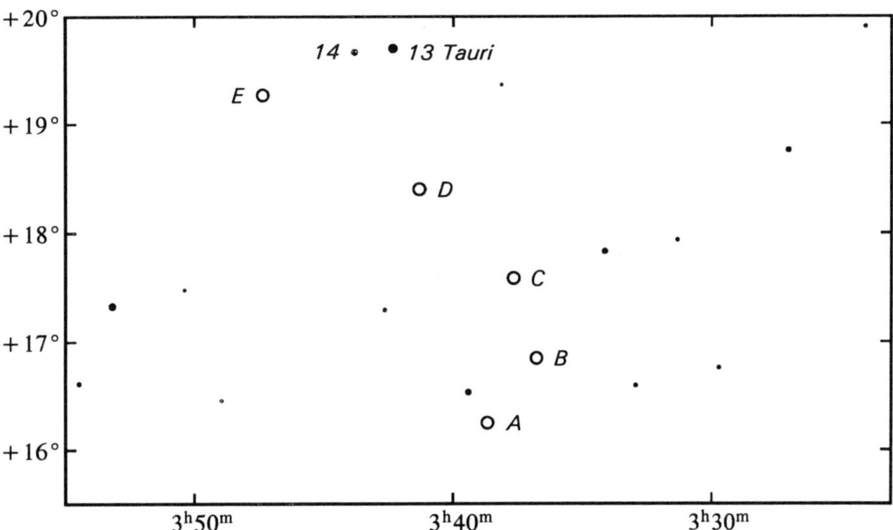

Fig. 47.d : *The positions A to E are those of Ceres at 0h Universal Time on 1801 January 1, 11, 21, 31, and February 10, respectively.*

TABLE 47.A

Ephemeris of Pluto — Positions are for equinox 2000.0

0h UT	α	δ	r	Δ
	h　m　s	°　′　″	AU	AU
1930 Jan. 20	7 23 23.7	+21 48 52	41.316	40.349
24	7 23 02.6	+21 49 49	41.313	40.362
28	7 22 42.0	+21 50 46	41.311	40.379
Feb. 1	7 22 22.0	+21 51 42	41.308	40.402
5	7 22 02.7	+21 52 35	41.306	40.428
9	7 21 44.2	+21 53 28	41.303	40.460
13	7 21 26.7	+21 54 18	41.301	40.495
17	7 21 10.3	+21 55 06	41.298	40.534
21	7 20 55.0	+21 55 52	41.295	40.577

Ceres, minor planet No. 1, was discovered by Giuseppe Piazzi (1746–1826), at Palermo, Sicily, on the evening of 1801 January 1. The new body was situated in the constellation Taurus, eight degrees southsouthwest of the Peiades. Edwin Goffin has calculated the following orbital elements (personal communication to the author): Epoch = JDE 2378900.5 = 1801 February 9.0 Dynamical Time; semimajor axis a = 2.7662074 AU; eccentricity e = 0.0805735; inclination i = 10°.63155; argument of perihelion ω = 65°.59355; longitude of ascending node Ω = 82°.96618 (equinox 1950.0); mean anomaly M = 299°.38975. From these elements, the ephemeris given in Table 47.B have been deduced. As before, the distances to the Sun (r) and to the Earth (Δ) are in astronomical units.

TABLE 47.B

Ephemeris of Ceres — Positions are for equinox 2000.0

0h UT	α	δ	r	Δ	Elon- gation	Visual Magn.
	h　m	°　′	AU	AU	°	
1801 Jan. 1	3 38.7	+16 15	2.702	1.933	133	7.8
6	3 37.4	+16 32	2.698	1.981	128	7.9
11	3 36.8	+16 51	2.694	2.032	123	8.0
16	3 36.9	+17 12	2.690	2.087	118	8.1
21	3 37.7	+17 35	2.686	2.145	113	8.2
26	3 39.2	+17 59	2.682	2.205	108	8.3
31	3 41.3	+18 24	2.678	2.266	104	8.3
Feb. 5	3 44.0	+18 49	2.674	2.329	99	8.4
10	3 47.3	+19 16	2.670	2.392	95	8.5

48. *Ecliptic and galactic equator*

The *solstitial points* are the points of the ecliptic at celestial longitude 90° and 270°, or at right ascension 6^h00^m and 18^h00^m. The center of the solar disk reaches these points at the times of summer and winter solstices.

The *galactic equator* is the great circle on the celestial sphere that passes through the plane of the Milky Way. (The Milky Way system, or Galaxy, is the huge star system of which the Sun is a member).

In some modern star atlases, such as Tirion's *Sky Atlas 2000.0* (Sky Publishing Corporation), and on charts 136 and 339 of *Uranometria 2000.0* by Tirion, Rappaport and Lovi (Willmann-Bell, ed.), we note that the galactic equator passes almost exactly through the two solstitial points; see also Figure 48.*a*. By reason of the precession — the slow, gradual turning of the Earth's axis around the poles of the ecliptic — the celestial longitude of every star increases by 50.3 arcseconds per year, or one degree every 72 years. Of course, this holds not only for stars, but also for mathematically defined points such as the intersections of the galactic equator and the ecliptic. Consequently, these points of intersection slowly move eastward with respect to the vernal equinox. When is the galactic equator passing *exactly* through the solstitial points of the ecliptic?

The current system of galactic coordinates has been defined by the International Astronomical Union in 1959. In the standard equatorial system of B1950.0, the galactic North Pole has the coordinates

$$\alpha_{1950} = 12^h49^m = 192°15', \qquad \delta_{1950} = +27°24'.$$

These values have been fixed *conventionally* and therefore must be considered as *exact* for the mentioned equinox of B1950.0. The 'B' here means that the epoch is based on the so-called Besselian year; see for instance page 125 of my *Astronomical Algorithms* (Willmann-Bell, ed.). The epoch B1950.0 corresponds to Julian Ephemeris Day 2433282.4235 = 1950 January 0.9235 Dynamical Time. The (mean) obliquity of the ecliptic at this epoch is $\varepsilon = 23°.4457889$. Using this value of ε, we can convert α_{1950} and δ_{1950} into the ecliptical coordinates (celestial longitude and latitude) by means of the classical transformation formulae. The results are

$$\lambda_{1950} = 179°.320948, \qquad \beta_{1950} = +29°.811954.$$

Using the precession formulae in the ecliptical system (see page 128 of my *Algorithms*), we may now calculate the coordinates of the galactic North Pole for other epochs. I did it for five instants separated by 18262.5 days (50 Julian years). The results are given in the table at the top of page 303. Here, λ and β are the ecliptical longitude and latitude of the galactic North Pole referred to the mean equinox of the date mentioned in the first column. Because the galactic equator intersects the ecliptic at the two points with longitudes $\lambda - 90°$ and $\lambda + 90°$, it will pass exactly through the solstitial points when λ is equal to 180° (or to 360°).

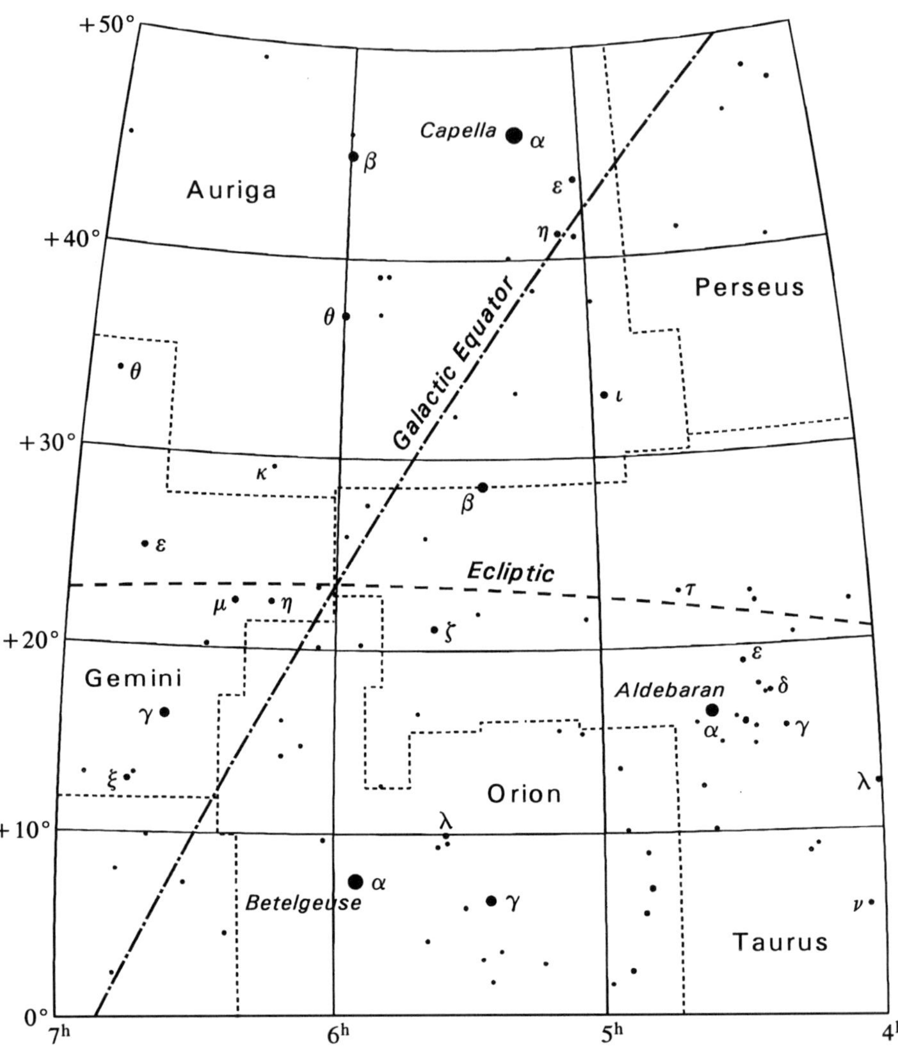

Fig. 48.a : *In this star chart, the declination circles are drawn at intervals of 10 degrees. The horizontal line at the bottom coincides with the celestial equator; the top line is at 50° north declination. The lines of right ascension are drawn at intervals of one hour, or 15 degrees. Stars brighter than visual magnitude 5.0 are plotted. The dotted lines are the official (IAU) boundaries of the constellations. Positions are for epoch and equinox 2000.0. The galactic equator intersects the ecliptic very near right ascension 6ʰ, or longitiude 90°. The exact time of coincidence is May 1998. It is only by chance that the Gemini-Taurus border passes very close to that point too !*

Dynamical Time	Julian Ephem. Day	λ	β
1900 Jan. 0.5	2415 020.0	178°.61897	+29°.81250
1950 Jan. 1.0	2433 282.5	179.32095	+29.81195
2000 Jan. 1.5	2451 545.0	180.02309	+29.81138
2050 Jan. 1.0	2469 807.5	180.72537	+29.81078
2100 Jan. 1.5	2488 070.0	181.42781	+29.81016

By inverse interpolation of the values of λ given in the table, one finds that the value $\lambda = 180°$ is reached in May 1998. Hence, in May 1998 both solstitial points (the summer and the winter ones) will be situated exactly on the 'official' galactic equator.

49. The equinoctial and solstitial points and the constellations

It is a well-known fact that the so-called 'signs' of the zodiac — those empty, rectangular areas of the astrologers — no longer coincide with the constellations of the same name. The reason is the precession, which I already mentioned in the previous chapter. Due to the precession, the shift is approximately one sign every 2000 years. Nowadays, the difference is almost one sign. On September 23, for instance, when the Sun enters the 'sign' of Scorpius (which the astrologers call Scorpio), it is still in the *constellation* Virgo, and about to enter that of Libra.

Two thousand years ago the constellations and the signs did match approximately, but in fact exact coincidence can never occur. The reason is obvious. While by definition each sign is 30 degrees long, the actual constellations occupy different lengths along the ecliptic. The official constellation boundaries adopted by the International Astronomical Union show that the Sun travels 44 degrees in Virgo, but only seven degrees in Scorpius. Moreover, each year the Sun spends 18 days in Ophiuchus, a constellation for which there is no corresponding sign!

So it is clear that no exact instant can be quoted when signs and constellations would have coincided. But of course it is possible to calculate the times at which the March equinox enters successive constellations along the ecliptic. The March (vernal) equinox is one of the two intersections of the ecliptic and the celestial equator; its celestial longitude is $0°$. Therefore, the vernal equinox defines the beginning of the 'sign' of Aries.

I have made the calculations taking into account the variable speed of the precession (50.2687 seconds of arc per year in the year 1900, but growing to 50.2910 by 2000) and the slow rotation (47″ per century) of the plane of the ecliptic itself. In fact, use was made of the precession formulae by J. H. Lieske *e.a.* (1977) which presently are officially adopted by the International Astronomical Union.

In the year −1865 the March equinox passed from Taurus into Aries, and in −67 it entered the constellation Pisces. That year, the beginning of the sign of Aries coincided, on the ecliptic, with the western boundary of the like-named constellation. Not until A.D. 2597 will the March equinox cross the boundary into Aquarius. In A.D. 4312 this equinox will enter Capricornus. See Figure 49.*a*.

The September equinox — the other ecliptic-equator intersection — passed from Libra into Virgo in the year −729, and it remains there until A.D. 2439, when it will enter Leo.

Now in Sagittarius (since the year −130), the December solstice will cross into Ophiuchus in the year 2269, and into Scorpius in 3597. The June solstice (longitude 90°) passed in −1458 from Leo into Cancer, in −10 into Gemini, and in December 1989 into Taurus, as inspection of any good star atlas shows — see also Figure 48.*a*. In A.D. 4609 it will pass from Taurus into Aries.

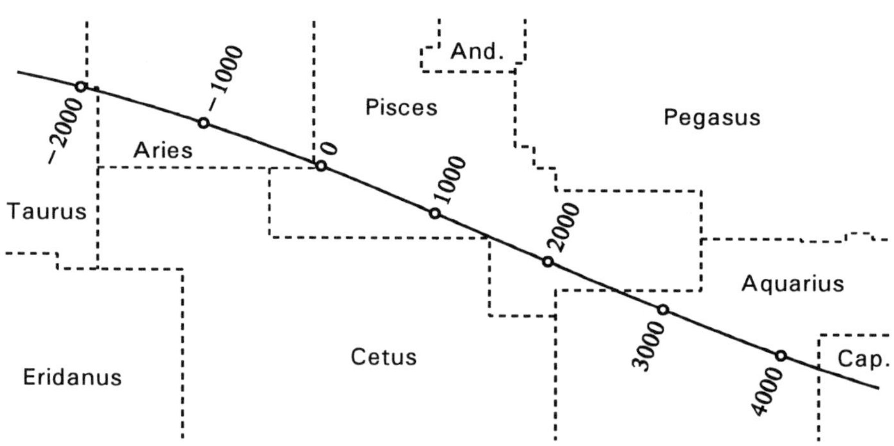

Fig. 49.a : *The successive positions of the vernal (March) equinox with respect to the official (IAU) constellation boundaries (the dashed lines). These boundaries are defined as arcs of equal right ascension or declination with respect to the celestial equator and the equinox of the epoch 1875.0. The position of the vernal equinox is indicated at intervals of 1000 years. In A.D. 1489 the equinox passed only 0°10′ from the 'corner' of Cetus, the Whale, but it did not enter this constellation.*

50. *The declination of Polaris*

Polaris, the Pole Star, or α Ursae Minoris, is not situated exactly at the northern celestial pole. As a result of the precession, the star is now drawing closer to the pole — or, more precisely, the pole is approaching the star.

Table 50.A gives the right ascension and declination of Polaris for some epochs, referred to the mean equator and equinox of that epoch. In the last column the star's north polar distance, 90° minus the declination, is given in arcseconds. In the first column, the decimal zero after the year, for instance 1850.0, means that the epoch is the beginning of that year. More precisely, the unit of time used here is the Julian year of 365.25 days, and the reference epoch is

$$2000.0 \; = \; \text{Julian Day } 2451545.0 \; = \; 2000 \text{ January } 1.5.$$

For instance, 2110.0 means 110×365.25 days after 2000.0, that is, Julian Day = $2451545.0 + (110 \times 365.25) = 2491722.5$, corresponding to 2110 January 2.0. Similarly, the epoch 1800.0 is found to be actually 1799 December 30.5, not the beginning of January 1st of A.D. 1800.

From the table, we note that the right ascension of Polaris is now increasing at an ever increasing rate, and that the declination will reach a maximum around A.D. 2100, when the star will be closest to the pole. An accurate calculation shows that the greatest declination, $+89°32'23''$, will be reached in February 2102. Then the star's north polar distance will be $27'37''$, or $1657''$.

In March 1944 the declination crossed the value $\delta = +89°00'00''$. Figure 50.*a* shows the motion of α UMi with respect to the northern celestial pole.

Apparent places

The values mentioned above and in the table refer to the *mean* coordinates of Polaris, that is, the positions *not* affected by the nutation and by the annual aberration.

The nutation is a periodic oscillation of the rotational axis of the Earth around its 'mean' position. Due to the nutation, the instantaneous pole of rotation of the Earth oscillates around a mean pole which itself moves by the precession around the pole of the ecliptic. The nutation is due principally to the action of the Moon, and can be described by a sum of periodic terms. The most important term has a period of 6798.4 days (18.6 years), the same as the regression of the nodes of the lunar orbit, but some other terms are of very short period (less than 10 days).

The aberration of starlight is an apparent displacement of a star's position due to the finite velocity of light combined with the orbital motion of the Earth around the Sun. At the end of a sidereal year, a star returns to its original place, so far as aberration is concerned.

TABLE 50.A   —   *Mean coordinates of α Ursae Minoris*

Epoch and equinox	Right ascension	Declination	Polar distance
	h m s	° ′ ″	″
1800.0	0 52 26	+88 14 24	6336
1850.0	1 05 02	+88 30 35	5365
1900.0	1 22 34	+88 46 26	4414
1950.0	1 48 48	+89 01 43	3497
1960.0	1 55 41	+89 04 40	3320
1970.0	2 03 18	+89 07 34	3146
1980.0	2 11 46	+89 10 24	2976
1990.0	2 21 14	+89 13 10	2810
2000.0	2 31 49	+89 15 51	2649
2010.0	2 43 42	+89 18 25	2495
2020.0	2 57 06	+89 20 53	2347
2030.0	3 12 11	+89 23 11	2209
2040.0	3 29 11	+89 25 20	2080
2050.0	3 48 16	+89 27 15	1965
2060.0	4 09 33	+89 28 56	1864
2070.0	4 33 00	+89 30 20	1780
2080.0	4 58 28	+89 31 23	1717
2090.0	5 25 30	+89 32 05	1675
2100.0	5 53 29	+89 32 22	1658
2110.0	6 21 39	+89 32 15	1665

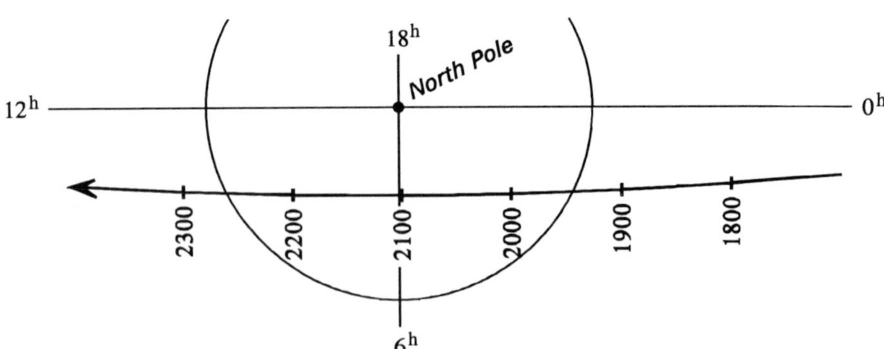

Fig. 50.a : *The path of Polaris with respect to the celestial North Pole. Actually, it is the Pole which moves with respect to the 'fixed' Pole Star. The lines of right ascension 0, 6, 12 and 18 hours are drawn, and the circle has a radius of one degree.*

So, due to the combined effects of nutation and aberration, a star's so-called *apparent* position is continuously changing around its mean position, and this mean position itself is moving due to the precession.

For instance, the apparent declination of Polaris reaches the following extreme values. See also Figure 50.*b*.

1997 February 7	+89°15′19″	maximum
1997 July 16	+89°14′46″	minimum
1998 January 31	+89°15′34″	maximum
1998 July 19	+89°15′01″	minimum
1999 February 2	+89°15′49″	maximum
etc . . .		

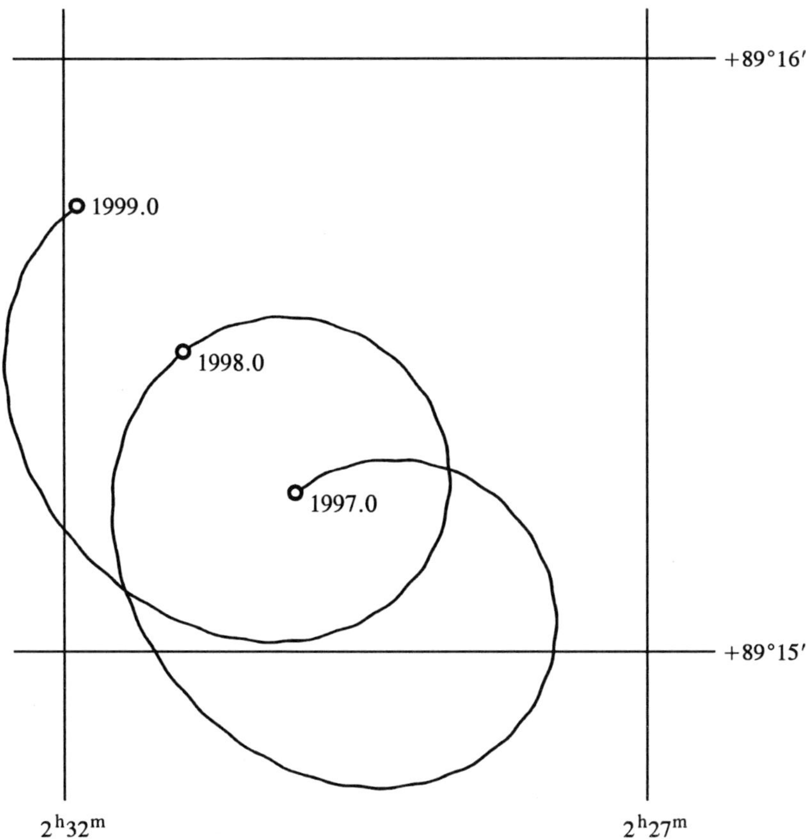

Fig. 50.b : *Variation of the apparent position of α Ursae Minoris from 1997.0 to 1999.0. The loops are due to the annual aberration. The general displacement to upper left is a result of the precession. The very small, wavy irregularities of the curve are due to short-periodic terms in the nutation.*

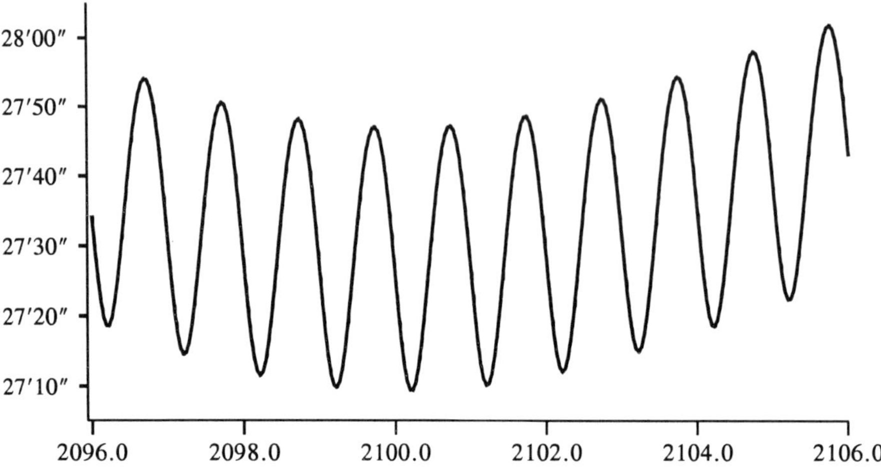

Fig. 50.c *: The angular distance of Polaris to the celestial North Pole,*
A.D. 2096 to 2105, taking into account the effects of nutation and
aberration. The ups-and-downs are due to the annual aberration.

The variation of the *apparent* north polar distance of Polaris from the
beginning of A.D. 2096 to the beginning of A.D. 2106 is shown in the figure
above. The effect of the annual aberration is readily seen. An accurate calculation,
using the FK5 position and proper motion of Polaris, shows that the star will reach
its greatest *apparent* declination, $+89°32'50''.62$, and hence its least apparent polar
distance, $27'09''$, on 2100 March 24, almost two years before the *mean* declination
reaches its greatest value.

51. *Alpha is not always the brightest*

Castor and Pollux, the two brightest stars of Gemini, form a prominent pair,
4½ degrees apart. Castor, the northern star, is the slightly fainter of the two,
although it is designed by the Greek letter α.

Gemini is not the only 'exception'. There are no less than 34 constellations,
out of 88, where the brightest star does not bear the Greek-letter designation α.
Of course, here we are speaking of the *apparent* visual magnitudes of the stars. For
the magnitudes I made use of those given in the *Atlas Coeli Katalog* by Antonín
Bečvař (Prague, 1959); these magnitudes are those of the Revised Harvard
Photometry (RHP).

Those 34 'abnormal' constellations are the following.

● In 13 constellatons, Beta (β), and no other star, is brighter than Alpha (α): Aquarius, Ara, Camelopardalis, Cetus, Coma, Delphinus, Gemini, Hydrus, Libra, Lupus, Monoceros, Orion, and Triangulum.

Here we should note that β Mon is a well-known triple star. Its three components can be seen in small telescopes, but not with the naked eye. The three components *together* make β slightly brighter than α.

● In the southern constellation Mensa, star γ is brighter than α.

● In Cassiopeia, stars β and γ are brighter than α.

● In Crater and Sagitta, stars γ and δ are brighter than α.

● In Corvus, stars β, γ, δ and ε are brighter than α.

● In Ursa Major, stars ε and η are brighter than α.

● In Cancer, β, δ and ι are brighter than α.

● In Pegasus, star ε is slightly brighter than α.

● In Volans, stars β, γ, δ and ζ are brighter than α.

● In Draco, stars β, γ, δ, ζ, η and ι are brighter than α.

● In Microscopium, stars γ, ε and θ_1 are brighter than α.

● In Octans, stars β, γ_1, δ, ε, θ, ν and χ are brighter than α.

● In Pisces, stars γ, η, ι and even ω (the *last* letter of the Greek alphabet!) are brighter than α.

● In Vulpecula, a star that even has no Greek-letter designation, 13 Vul, is slightly brighter than α Vul.

● In Capricornus, α Cap is a wide, naked-eye pair. Its components are of visual magnitudes 3.77 and 4.53, resulting in a total magnitude of 3.33. Nevertheless, the stars β Cap and δ Cap are brighter than this combined magnitude.

● In Hercules too α is a double star. The fainter component is of magnitude 5.39. The brighter star is a semi-regular variable, of magnitude 3.0 to 4.0. Consequently, the 'total' magnitude of α varies between the extremes 2.9 and 3.7. Star β Her is always brighter than α, whereas δ, ζ, η, μ and π Her *now and then* are, depending on the brightness of the variable star.

● In Sagittarius, the two brightest stars are ε (Epsilon) and σ (Sigma). No less than thirteen stars in this constellation are brighter than α, namely γ, δ, ε, ζ, η, λ, ξ_2, o, π, ρ_1, σ, τ and φ. See Figure 51.*a*.

● Finally, there are four constellations *where no star Alpha at all is present*: Leo Minor, Norma, Puppis and Vela. For the latter two, the reason is well-known: the former, extensive constellation Argo (the Ship Argo) has been officially divided into the constellations Carina (the Ship's Keel), Puppis (the Ship's Stern), and Vela (the Ship's Sails). Star α of Argo, Canopus, then became α Carinae, and that's why Puppis and Vela lack an α star.

Here we should note that Antonín Bečvař erroneously designated the star *a* Puppis (lowercase Latin letter *a*) as α Puppis. This error occurs in his *Katalog*

(Prague, 1959) as well as in his *Atlas Coeli* itself. It concerns a star of magnitude 3.8 situated at $\alpha = 7^\mathrm{h}50^\mathrm{m}29^\mathrm{s}\!.8$, $\delta = -40°26'45''$ (1950.0 coordinates), or at $\alpha = 7^\mathrm{h}52^\mathrm{m}13^\mathrm{s}\!.0$, $\delta = -40°34'33''$ (2000.0 coordinates). This star also bears the designations HR 3080 (its number in the Harvard Revised Photometry, *Harvard Annals*, vol. 50; 1908), HD 64440 (its number in the Henry Draper Catalogue, *Harvard Annals*, vols. 93-99; 1918-1924), SAO 219082 (*Smithsonian Astrophysical Observatory Star Catalog*; 1966), and PPM 312247 (*Positions and Proper Motions*, Astronomisches Rechen-Institut Heidelberg; 1991-1993).

The reason why Leo Minor has no star α is unknown to the author of these lines. The southern constellation Norma, the Carpenter's Square, was conceived by the French astronomer Lacaille in 1752. The star which he called α Normae is now inside of the official (IAU) boundaries of Scorpius, where it bears the designation *N* Scorpii (= SAO 207732 = PPM 295143).

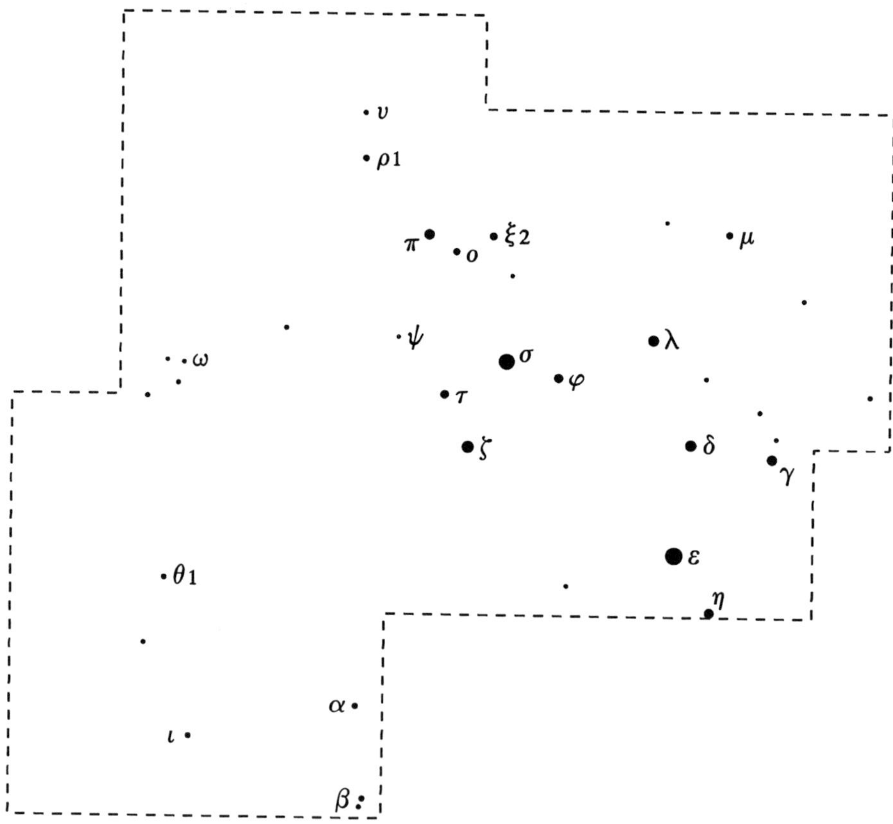

Fig. 51.a : *In Sagittarius, thirteen stars are brighter than* α *Sgr. The dashed lines are the official (IAU) boundaries of this constellation. North is up. Stars brighter than visual magnitude 5.0 are shown. From north to south, Sagittarius extends over 33½ degrees.*

STATISTICS, ETC.

52. *The mean frequency, yes, but ...*

Suppose that the hourly rate of a meteor stream is 60. This means that the *mean* number of meteors seen by a single observer watching a clear sky is 60 per hour, or one per minute. But we may not expect to observe exactly one meteor every minute! Sometimes none will be seen during several minutes, and another time two meteors will appear almost simultaneously. In practice, the conception of 'mean frequency' is meaningfull only when a sufficiently long period is considered.

In some cases, the events are indeed distributed almost uniformly. This is the case, for instance, for sunrise at a given place, or for the phases of the Moon. At Washington, D.C., there are 365 sunrises annually (366 in a bissextile year), so that the mean frequency is one per day. And actually there *is* just one sunrise every day, no more, no less.

But for other astronomical phenomena the distribution can be far from uniform. As we have seen in Chapter 25, during the period 1900–2100 there are fourteen occultations of stars brighter than visual magnitude 3.5 by a planet visible somewhere on the Earth's surface. This means *on an average* one case every 14 years. In fact these events do not happen exactly at intervals of 14 years: occultations of bright stars by planets are distributed irregularly in time. For example, four events took place from 1971 to 1984, but for the next one we have to wait till A.D. 2035.

For another example of a very irregular distribution, consider the total solar eclipses visible from a given place on the Earth's surface. For instance, the eclipse of 2142 May 25 will be the first total solar eclipse at Antwerp, Belgium, for at least seven centuries; but only nine years later, on 2151 June 14, there will be another total one visible from the same city.

At New York, N.Y., four total solar eclipses are visible within a period less than 300 years: on 1925 January 24, 2079 May 1, 2144 October 26, and 2200 April 14. This is much more than may be expected from the calculated mean frequency (see Chapter 13).

And sometimes two total solar eclipses are visible from the same place even within a period less than 18 months — see Chapter 14.

For some astronomical phenomena we have to pay special attention, however. They don't take place at regular intervals, yet their distribution is not arbitrary. At San Francisco, California, *five* occultations of the bright star Aldebaran (α Tauri) by the Moon were visible in 1980 (some of them in daylight), and another one occurred on 1981 February 12. The next one for San Francisco does not happen before 1997 April 11, however. Very strange? Not at all, because occultations of a given star by the Moon take place in series. One Aldebaran-series lasted from 1978 January 19 to 1981 April 8, and the next one started on 1996 August 8 (see Chapter 20), so it is not surprising that several occultations of this star were visible in 1980, and that none *could* have happened from 1982 to 1995.

From A.D. 1500 to 3000 there are 26 transits of Venus over the Sun's disk. This gives a mean frequency of one event every 58 years for this period. However, these phenomena occur in pairs, the two transits of such a pair being separated by eight years. For example, there were transits in 1874 and 1882, the next two not being until 2004 and 2012. Consequently, a person born in 1890 and who died in 1990 could not see any Venus transit during his 100-year long life, while somebody born in 2000 could have witnessed *two* transits before he reaches the age of twelve!

Finally, here are two examples taken from the realm of pure mathematics. The first concerns the distribution of the digits 0 to 9 in the decimal expression of the famous number π. The frequency with which each of these digits appears tends to the limit 1/10 as the number of decimal places increases beyond all bounds. However, the first digit zero does not occur before the 32nd place, although 'normally' one could have expected it to appear once in the ten first decimals:

$$\pi = 3.14159\ 26535\ 89793\ 23846\ 26433\ 83279\ 50\mathbf{2}\dots$$

A bigger surprise in the decimal expression of π is the unexpected occurrence of *six* successive digits 9, at the places 762 to 767.

In the *Journal of Recreational Mathematics*, Vol. 16, No. 2, page 138 (1983), Steven Verhezen posed the following problem. The smallest integers whose square and cubic roots start with a pandigital string of digits [that is, each of the digits 0 to 9 exactly once] are 1362 and 2017:

$$\sqrt{1362} = 36.90528417\dots \quad \text{and} \quad \sqrt[3]{2017} = 12.63480759\dots$$

What is the smallest integer having *both* its square and cubic roots starting pandigitally? The mean frequency of those integers is

$$(0.9 \times 0.8 \times 0.7 \times 0.6 \times 0.5 \times 0.4 \times 0.3 \times 0.2 \times 0.1)^2 = 1/7594058,$$

so we may expect one case every 7594058 numbers. However, a computer search performed for all integers from 1 to 25000000 provides *no* solution, although about three might be expected in this range. In fact, the first (and smallest) solution to the problem is the number 29265786. The next two solutions follow after rather short intervals, however. Here is the list of all solutions smaller than 100 million. There are fifteen of them, in good agreement with the mean frequency mentioned above:

29265786	67177815	92242910
32723928	67863673	92260201
35809137	69065623	92456375
55989332	69101140	94967371
57522048	91782482	96318633

Just as is the case for the appearance of meteors, these 15 numbers are irregularly distributed. Notice the gap between 35809137 and 55989332, and that between 69101140 and 91782482. On the other hand, three cases occur between 92242000 and 92457000.

53. *Statistics: danger!*

Examples are given of statistical investigations giving incorrect results because they are based on a too small number of data.

Phase	1980	UT
		h m
N.M.	Oct. 9	2 50
F.Q.	Oct. 17	3 47
F.M.	Oct. 23	20 52
L.Q.	Oct. 30	16 33
N.M.	Nov. 7	20 43
F.Q.	Nov. 15	15 47
F.M.	Nov. 22	6 39
L.Q.	Nov. 29	9 59
N.M.	Dec. 7	14 35
F.Q.	Dec. 15	1 47

The table at right gives the times (Universal Time) of the lunar phases (New Moon, etc.) from October 9 to December 15, 1980. Before you continue reading, do you notice something special about these instants?

The remarkable fact is that the *minutes* of the times of these ten successive lunar phases all lie in the second half of the hour, between the minutes 30 and 60. Of course, this peculiarity just happens by chance. Indeed, what have lunar phases to deal with the minutes of the hour at which they take place? Moreover, the subdivision of the day into 24 hours, and of the hour into 60 minutes, is a fully arbitrary choice. The day could have been divided into 10 or 36 hours as well, and the hour into 50 or 100 minutes, and in such a case the above-mentioned 'effect' for those ten lunar phases would have disappeared completely.

The probability that 10 successive lunar phases all take place within the 'same half hour' is equal to $1:2^9$, or $1:512$. This is a rather small probability, yet here we see such a case taking place. But we do know that it happens only by chance. It really becomes dangerous when similar 'effects' are found in some areas of scientific (or pseudoscientific?) investigation, especially when this is done by persons eagerly watching for 'strange effects'.

In 1975, solar expert W. Gleissberg [1] found an unexplained anomaly in the annual distribution of the epochs of the sunspot maxima: of the 21 maxima occurring during the period 1750–1970, thirteen (more than 60%) took place during the 4-month period February-May, while only four occurred during each of the periods June-September and October-January. This preponderance of the months February to May, if real, cannot be explained, neither by the variable Sun-Earth distance, nor by weather conditions, nor by the slight inclination of the Sun's equator on the plane of the Earth's orbit.

The effect became even stronger if only the eight highest sunspot maxima were taken into consideration. The distribution was then found as follows:

7 maxima in the period February to May,
1 maximum   —   —   June to September,
0 maximum   —   —   October to January.

Gleissberg noted that "the application of the calculus of probability leads to the result that, if 8 maxima were randomly distributed over the three 4-month periods, there would be only a chance of 1:129 that at least 7 maxima would fall into the same period."

Should we believe that the time of the year, and hence the position of the Earth in its orbit, influences the time of sunspot maximum? Is this an indication that there is an influence of the planetary positions on the frequency of sunspots?

Well, believe it or not, but the 'Gleissberg effect' was a result of bad luck. It was precisely that only one possibility out of the 129 that *did* happen. Statistics can play nasty tricks on us. Remember the still smaller probability of 1:512 concerning the above-mentioned 10 lunar phases!

Gleissberg's result was based on the sunspot relative numbers' smoothed means as calculated by the Zürich Observatory. This method of smoothing is as follows. First, calculate the arithmetical mean M_1 of the months Nos. 1 to 12, and the mean M_2 of the months Nos. 2 to 13. Then the mean M of M_1 and M_2 is taken as the smoothed mean for the middle month (No. 7).

One might think that the 'mean of means' calculated in this manner will provide a very smoothed mean. This is not so, however, because calculation shows that M is very nearly equal to simply the mean of the months 1 to 13. The Zürich method gives the same weight to all the months 2 to 12 (and half that weight to the extreme months 1 and 13). Moreover, the 'smoothed' curve thus obtained is not yet sufficiently regular.

For these reasons, I proposed a new smoothing formula [2], still based on 13 consecutive monthly means, but where a greater weight is given to the central months, namely

$$S_i = \frac{1}{81} (11\,N_i + 10\,N_{i\pm1} + 9\,N_{i\pm2} + 7\,N_{i\pm3} + 5\,N_{i\pm4} + 3\,N_{i\pm5} + N_{i\pm6})$$

Here, the smoothed mean S_i for month i is calculated from the monthly means of the 13 consecutive months $i-6$ to $i+6$. A weight of 11 is given to the central month i, a weight of 10 to the adjacent months $i-1$ and $i+1$, etc. The constant 81 in the denominator is the sum of the weights.

If we now use *this* formula to calculate the smoothed monthly means of the Zürich sunspot numbers, and when we adopt by definition that the month having the highest smoothed mean is the epoch of the sunspot maximum, then understandably we will obtain times which differ a little from the Zürich values [3]. The annual distribution of the sunspot maxima for the period 1750–1970 is now as follows [4]:

February–May   6 cases (1750, 1778, 1816, 1837, 1937, 1969)
June–September   8 cases (1761, 1829, 1860, 1870, 1893, 1917, 1928, 1947)
October–January   7 cases (1769, 1787, 1804, 1847, 1884, 1905, 1957)

The 'anomaly' found by Gleissberg now has completely disappeared. The three four-month periods have practically the same number of cases. We may deduce that the 'unexpected effect' found by Gleissberg was just accidental, since this anomaly disappears when the Zürich method of smoothing is replaced by another method. On the other hand, we might not deduce that the 'anomaly' resulted from the special method of smoothing used at Zürich. Indeed, even when a 'wrong' smoothing formula is used, there is no reason why some months of the year should be favored, since the same formula is used continually for all months. I showed, however, that the effect is removed simply by using another (and better) formula for the calculation of the smoothed means.

Of course, the 'effect' would disappear also when a larger number of sunspot maxima is considered. But this cannot be performed before one or two centuries, because only one sunspot maximum occurs every 11 years or so, so that our observational material increases only very slowly. (Since the publication of Gleissberg's article, two more sunspot maxima have occurred: November 1979 and October 1989, both contradicting the 'effect').

The flaw in the Gleissberg investigation was that this study was based on *too small number of cases*, and an unlucky mischance did deceive him. The same holds in the case of the ten successive lunar phases mentioned at the beginning of this chapter: if we consider a much larger number of phases, say 100 or 200, then we would find that the phases do not occur more often during a certain part of an hour than in other parts, statistically speaking.

In 1936, F. Sanford [5] thought he had found the proof that planetary positions influence the formation of sunspots. He noted that there are more sunspots visible on the Sun's disk when Venus is in superior conjunction with the Sun than when it is at inferior conjunction. This discovery was based on observations of sunspots made from 1917 to 1932.

However, during this period there were only *ten* superior conjunctions and ten inferior conjunctions of Venus, a number too small to yield any meaningful statistical result. By considering 40 superior and 40 inferior conjunctions of Venus in the longer period 1902–1965, no 'Venus effect' at all is found [6].

Sometimes even a statistical test can fool us, as shown by the following example. In 1948, Panofsky and Hess published a paper [7] suggesting that there was a correlation (*) of 0.62 between the wind on Jupiter (as indicated by the motion of the Great Red Spot) and mean zonal winds at the surface of the Earth.

(*) The correlation coefficient is a statistical measure of the degree to which two variables are related to each other. This coefficient is always between $+1$ and -1. A value of $+1$ or -1 would indicate that the two variables are totally correlated. A value of zero indicates that there is no relationship between the two variables. In practice, however, when there is no relationship, one may find that the correlation coefficient is not exactly zero, due to the fortuitous coincidences that generally occur except for an infinite number of points.

Thirty years later, in 1978, another article appeared in the same journal, by the same Panofsky and F. J. Lucadamo [8]. Here we read that additional observations made since 1948 (the publication of the first article) now "indicate a correlation of −0.12, suggesting that the original correlation was spurious, even though it appeared statistically significant according to standard tests."

So, Panofsky and Hess too had been victims of statistics based on a too small number of data. Might this be a good lesson to those who hope to find a relation between sunspot activity or the lunar phases and the weather.

It is even worse when occurrences close in time are inferred to be causally connected. The very hot summer of 1947 in Western Europe occurred in a year of high sunspot maximum, and this led some persons to deduce that the weather depends on solar activity. However, when a sufficiently number of years is considered, it appears that there is not the slightest relation between the weather at a given place and sunspot activity — see the next chapter.

One single accidental hit proves nothing. If you are sick precisely on the day of a Full Moon, this doesn't prove that there is an influence of the Moon on our health. Consider a sufficiently large number of Full Moons, and you shall notice no effect at all on illness or on other events in our lives. A dream followed by an experience that can be perceived as correlative may seem precognitive. But *thousands* of other dreams do not come true!

It becomes real fraud when the many cases, where the expected 'effect' does not appear, simply are concealed. Martin Gardner writes in one of his books [9]:

> "Many of the published results of extrasensory-perception testing are examples of patterns that are inevitable in any long series of random results. When such patterns fail to turn up, ESP proponents are unlikely to publish the results, not having found evidence for ESP; when they do find such patterns, they publish. Perhaps if the total picture could be surveyed, the published patterns would be less surprising."

Of course, statistics *is* serious business. But the aim of this chapter is to draw the reader's attention on the need for prudence and critical mind, especially when it concerns provisional results, or results based on a too small number of data. Critical mind is as important as open mind. Wait and see!

References

1. W. Gleissberg: "An unexpected anomaly in the annual distribution of the maxima in the 11-year sunspot cycle", *Journal of Interdisciplinary Cycle Research*, Vol. 6, No. 1, pages 37-40 (1975).

2. J. Meeus: "Une formule d'adoucissement pour l'activité solaire", *Ciel et Terre* (Belgium), Vol. 74, No. 6, pages 445-449 (November-December 1958).

3. J. Meeus: *Astronomical Tables of the Sun, Moon and Planets*, 2nd ed., pages 370-371 (Willmann-Bell, ed.; 1995).

4. J. Meeus: "On an 'unexpected anomaly' in the annual distribution of the maxima of the 11-year sunspot cycle, *Journal of Interdisciplinary Cycle Research*, Vol. 8, Nos. 3-4, pages 205-206 (1977).

5. F. Sanford: "Influence of planetary configurations upon the frequency of visible sun spots", *Smithsonian Miscell. Collections*, Vol. 95, No. 11, pages 1-5 (1936).

6. J. Meeus: "Comments on 'The Jupiter Effect'", *Icarus*, Vol. 26, pages 257-267 (1975).

7. H. A. Panofsky and S. L. Hess: "Zonal index and the motion of the Red Spot on Jupiter", *Bulletin of the American Meteorological Society*, Vol. 29, pages 426-428 (1948).

8. F. J. Lucadamo and H. A. Panofsky: "On a relation between winds on the Earth and Jupiter", *ibid.*, Vol. 59, pages 700-701 (1978).

9. Martin Gardner: *Mathematical Carnival*, page 165 (London, 1976).

54. Sunspots and the weather

Although many papers have been published about a possible relation between sunspot activity and the weather on Earth, no such relation has ever been proved once and for all. By 'weather' I mean *the actual weather of every day*, that which is seen by the man-in-the-street, not such things as the winds in the stratosphere.

Either the 'correlation' was spurious and could not be confirmed, or it did not concern the actual *weather* that interests everybody: will next summer be hot and dry, or will there be much rain next year?

For example, in 1991 Danish researchers Eigil Friis-Christensen and Knud Lassen found that the *duration* of a sunspot cycle seemed related to the temperatures in the northern hemisphere (*Sky and Telescope*, November 1995, pages 12-13). However, first, the total change in the mean temperature was just less than one degree centigrade, a variation that is not appreciable by ordinary means; secondly, the hot or cold weather (for instance during a summer in a given country) has nothing to do with slight changes of the *mean* temperature in a *hemisphere*, but it mainly depends on the existence and positions of centers of high pressure on the Earth's surface; and thirdly, the effect is not related at all to the *phase* in the 11-year sunspot cycle but, if real, it is a variation of *long* period.

In another study, a correlation was announced between the phase of the solar cycle and the winds in the upper atmosphere. However, the latter is not the 'weather' as we call it in the life of every day!

Of course, we *cannot prove* that sunspot activity doesn't influence the weather. But here, as elsewhere, the burden of proof is on the claimant. As long as that burden is not met, the examples which follow can be taken as indication that there is no relation between the phase of the sunspot cycle and the weather on Earth.

Currie [1] found that during a period of 100 years the air temperature in North America showed an oscillation with a period of 10.6 ± 0.3 years, but with an amplitude of only 0.1 degree centigrade. Even if this 'correlation' appears to be *statistically* significant, it is so extremely weak that it has no practical consequence: absolutely *nothing* can be deduced from it for weather forecast in a given year. The weather fluctuates too much by other causes.

On the other hand, many others have failed to detect a correlation between the 11-year sunspot cycle and such surface phenomena as temperature, air pressure and precipitation. This was the case for Gerety *et al.* [2]. They found that "cross-spectral computations using the time series of Zurich sunspot numbers and seasonal mean temperature and precipitation totals indicate that these series are uncorrelated at individual stations and grouped together into latitude bands."

Others searched for the influence of sunspot activity on tree rings. But the results so obtained [3] are far from convincing, and the 'relation' has been called in question by — among others — Fritts and LaMarche [4].

The influence of solar activity on the water level in African lakes also is extremely doubtful, and Ratcliffe [5] pointed out that "this correlation disappeared soon after its discovery"!

If the sunspot activity had an influence on the weather, this effect would be *periodic*, and hence would have been proved for a long time past, for instance by performing harmonic analysis on any weather element. (Harmonic analysis is the process of expressing functions in a series of sines and cosines).

For instance, it has been claimed that there is a correlation between the annual amount of rainfall and sunspot activity. In Table 54.A, the second column gives the mean R of the definitive Zürich sunspot numbers for the years 1901 to 1995. The next column gives the total annual amount of precipitation in millimeters P at the Uccle Observatory, near Brussels, Belgium. Calculation shows that the correlation coefficient between these two variables is -0.060, which indicates that there is no significant correlation between R and P. For the years 1901–1988, the correlation even drops to -0.027. (About the correlation coefficient, see the note on page 317).

The 95 values of P are plotted in Figure 54.*a*. Here also no periodicity of 11 years is seen.

One might object that nevertheless there may exist a correlation between the phase of the sunspot cycle and the yearly amount of precipitation, but that the maxima of the two periodicities may not coincide, that is, the two periodicities may be out of phase. For example, a moderate sunspot activity on the *ascending* part of the curve might be correlated to much precipitation, while a similar moderate sunspot activity on the *descending* branch could be associated with drier weather. In such a case, indeed, the correlation coefficient could be small even in the presence of an actual Sun-weather correlation.

For this reason, I constructed Figure 54.*b*. Here, each of the 95 years in the period 1901-1995 is represented by a dot, determined vertically by the annual amount of precipitation at Uccle, and horizontally by the number of years elapsed

TABLE 54.A

Annual means of the relative Zürich sunspot numbers (R), total annual amount of precipitation in mm at Uccle (P), annual frost days at Uccle (F), annual hot days at Uccle (H), and summer index at Manchester [references 6, 7, 8]

Year	R	P	F	H	Index at Manchester	Years since sunspot minimum
1901	2.7	700			249	0
1902	5.0	762			195	1
1903	24.4	854			209	2
1904	42.0	663			212	3
1905	63.5	912			223	4
1906	53.8	821			214	5
1907	62.0	622			147	6
1908	48.5	678			220	7
1909	43.9	842			171	8
1910	18.6	990			190	9
1911	5.7	741			274	10
1912	3.6	941			156	11
1913	1.4	801			205	0
1914	9.6	877			222	1
1915	47.4	910			196	2
1916	57.1	1054			188	3
1917	103.9	851			228	4
1918	80.6	848			200	5
1919	63.6	980			203	6
1920	37.6	760			174	7
1921	26.1	417			249	8
1922	14.2	938			178	9
1923	5.8	917			174	0
1924	16.7	849			158	1
1925	44.3	1075			246	2
1926	63.9	896			227	3
1927	69.0	837			175	4
1928	77.8	882			197	5
1929	64.9	688			211	6
1930	35.7	953			199	7

Table continues on next page

The values of F and H for the years 1901 to 1930 are not available to the author.

TABLE 54.A   (Cont.)

Year	R	P	F	H	Index at Manchester	Years since sunspot minimum
1931	21.2	858	74	1	173	8
1932	11.1	858	61	8	223	9
1933	5.7	738	77	4	251	0
1934	8.7	707	51	4	238	1
1935	36.1	916	58	12	243	2
1936	79.7	763	54	4	190	3
1937	114.4	900	55	6	213	4
1938	109.6	711	42	8	177	5
1939	88.8	928	80	3	213	6
1940	67.8	837	78	1	238	7
1941	47.5	744	92	12	236	8
1942	30.6	841	42	8	214	9
1943	16.3	738	71	9	209	10
1944	9.6	766	50	7	200	0
1945	33.2	745	62	11	223	1
1946	92.6	861	77	6	170	2
1947	151.6	640	46	28	255	3
1948	136.3	792	63	6	176	4
1949	134.7	521	64	9	267	5
1950	83.9	951	71	9	216	6
1951	69.4	878	68	1	201	7
1952	31.5	926	88	6	198	8
1953	13.9	557	49	2	193	9
1954	4.4	741	83	3	143	0
1955	38.0	616	85	3	277	1
1956	141.7	795	47	1	155	2
1957	190.2	801	75	9	216	3
1958	184.8	834	57	2	184	4
1959	159.0	560	47	8	269	5
1960	112.3	962	32	1	217	6

Table continues on next page

The number F of frost days given for 1931 are those of the winter 1931–1932, and so on for the other years.

TABLE 54.A   (Cont.)

Year	R	P	F	H	Index at Manchester	Years since sunspot minimum
1961	53.9	903	74	6	203	7
1962	37.5	862	95	1	197	8
1963	27.9	713	73	1	194	9
1964	10.2	785	62	9	197	0
1965	15.1	1073	54	0	189	1
1966	47.0	1054	43	5	192	2
1967	93.8	707	75	4	223	3
1968	105.9	776	72	2	215	4
1969	105.5	776	84	7	234	5
1970	104.5	727	38	2	235	6
1971	66.6	691	36	2	205	7
1972	68.9	710	53	4	185	8
1973	38.0	690	34	11	234	9
1974	34.5	1039	31	2	199	10
1975	15.5	734	61	8	268	11
1976	12.6	541	54	25	301	0
1977	27.5	855	53	1	223	1
1978	92.5	767	86	2	173	2
1979	155.4	839	44	0	199	3
1980	154.6	913	67	2	173	4
1981	140.5	1016	57	4	196	5
1982	115.9	800	40	5	203	6
1983	66.6	689	45	6	278	7
1984	45.9	931	65	3		8
1985	17.9	758	81	0		9
1986	13.4	946	63	1		0
1987	29.2	908	29	2		1
1988	100.2	1005	24	0		2
1989	157.6	639	29	3		3
1990	142.6	759	54	7		4
1991	145.7	794	42	2		5
1992	94.3	916	44	2		6
1993	54.6	857	38	0		7
1994	29.9	894	27	9		8
1995	17.5	763	83	12		9

since the last sunspot minimum (last column of Table 54.A). These points show a considerable scattering, so here again there appears to exist no correlation between the phase of the solar cycle and the amount of precipitation at Uccle. If there was a *clear* correlation, it certainly would have become recognizable on the basis of observations made during a period of 95 years.

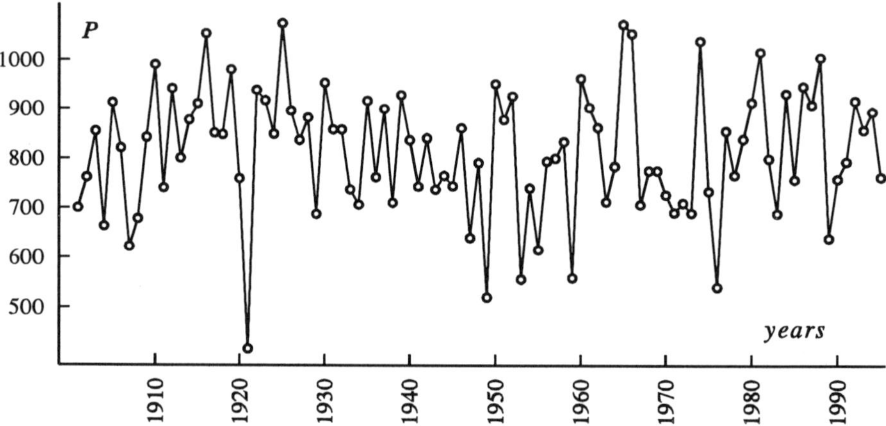

Fig. 54.a : *The annual amount of precipitation in millimeters, at Uccle, Belgium, for the years 1901 to 1995.*

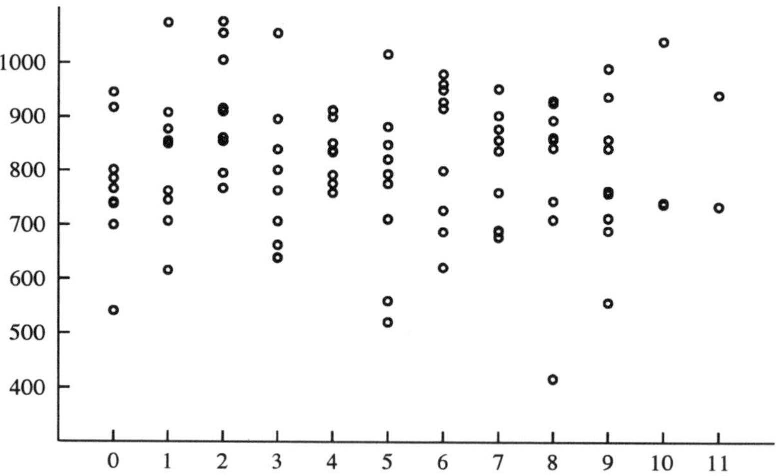

Fig. 54.b : *Distribution of the years 1901 to 1995 as a function of the amount of precipitation at Uccle (vertically) and the number of years since the last sunspot minimum (horizontally). No correlation is seen.*

Maybe the preceding analysis is still too rough, and should we use more 'sophisticated' statistical methods? Well, if one really needs to make appeal to advanced methods, or to 'manipulate' the observational data in order to finally obtain a vague correlation, then this is in itself an indication that the relation between sunspot activity and terrestrial weather is very weak, if not illusory.

Here we must note that if the relative Zürich sunspot number (the Wolf number) is not a universal panacea to define solar activity, it is nevertheless an excellent indicator for this activity: a calm Sun always corresponds to a small sunspot number, while a large number corresponds to an active Sun.

Another claim is that long and cold winters would tend to occur near the epochs of sunspot maxima. At least for Western Europe, this certainly is not the case — see Figure 54.c.

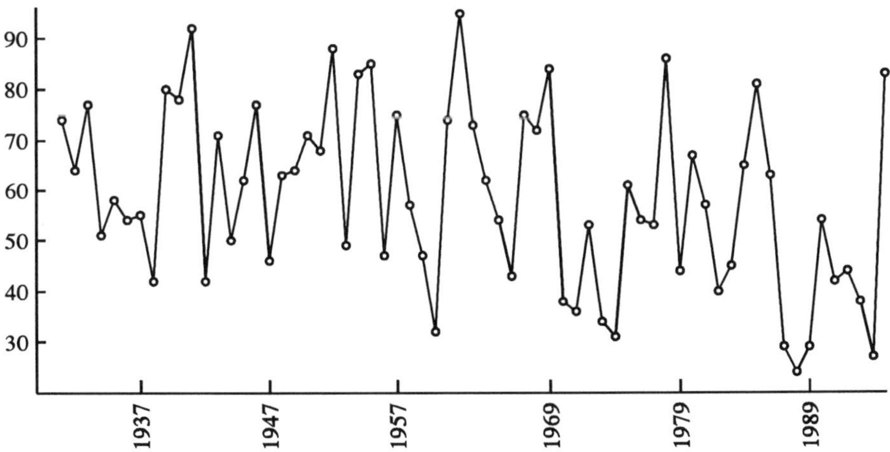

Fig. 54.c : *The number F of frost days at Uccle, Belgium, during the winters from 1931 to 1995. No periodicity is seen. The years labeled along the horizontal axis are those of sunspot maxima.*

Similarly, there appears to be no relation between the number of hot days in Belgium and the level of sunspot activity. Table 54.A gives the number H of hot days at Uccle for the years 1931 to 1995. These are the days when the maximum air temperature exceeded 30.0 degrees centigrade. We leave it as an exercise to the reader to draw a graph with the years plotted as a function of R and H. It will appear that the points are disorderly distributed in the R–H diagram.

And when we calculate the correlation coefficient for the 65 R–H points of the period 1931-1995, we obtain a value as small as -0.025. This value is so close to zero that we may state without hesitation that there is not the slightest relation between the yearly mean of the Zürich sunspot numbers and the yearly number of hot days at Uccle. (If we drop the last value, that for 1995, the correlation coefficient is even closer to zero, namely -0.004).

Hughes [8] presented a 'summer index' for the 83 summers 1901 to 1983 at Manchester, England. In meteorology, summer is defined as the period June 1 to August 31, and Hughes defined his index by the formula

$$\text{Index} \ = \ 10 \times (T + \frac{S}{67} - \frac{D}{8})$$

where T is the mean maximum temperature in degrees centigrade, S is the total hours of sunshine, and D is the number of days with rain. Hughes' indices are reproduced in the sixth column of Table 54.A.

Figure 54.d shows the distribution of those 83 summers as a function of the mean relative sunspot number R of that year (horizontally), and of the Hughes index (vertically). We see at once that there is no functional relationship between the two variables. The 'best' straight line through the 83 points, calculated by means of the method of least squares, is

$$\text{index} \ = \ 210.45 \ - \ 0.0167\,R.$$

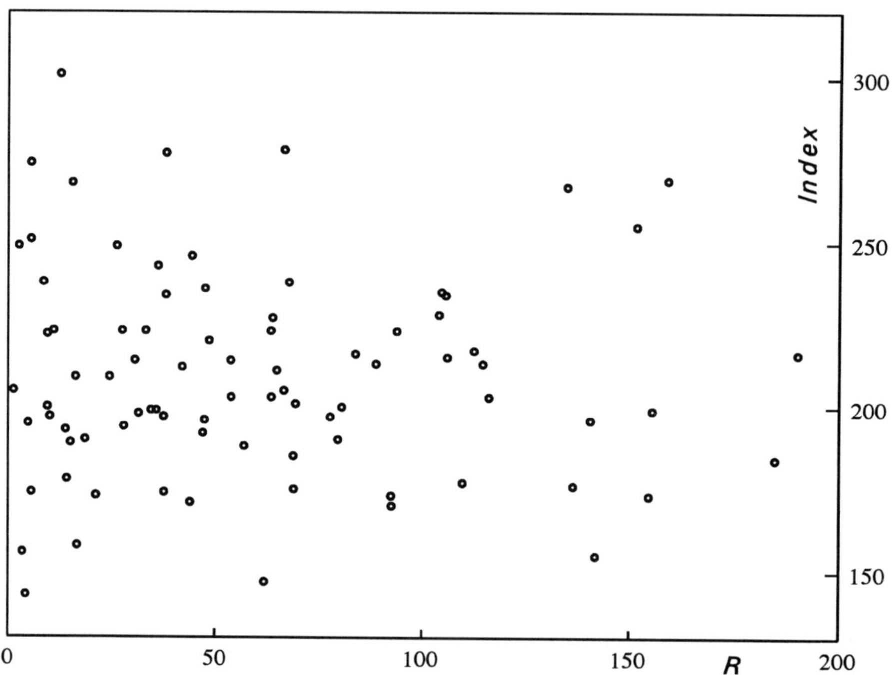

Fig. 54.d : *Distribution of the summers 1901 to 1983 at Manchester, England, as a function of the yearly mean relative Zürich sunspot number R (horizontally) and the Hughes index (vertically). No correlation is seen.*

This line (not drawn in the figure) is almost horizontal. The very small coefficient of R in this formula, as well as the very small value of the coefficient of correlation (-0.025), are consistent with the hypothesis that there is no functional relationship between the summer weather at Manchester and the level of sunspot activity.

Yes, solar disturbances *do* influence phenomena on Earth. For example, some radio fadings are caused by solar flares, eruptions on the Sun. Charged particles ejected from the Sun and entering the Earth's upper atmosphere collide with and ionize molecules of the air to cause aurorae. Geomagnetism also is very sensible to solar activity.

The density of the *upper* atmosphere, above 200 kilometers, is highly variable, depending on the sunspot activity. When there are many spots on the Sun, as in 1957 and 1969, this density is greater than in years when the Sun is calm (1964, 1976). This relation was discovered shortly after the launch of the first artificial satellites, from the decrease of the revolution periods of these objects circling the Earth.

But, once more, all this has nothing to do with the *weather*. And while there might indeed be a link between *global* climate (*) and solar minima of *long* duration such as the Maunder minimum (**), all happens as if there is not the slightest relation between the 11-year sunspot cycle and weather phenomena. Here, as elsewhere, one single coincidence proves nothing. While the hot and dry summer of 1947 in Western Europe coincided with a very high sunspot maximum, the similar hot summer of 1976 took place at the time of a *minimum* of sunspot activity!

Even if one or another correlation were confirmed without any possible doubt, it would be grossly exaggerated to assert that "the weather depends on solar activity". Such a correlation could be only very weak; perhaps it would be of theoretical interest, but for actual weather forecast it would be worthless.

(*)   The word 'climate' refers to *long-term* (many years) mean weather characteristics.

(**) The period, from about 1645 to 1715, when almost no sunspots were seen. This was pointed out in the 1890's by G. Spörer and E.W. Maunder. See also [9].

References

1. R. G. Currie : "Solar cycle in surface air temperature", *Journal of Geophysical Research*, Vol. 79, pages 5657-5660 (1974).
2. E. J. Gerety, J. M. Wallace, and C. S. Zerefos : *Journal of the Atmospheric Sciences*, Vol. 34, pages 673-678 (April 1977).
3. See for instance the curve by Douglas (1919) in *Astronomy and Astrophysics*, Vol. 54, page 860 (1977).
4. H. C. Fritts and F. N. LaMarche : *Tree Ring Bulletin*, Vol. 32, page 21 (1972).

5. R. A. S. Ratcliffe: *Meteorological Magazine*, year 1976, page 394.

6. *Observations Climatologiques*, bulletin mensuel de l'Institut Royal Météorologique (Brussels).

7. L. Poncelet and H. Martin: *Esquisse Climatologique de la Belgique*; Mémoires de l'I.R.M., Vol. XXVII (1947).

8. G. H. Hughes: "Summer weather at Manchester, 1901 to 1983", *Weather*, Vol. 41, No. 9, pages 302-304 (September 1986).

9. J. A. Eddy: "The Maunder minimum", *Science*, Vol. 192, pages 1189-1202 (1976 June 18).

55. *Solar activity and the brightness of lunar eclipses*

The Danjon scale is often used to estimate how dark a lunar eclipse appears. This five-step scale, named after the French astronomer André Danjon (1890–1967), ranges from $L = 0$ (very dark) to $L = 4$ (very bright).

From observations of this type, Danjon (*Comptes Rendus de l'Académie des Sciences*, Vol. 171, page 1127; 1920) deduced that the brightness of the totally eclipsed Moon is a function of the phase of the solar activity in its 11-year cycle. During the first two years after a sunspot *minimum*, the eclipsed Moon is rather dark ($L \approx 1$). The next years, the brightness of the eclipsed Moon gradually increases till the next solar minimum. Then the eclipse luminosity suddenly drops, and the cycle repeats.

According to this 'Danjon law', the instant of the sunspot maximum provokes no peculiarity at all, while the passage through the minimum causes a sudden drop in the brightness. This discontinuity is illustrated in Figure 55.*a*.

Later, G. de Vaucouleurs (1918–1995) found that the observations can be better represented by the *linear relation*

$$L = 1.3 + 2.4\,\Phi \tag{1}$$

where Φ is the phase of the solar cycle expressed as a fraction of the period. According to this formula, L has a mean value of 1.3 at the start of the sunspot cycle, increasing to 3.7 towards the end, just before the next sunspot minimum.

Table 55.A lists the 47 lunar eclipses considered by de Vaucouleurs (*Comptes Rendus de l'Académie des Sciences*, Vol. 218, pages 655-656; 1944). The second column gives the eclipse luminosity L, and the next column the phase Φ of the sunspot cycle, according to the same reference. In the fourth column, I added the magnitude of the eclipse.

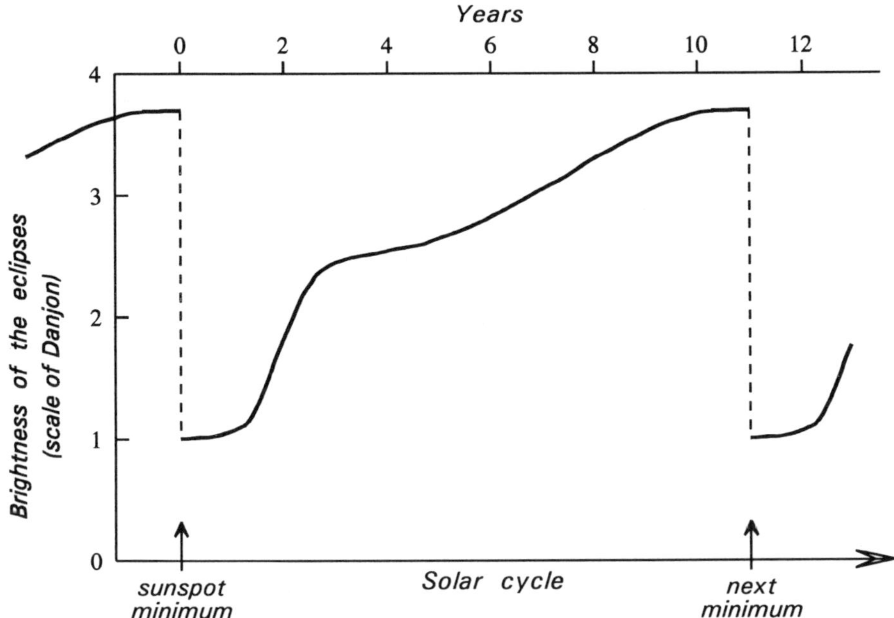

Fig. 55.a : *Mean brightness of the total eclipses of the Moon
in the course of the sunspot activity cycle, according to Danjon.*

Those 47 eclipses are represented in Figure 55.*b*, where each dot is defined
by its value of Φ (horizontally) and that of L (vertically). We clearly see that, on the
average, the value of L increases with the phase Φ of the sunspot cycle, in
accordance with Danjon's law. In particular, all twelve eclipses for which L is
larger than 3.0 occur for $\Phi > 0.64$. The 'best' straight line through the 47 points,
calculated by means of the method of least squares, is defined by the equation $L =
1.2 + 2.3\,\Phi$, which is almost identical to formula (1) found by de Vaucouleurs.

Unfortunately, there are problems. Firstly, several eclipses were abnormally
dark, being affected by aerosols spewed into the Earth's stratosphere by large
volcanic eruptions: Santa Maria, Guatemala, in October 1902; and Katmai, Alaska,
in June 1912. These cases are indicated by an asterisk in Table 55.A. Secondly, the
values of L placed between parentheses are uncertain for several reasons (few
estimations, or observations made during twilight), according to de Vaucouleurs
himself.

And, thirdly, of the 47 lunar eclipses considered by de Vaucouleurs *no less
than 21 were partial eclipses*. Nine of them were of magnitude less than 0.50, and
even six had a magnitude smaller than 0.25. The 'best' values of L according to de
Vaucouleurs are indicated by an exclamation mark in Table 55.A. Yet three of them
concerned eclipses with magnitudes 0.37, 0.22, and even as small as 0.18. To me

TABLE 55.A

The 47 lunar eclipses considered by de Vaucouleurs

Date of the eclipse	L	Φ	Magni-tude	Date of the eclipse	L	Φ	Magni-tude
1894 Sep. 15	1.7 !	0.45	0.22	1917 Jan. 8	(2.0)	0.34	1.36
1895 Mar. 11	2.0 !	0.50	1.62	1917 July 4	2.0 !	0.39	1.62
1895 Sep. 4	(1.5)	0.55	1.55	1919 Nov. 7	3	0.62	0.18
1896 Feb. 28	2.6 !	0.59	0.87	1920 May 3	3.3 !	0.67	1.22
1898 Jan. 8	(2.1)	0.75	0.15	1921 Oct. 16	3.7 !	0.81	0.93
1898 July 3	3.0 !	0.80	0.93	1923 Mar. 3	3.8 !	0.95	0.37
1898 Dec. 27	2.9 !	0.85	1.38	1924 Feb. 20	1.0 !	0.06	1.60
1899 Dec. 17	3.4 !	0.93	0.99	1924 Aug. 14	2.4 !	0.11	1.65
1902 Apr. 22	2.3 !	0.05	1.33	1925 Feb. 8	2.6 !	0.16	0.73
1902 Oct. 17	(2.5)	0.09	1.46	1927 Dec. 8	2.7 !	0.44	1.35
1903 Apr. 12	0.4 *	0.13	0.97	1931 Apr. 2	3.4 !	0.76	1.50
1903 Oct. 6	0 *	0.18	0.86	1931 Sep. 26	3.5 !	0.80	1.32
1905 Feb. 19	1.3 *	0.29	0.41	1932 Sep. 14	3.4 !	0.90	0.97
1905 Aug. 15	1.5 *	0.33	0.29	1934 Jan. 30	(0.5)	0.04	0.11
1906 Feb. 9	2.5 !	0.37	1.63	1935 Jan. 19	0.8 !	0.14	1.35
1906 Aug. 4	2	0.41	1.78	1935 July 16	(1)	0.18	1.75
1907 July 25	3	0.49	0.62	1936 Jan. 8	1.6 !	0.23	1.02
1909 June 4	3.5 !	0.65	1.16	1937 Nov. 18	(1.5)	0.32	0.15
1909 Nov. 27	(3)	0.69	1.37	1938 Nov. 7	2.6 !	0.42	1.35
1910 Nov. 17	3.4 !	0.77	1.13	1942 Mar. 3	2.8 !	0.84	1.56
1912 Apr. 1	3.7 !	0.89	0.18	1942 Aug. 26	2.8 !	0.88	1.53
1913 Mar. 22	0.7 *	0.97	1.57	1943 Feb. 20	3.6 !	0.93	0.76
1913 Sep. 15	(1) *	0.01	1.43	1943 Aug. 15	3.5 !	0.98	0.87
1914 Mar. 12	2.0 !	0.06	0.91				

it seems dubious that trustful values of L could be estimated in the case of such small partial eclipses.

If one suppresses the partial eclipses with a magnitude less than 0.50, and those which according to de Vaucouleurs are dubious or marked by an asterisk in the table, then only 29 eclipses out of the 47 remain. In the $\Phi-L$ diagram, these 29 eclipses are distributed as shown in Figure 55.c. The 'law of Danjon' now no longer is so evident, and the best straight line through the 29 points has the equation $L = 1.6 + 2.0\,\Phi$. We note that the coefficient of Φ has decreased a little. According to this new formula, L has a mean value of 1.6 at the beginning of a sunspot cycle, increasing to 3.6 towards the end of that cycle, or a variation in brightness a little smaller than the limits 1.3 and 3.7 found by de Vaucouleurs.

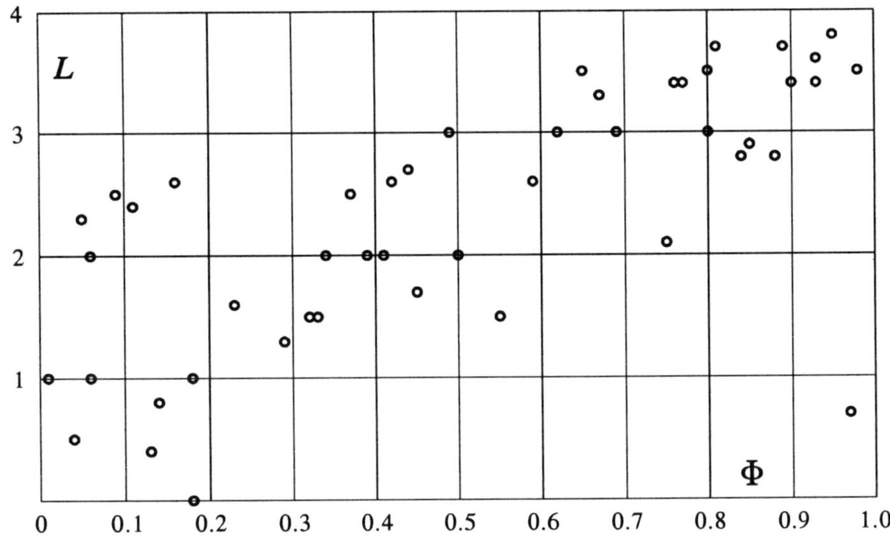

Fig. 55.b

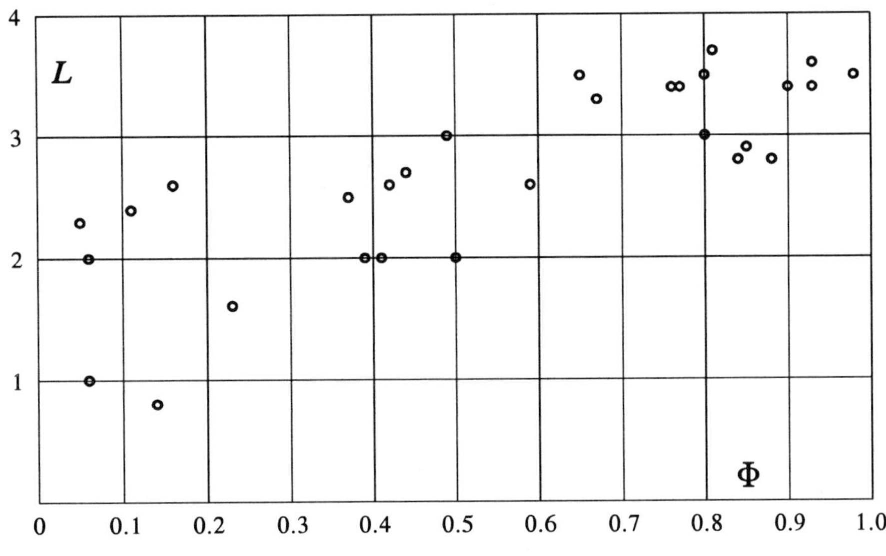

Fig. 55.c

I have made a similar analysis for the lunar eclipses which took place from 1960 to mid-1996. This time, only total eclipses were taken into consideration. The data are given in Table 55.B. Instead of the phase Φ of the sunspot cycle, the number m of months elapsed since the time of the last sunspot minimum is given. The value of the Moon's luminosity L in the Danjon scale has been taken from *Sky and Telescope*, and the reference is mentioned in the last column of the table: year, month, and page of the issue.

TABLE 55.B

Date of the eclipse	L	m	Reference (Sky and Telescope)
1960 Mar. 13	1.9	71	1968 June / 352
1960 Sep. 5	1.8	77	1968 June / 352
1963 Dec. 30	0.2	116	1964 March / 143, 1965 Feb. / 76
1964 June 25	0.0	122	1964 Aug. / 105
1964 Dec. 19	1.6	4	1965 Feb. / 76
1967 Apr. 24	2.0 ?	32	1967 July / 53, 1968 June / 352
1967 Oct. 18	3.2	38	1967 Dec. / 408
1968 Apr. 13	2.3	44	1968 June / 352
1968 Oct. 6	1.6	50	1968 Dec. / 413
1971 Feb. 10	2.7	78	1971 May / 274
1971 Aug. 6	2.0 ?	84	1971 Oct. / 243
1972 Jan. 30	2.9	89	1972 Apr. / 260
1974 Nov. 29	2.5 ?	123	1975 Feb. / 130
1975 May 25	1.6	129	1975 Oct. / 222
1975 Nov. 18	2.9	135	1976 Feb. / 77
1978 Mar. 24	2.1	24	1978 Aug. / 168
1978 Sep. 16	2.5	30	1978 Dec. / 578
1979 Sep. 6	3	42	1980 Jan. / 31
1982 Jan. 9	3	70	1982 Apr. / 426
1982 July 6	0.5	76	1982 Oct. / 390
1982 Dec. 30	0	81	1983 Mar. / 288
1986 Oct. 17	2	1	1987 Feb. / 224
1989 Aug. 17	2	35	1989 Nov. / 549
1992 Dec. 9	0.5	75	1996 Sep. / 101
1993 June 4	3	81	1993 Nov. / 102
1993 Nov. 29	1.7	86	1996 Sep. / 101
1996 Apr. 4	2.2	115	1996 Sep. / 101

At the eclipse of 1982 July 6, because of the wide variation of illumination across the Moon at mid-totality, this event was rather difficult to rate. The majority of reports were similar to that of Robert Hays, who gave the northern half of the lunar disk a rating of 0, and the southern half a 1. In the table, I adopted for this eclipse the mean value 0.5 for L.

Five total eclipses in the period 1985-1990 are not mentioned in the list, because for them no value of L is known to me.

The 27 eclipses of Table 55.B are represented in Figure 55.*d*. Here we no longer see any indication in favor of Danjon's law!

Five of these eclipses were very dark, with L-values ranging from 0 to 0.5. These eclipses occurred shortly after the large volcanic eruptions of Mount Agung, Bali, Indonesia (March 1963), of El Chichón, Mexico (April 1982), and of Mount Pinatubo, Philippines (June 1991). If we discard these five abnormally dark eclipses, we find by the method of least squares that the best straight line trough the remaining 22 points has the equation $L = 2.16 + 0.00202\,m$. This corresponds to $L = 2.16 + 0.27\,\Phi$ where, as before, Φ is the phase of the current sunspot cycle. So here we find for the coefficient of Φ a value *much* smaller than that obtained by de Vaucouleurs. The best straight line, not represented in the diagram, is now nearly horizontal. From the observational data of the years 1960-1996, we thus are lead to conclude that the brightness of the eclipsed Moon does *not* depend on the sunspot activity, or that in any case the relation found by Danjon is very dubious.

Of course, it would be necessary to continue to estimate the brightness of the totally eclipsed Moon during many more years, so that a really definitive conclusion can be drawn.

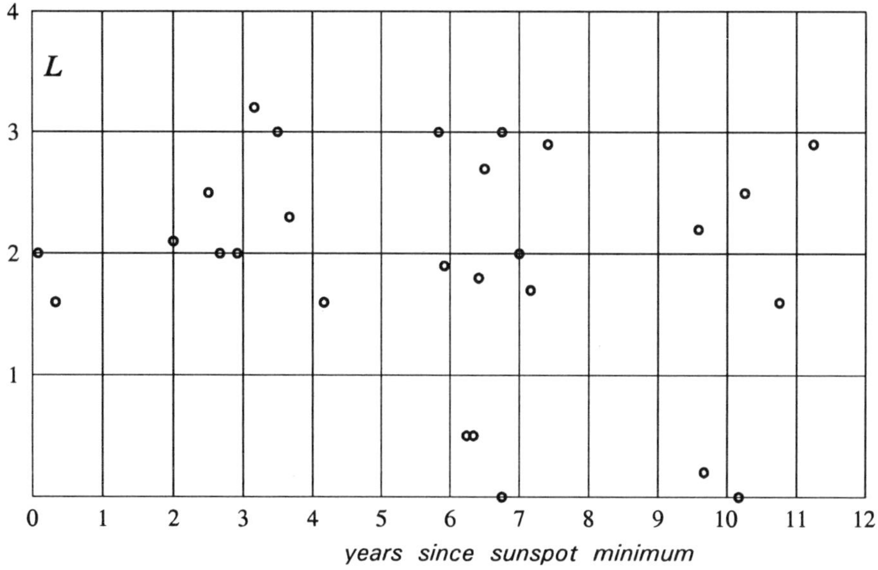

Fig. 55.d

The brightness estimates of the lunar eclipses 1960–1996 give no indication in favor of the relation stated by Danjon.

VARIA

56. *The equation of time*

Former versions of the following text were published in the Dutch journal *Hemel en Dampkring*, Vol. 68, No. 2, pages 21-27 (February 1970) and in *L'Astronomie* (Société Astronomique de France), Vol. 109, pages 188-193 (June 1995).

In ancient astronomy, the word 'equation' designated a correction to be added algebraically to a mean value in order to obtain a true value. For example, Ptolemy as well as Copernicus used an 'equation of the center', a quantity to be added to the mean longitude of a planet in order to obtain its true longitude, and this expression is still used in modern astronomy. Hence, in ancient astronomy, the equation of time was the correction to be applied to the mean solar time to obtain the true solar time.

Definition

The time interval between two successive upper transits of the Sun's center through the meridian of a given place is called a true solar day. This interval is not constant, but varies between the extremes 23 hours 59 minutes 39 seconds and 24 hours 00 minute 30 seconds (in units of mean time).

For example, on 1995 December 25 at the Paris Observatory the transit of the Sun took place at $11^h50^m32^s$ UT. The next day, it occurred at $11^h51^m02^s$, so the true solar day had a length of 24 hours 00 minute 30 seconds. But let us now consider the transits of the Sun on 1995 August 15 and 16, again for Paris: the instants were $11^h55^m11^s$ and $11^h54^m59^s$, so the true solar day had a length of 23 hours 59 minutes 48 seconds in this case.

The daily increase of the right ascension of the Sun is not constant for the following two reasons:

1. The Sun does not move along the celestial equator, but along the ecliptic. From this it results that the daily increase of the Sun's right ascension is the *projection* on the equator of the increase of the Sun's celestial longitude.

Suppose, for the sake of simplification, that the Sun moves every day over exactly 1 degree along the ecliptic. When the Sun is close to the vernal equinox, the projection $A'B'$ of its displacement AB is smaller than $1°$ — see Figure 56.*a*. On the other hand, near the solstitial points the projection $C'D'$ of the displacement CD is larger than $1°$. As a comparison: on the Earth's globe an arc of one degree in longitude along the equator corresponds to a distance of 111 kilometers, but if in Alaska you travel 111 km towards the east or the west (which corresponds to one degree on a *great circle*), then your longitude will increase more than 1 degree.

2. The Sun's speed along the ecliptic is not constant. This speed is largest in early January, when the Earth is near perihelion.

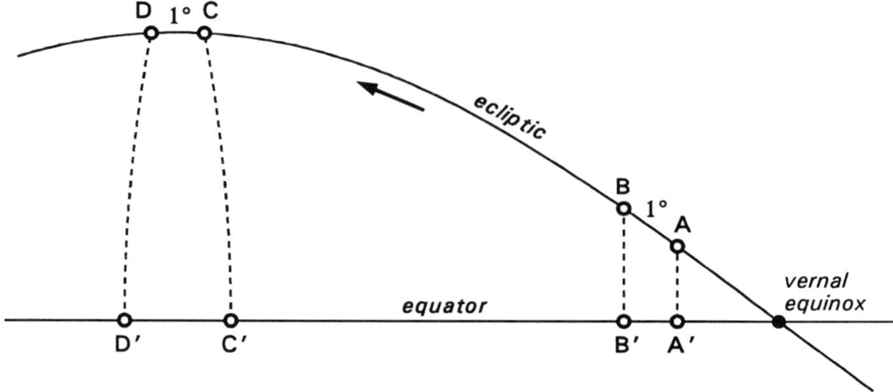

Fig. 56.a : *When the ecliptical longitude of the Sun increases by
one degree, the right ascension increases by a variable amount.*

To avoid the inconveniences resulting from the inequal true solar days, the
mean time based on the consideration of the mean solar day has been conceived.

Consider a first fictitious Sun travelling along the *ecliptic* at a constant speed
and coinciding with the true Sun at the perigee and the apogee (when the Earth is
in perihelion and aphelion, respectively). Then consider a second fictitious Sun
travelling along the *celestial equator* at a constant speed and coinciding with the first
fictitious Sun at the equinoxes. This second fictitious Sun is called the *mean Sun*,
and by definition its right ascension is equal to the longitude of the first fictitious
Sun. From this it results that the right ascension of the mean Sun increases at a
uniform rate.

The interval between two successive meridian transits of the mean Sun will
also be constant. This interval is called the *mean solar day.*

For a given place, the true (or apparent) solar time at a given instant is the
hour angle of the true Sun, while the mean time is the hour angle of the mean Sun,
the second fictitious Sun which has just been defined.

At a given instant, the *equation of time* is the difference between apparent
and mean solar time. In other words, it is the difference between the hour angles
of the true and the mean Sun.

The equation of time in the course of the year

The equation of time slowly varies from one day to the next, and this
variation repeats almost identically from one year to the next. It is the sum of two
components, corresponding to the two effects mentioned before :

— the first component, due to the fact that the Sun moves in the ecliptic and not

along the celestial equator, has a period of six months. It is zero four times a year: at the equinoxes and at the solstices. Its extreme values are -9.87 and $+9.87$ minutes (for the value of the obliquity of the ecliptic in A.D. 2000). This term is illustrated by curve *a* in Figure 56.*b*;
— the second component, due to the fact that the Sun's motion along the ecliptic is not uniform, has a period of one year. It is zero twice a year, namely when the Earth is in perihelion and in aphelion. Its extreme values are -7.66 and $+7.66$ minutes (for the value of the eccentricity of the Earth's orbit in A.D. 2000). This component is illustrated by curve *b* in Figure 56.*b*.

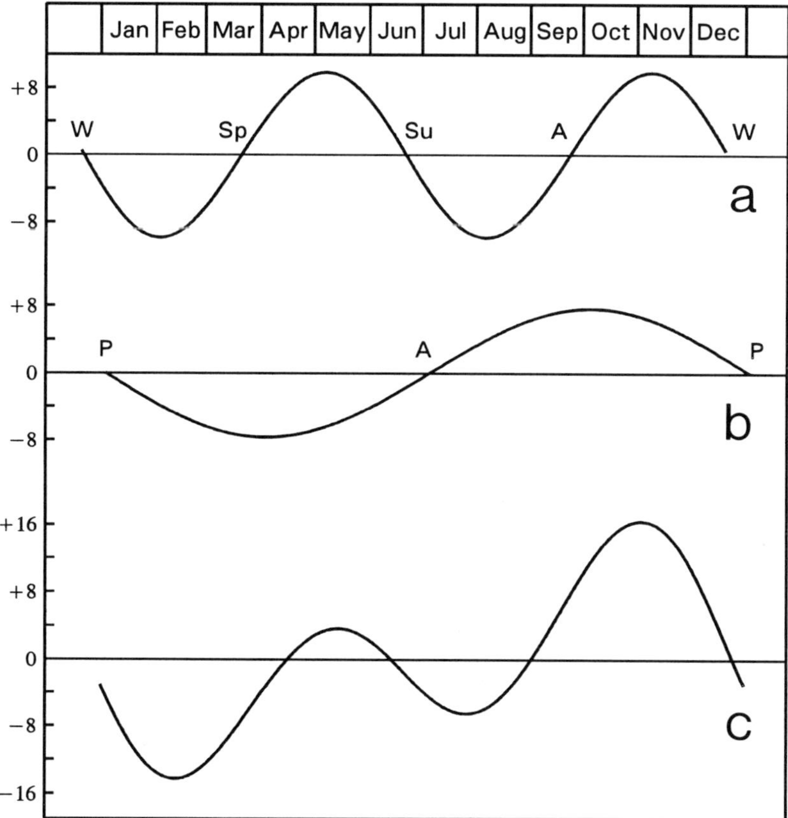

Fig. 56.b : *The equation of time in the course of the year. The upper curve (a) shows the effect of the obliquity of the ecliptic on the equator; W, Sp, Su and A are the instants of the beginning of the astronomical seasons (winter, spring, summer, and autumn). The middle curve (b) shows the effect of the eccentricity of the Earth's orbit; P = perihelion, A = aphelion. The lower curve (c) is the combined effect, which is the equation of time. The scale at left is in minutes of time.*

The two undulating curves a and b are symmetric, but their amplitudes and their periods differ. Moreover, they are out of phase: near A.D. 2000, curve b becomes zero (in the perihelion) two weeks after curve a (at the winter solstice). Consequently, their sum (curve c) is not symmetric with respect to the 'zero' line.

The equation of time is zero four times a year. And four times a year it reaches a maximum or a minimum value. Nowadays these particular values occur on the dates mentioned in the following table.

TABLE 56.A

*Zero and extreme values of the equation of time
near A.D. 2000*

minimum	$-$ 14 *min.* 15 *sec.*	February 11
zero	0	April 15
maximum	$+$ 3 *min.* 41 *sec.*	May 14
zero	0	June 13
minimum	$-$ 6 *min.* 30 *sec.*	July 26
zero	0	September 1
maximum	$+$ 16 *min.* 25 *sec.*	November 3
zero	0	December 25

We note that the most important component of the equation of time is that which is due to the obliquity of the axis of rotation of the Earth. Consequently, even if the Earth's orbit was exactly circular, the equation of time would still exist: then there still would be four extreme values each year, and the equation of time would be zero exactly at the equinoxes and at the solstices.

Calculation of the equation of time

Because the elements of the Earth's orbit slowly vary, the curve representing the yearly variation of the equation of time will take different shapes in the course of the centuries. How can such a curve be calculated for a given year? Here we shall neglect the small irregularities due to the nutation and the planetary perturbations.

From what has been said, the equation of time E is obtained by subtracting the right ascension α of the true Sun from the right ascension of the mean Sun, that is, from the longitude L of the first fictitious Sun: $E = L - \alpha$.

Let λ be the ecliptical (celestial) longitude of the true Sun, and C the equation of the *center* of the Sun. The equation of the center is the difference between the longitude λ of the true Sun and the longitude L of the mean Sun, that is, $C = \lambda - L$. Consequently, we have

$$E = L - \alpha = \lambda - C - \alpha = (\lambda - \alpha) - C$$

When the elements of the orbit of the Earth are known, it is possible to calculate the variation of the equation of time in the course of the year as follows. Let ε be the obliquity of the ecliptic (that is, the inclination of the axis of rotation of the Earth), e the eccentricity of the Earth's orbit, and π the longitude of the Sun's perigee measured from the vernal equinox, or the longitude of the Earth's perihelion increased by 180 degrees.

Let us take a sequence of arbitrary values for the mean anomaly M of the Sun, for instance at intervals of $5°$. The mean anomaly is the difference between the mean longitude of the Sun and the longitude of the perigee, that is, $M = L - \pi$. Let us write, for short,

$$U_1 = 2e - e^3/4 \qquad U_2 = 5e^2/4 \qquad U_3 = 13e^3/12$$

Then the equation of the center can be expressed as a series in M:

$$C = U_1 \sin M + U_2 \sin 2M + U_3 \sin 3M + \text{negligible terms.}$$

The result is expressed in *radians*. It can be converted into degrees by multiplying it by $180/\pi$, or 57.29578. Then the true longitude of the Sun is $\lambda = \pi + M + C$.

The right ascension α of the Sun can be calculated by means of the formula $\tan \alpha = \cos \varepsilon \tan \lambda$, but in the present case it is easier to calculate directly the difference $\lambda - \alpha$ by another series expression. Let $y = (\tan \varepsilon/2)^2$. Then we have

$$\lambda - \alpha = y \sin 2\lambda - \frac{1}{2} y^2 \sin 4\lambda + \frac{1}{3} y^3 \sin 6\lambda + \text{negligible terms}$$

where, again, the result is expressed in radians. And finally, as has been said above, $E = (\lambda - \alpha) - C$. If this result is expressed in degrees, it should be multiplied by 4 to convert it into minutes of time.

Performing this calculation for several values of the mean anomaly M from 0 to 360 degrees, we obtain the variation of E in the course of the year.

In some books, the development of E is restricted to the first two terms only, namely $y \sin 2\lambda - 2e \sin M$. However, in the series expression for $\lambda - \alpha$, the coefficient of $\frac{1}{2}y^2$ has the value 12.7 seconds of time, which cannot be neglected. And in the series for C, the term $5e^2/4$ is equal to 4.8 seconds. Hence, when the equation of time in calculated by this method, one should at least take into account the terms of the second order in the series for C and $\lambda - \alpha$ when one wishes to obtain a result accurate to the second of time.

A last question must be solved, however. To what dates of the year do the various values of M correspond? One should note that M varies linearly with time, that is, it has a uniform motion in the course of the year, as it increases every day by 0.985 600 282 degree.

The concordance with the dates can be found by considering that $\lambda = 0°$ at the instant of the March equinox. However, for the sake of uniformity, I have assumed in the calculation that the beginning of the year corresponds to instant when $M = 280° - \pi$. This is the instant when the right ascension of the second fictitious Sun is equal to 280°, which, by definition is the beginning of the so-called Besselian year, or the beginning of the 'tropical year'. This instant differs little from the beginning of our Gregorian years, and it could be considered as the start of the year in an 'ideal' calendar.

Table 56.B gives the values of ε, π, e and $280° - \pi$ for the beginning of the years -2000 to $+5000$, at intervals of 1000 years. The values of ε have been calculated by means of Laskar's formula (*Astronomy and Astrophysics*, Vol. 157, page 68; 1986), while those of π and e have been found by using the formulae by Simon *et al.* (*Astronomy and Astrophysics*, Vol. 282, page 678; 1994).

We note that presently both the obliquity of the ecliptic and the orbital eccentricity are decreasing. Consequently, on the average the equation of time is decreasing in the course of the centuries. This change is not big, however: in the year $+5000$ ε will still be equal to 96.4 percent of the value it had in the year -2000, while for e it is 84.4 percent.

TABLE 56.B

Values of the parameters needed for the calculation of the equation of time

Year	ε (degrees)	π (degrees)	e	$280° - \pi$ = value of M on January 1.0 in the 'ideal' year
-2000	23.92409	214.8819	0.0181777	65.1181
-1000	23.81438	231.7623	0.0178517	48.2377
0	23.69488	248.7305	0.0174975	31.2695
$+1000$	23.56876	265.7884	0.0171162	14.2116
$+2000$	23.43929	282.9373	0.0167086	357.0627
$+3000$	23.30981	300.1776	0.0162757	339.8224
$+4000$	23.18362	317.5086	0.0158184	322.4914
$+5000$	23.06388	334.9294	0.0153374	305.0706

But it is for the longitude of the perigee that the changes are greatest. During the period of 70 centuries considered here, π increases by 120 degrees. Now, a change of the value of π results in a shift of point P with respect to point W in Figure 56.*b*, with the consequence that the combined effect (curve *c*) will take various shapes over time. *It is the shift of the Sun's perigee with respect to the vernal equinox which is the main cause of the evolution of the curve of the equation of time in the course of the centuries.*

I leave it to the interested reader to make the calculation. My results are given in Table 56.C. The values of M are in degrees. Use is made of the 'ideal' year described above. For this reason, in the table no discontinuity is seen when passing from the Julian to the Gregorian calendar in 1582.

TABLE 56.C

The extreme values of the equation of time through the centuries

Year	Minimum	Maximum	Minimum	Maximum
-2000	$-18\ m\ 33\ s$ $M = 95.4$ January 31	$+12\ m\ 45\ s$ $M = 203.0$ May 20	$-2\ m\ 06\ s$ $M = 283.3$ August 10	$+9\ m\ 30\ s$ $M = 359.3$ October 26
-1000	$-18\ m\ 18\ s$ $M = 81.4$ February 3	$+10\ m\ 14\ s$ $M = 186.3$ May 21	$-2\ m\ 06\ s$ $M = 262.2$ August 6	$+11\ m\ 45\ s$ $M = 343.6$ October 27
0	$-17\ m\ 27\ s$ $M = 67.3$ February 6	$+7\ m\ 44\ s$ $M = 168.5$ May 20	$-2\ m\ 57\ s$ $M = 241.0$ August 1	$+13\ m\ 45\ s$ $M = 328.6$ October 29
$+1000$	$-16\ m\ 04\ s$ $M = 53.1$ February 9	$+5\ m\ 27\ s$ $M = 149.3$ May 18	$-4\ m\ 30\ s$ $M = 220.5$ July 29	$+15\ m\ 20\ s$ $M = 314.0$ November 1
$+2000$	$-14\ m\ 15\ s$ $M = 38.4$ February 11	$+3\ m\ 41\ s$ $M = 129.0$ May 14	$-6\ m\ 30\ s$ $M = 201.0$ July 26	$+16\ m\ 25\ s$ $M = 299.6$ November 3
$+3000$	$-12\ m\ 08\ s$ $M = 23.2$ February 14	$+2\ m\ 37\ s$ $M = 107.8$ May 10	$-8\ m\ 41\ s$ $M = 182.6$ July 25	$+16\ m\ 57\ s$ $M = 285.2$ November 6
$+4000$	$-9\ m\ 52\ s$ $M = 7.1$ February 15	$+2\ m\ 24\ s$ $M = 86.4$ May 6	$-10\ m\ 48\ s$ $M = 165.3$ July 25	$+16\ m\ 54\ s$ $M = 270.6$ November 9
$+5000$	$-7\ m\ 38\ s$ $M = 349.9$ February 15	$+3\ m\ 00\ s$ $M = 65.4$ May 3	$-12\ m\ 38\ s$ $M = 148.8$ July 26	$+16\ m\ 17\ s$ $M = 255.7$ November 12

Fig. 56.c : *The curve of the equation of time at intervals of 1000 years, from −2000 to +5000. For each of the curves, the scale is given at left, in minutes of time. The horizontal line corresponds to E = 0. The tics on this axis divide the year in four quarters: January 1 is at left, and December 31 at right.*

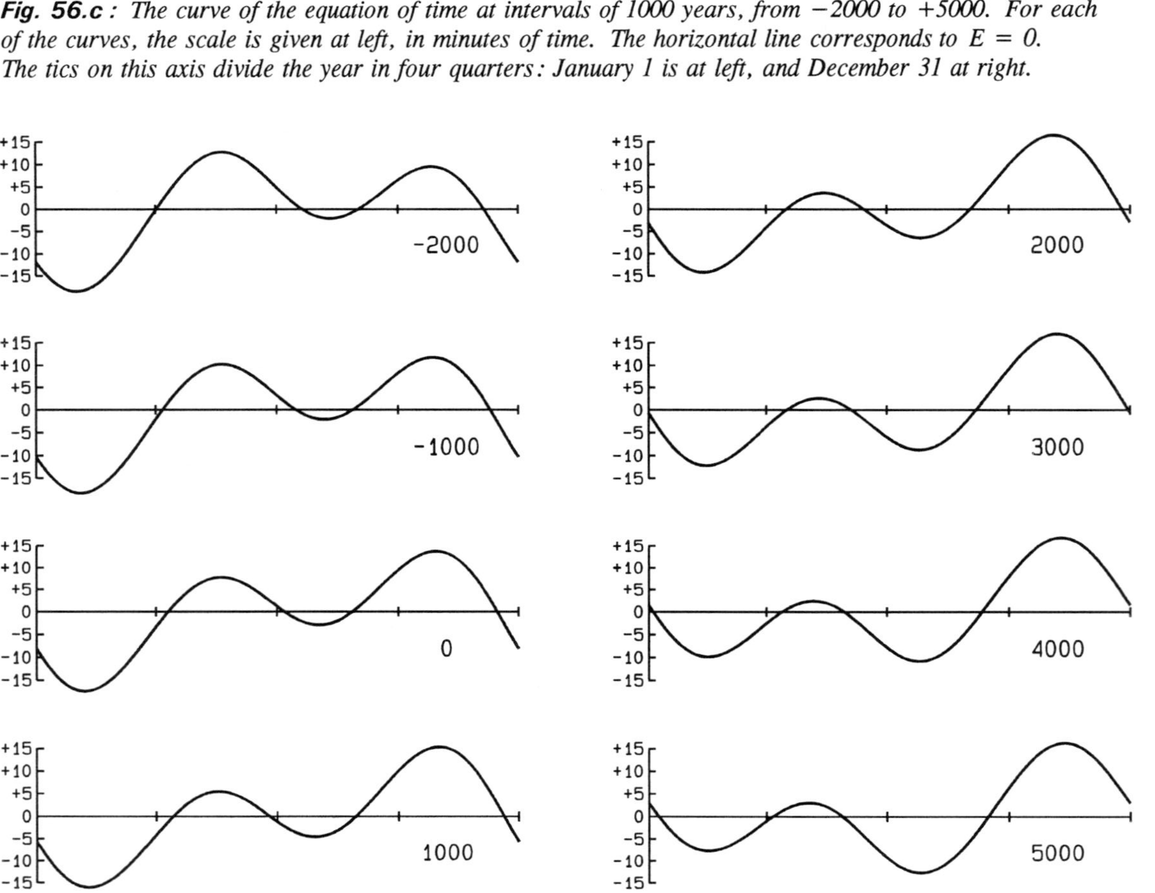

My results are illustrated in Figure 56.*c*. Before the year −1320, the maximum of May was higher than that of October-November. The latter will continue to increase a little, in spite of the decrease of both the obliquity of the ecliptic and the orbital eccentricity, to reach its largest value about A.D. 3400, namely + 17 minutes 00 second. The minimum of July-August, which around −1500 was equal to only − 1 minute 59 seconds, is becoming more pronounced.

In the year +1246 we had $\pi = 270°$, so the perigee *P* of the Sun coincided with the winter solstitial point *W* of the ecliptic. At that epoch, the curve of the equation of time was exactly symmetric. The minimum of February (− 15 minutes 39 seconds) was exactly equal, in absolute value, to the maximum of November (+ 15 minutes 39 seconds), and the maximum of May (+ 4 minutes 58 seconds) was equal to the July minimum (− 4 minutes 58 seconds).

On 1996 November 2, the equation of time was equal to + 16 minutes 26.3 seconds, its greatest value since more than 1000 years.

During the period of 70 centuries considered here, the equation of time reaches each year *two* maximum and *two* minimum values. This is not always the case, however. The obliquity of the ecliptic and the eccentricity of the Earth's orbit cannot have any values over time; they remain between limits they cannot exceed. According to A. Berger ("Milankovitch theory and climate", *Review of Geophysics*, Vol. 26, No. 4, pages 624-657) these limits are 21°59′ and 24°36′ for the obliquity, and 0 and 0.06939 for the eccentricity. By means of the method of calculation described above, one finds that even for the largest possible value of the obliquity, 24°36′, the equation of time reaches during the year only one maximum and one minimum as soon as the orbital eccentricity of the Earth is larger than 0.047, but only for some values of π. Two examples are shown in Figure 56.*d*.

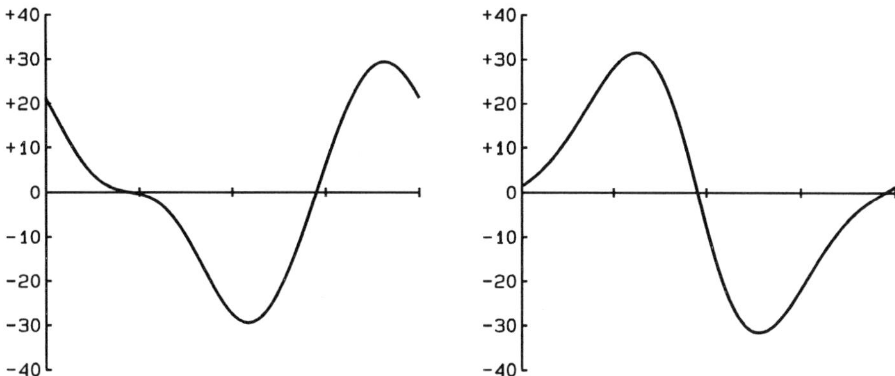

Fig. 56.d : *When the eccentricity of the orbit of the Earth is larger than 0.047, for some values of the longitude of the Sun's perigee π the equation of time may reach only one maximum and one minimum in the course of the year. Two examples are given here. At left is the curve of the equation of time for $\varepsilon = 24°00′$, $e = 0.05$, and $\pi = 0°$. At right, the curve for $\varepsilon = 23°00′$, $e = 0.06$, and $\pi = 90°$.*

The duration of the true solar day

As we have seen, the variation of the equation of time causes a variation of the length of the true solar day. With the present values of the obliquity of the ecliptic, the eccentricity of the orbit of the Earth and the longitude of the perigee of the Sun, the true solar day can exceed the mean solar day by up to 30 seconds. This happens near the time of the December solstice. On the other hand, near the equinoxes the true solar day is smaller than the mean one by about 20 seconds.

Four times a year, the length of the true solar day is exactly 24 hours of mean time. This does *not* occur when the equation of time becomes zero, but rather when the *variation* of the equation of time is zero. For the year 1998, these particular values of the length of the true solar day are listed in the following table.

TABLE 56.D

Variation of the length of the true solar day in the course of 1998

zero	24 hours exactly	February 11
minimum	24 hours − 18.1 seconds	March 26
zero	24 hours exactly	May 14
maximum	24 hours + 13.1 seconds	June 19
zero	24 hours exactly	July 26
minimum	24 hours − 21.3 seconds	September 16
zero	24 hours exactly	November 3
maximum	24 hours + 29.9 seconds	December 22

57. *About the equinoxes and the solstices*

Many people think that by definition spring begins on March 21, summer on June 21, autumn on September 21, and winter on December 21, just as by definition Christmas falls on December 25, or the day of Saint Germanus is May 28.

In fact, when UT dates are considered, the astronomical spring can start as well on March 20 (as in 1997 and 1998) as on March 21 (in 1999), the autumn on September 23 (1998 and 1999), etc., as we shall see.

By definition, the times of the equinoxes and solstices, and hence of the beginning of the astronomical seasons, are the instants when the apparent geocentric longitude of the Sun (that is, calculated by including the effects of aberration and nutation) is an integer multiple of 90°. Because the latitude of the Sun, though very small, is not zero, the declination of the center of the solar disk is not exactly zero at the time of an equinox.

The displacement after 1, 4, 400 years

The duration of the tropical year is 365.24219 days, or 365 days 5 hours 49 minutes. This is the time needed for the mean Sun (as defined in the preceding chapter) to increase its longitude by 360°. Consequently, after one common year of 365 days the seasons begin, on the average, 5 hours 49 minutes later in the year. To correct for this shift, an additional day (bissextile or leap day, February 29) is introduced every four years. Hence, after four years there is again a jump of one day backwards.

However, the tropical year is a little shorter than 365.25 days. While four calendar years have a total duration of 365 + 365 + 365 + 366 = 1461 days, four tropical years have 4 × 365.24219 = 1460.9688 days. Hence, after four years the equinoxes and the solstices occur approximately 0.0312 day, or 45 minutes, earlier.

The shift of +6 hours after one year, and that of approximately −45 minutes after four years, are illustrated in Figure 57.*a* and in Table 57.A.

The shift of −0.0312 day after four years would add up to a shift of −3.12 days after 400 years. To correct for this, the well-known rule has been introduced at the Gregorian calendar reform: the century years exactly divisible by 400 (such as 1600 and 2000) shall be leap years and contain 366 days, while the other century years (1700, 1800, 1900, 2100, ...) shall be common years of 365 days.

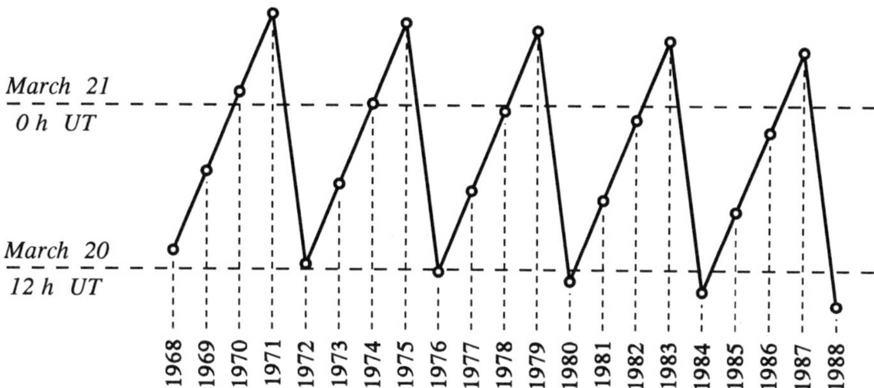

Fig. 57.a : *The instant of the March equinox from 1968 to 1988.*
The jump every four years is due to the bissextile day, February 29.

TABLE 57.A

Instants of the March equinox, 1980 to 1988

Year	March	Universal Time	Difference = 365 days 5 hours +
		h m	
1980	20	11 10	53 minutes
1981	20	17 03	53
1982	20	22 56	43
1983	21	4 39	45
1984	20	10 24	50
1985	20	16 14	49
1986	20	22 03	49
1987	21	3 52	47
1988	20	9 39	

After 400 years there still remains an error of $-3.12 + 3 = -0.12$ day. Consequently, after four centuries the astronomical seasons start 3 hours earlier in the year. This is a mean value, however, as can be seen from the following example where the instants are given in the uniform time-scale known as Dynamical Time:

Year	Equinox March		Solstice June		Equinox Sept.		Solstice Dec.	
1598	20	21^h12^m	21	22^h26^m	23	9^h42^m	21	23^h11^m
1998	20	19 56	21	14 04	23	5 38	22	1 58
2398	20	18 38	21	5 32	23	0 33	22	3 46
2798	20	18 07	20	21 33	22	18 40	22	5 22
Mean shift:	-1^h02^m		-8^h18^m		-5^h01^m		$+2^h04^m$	

We see that after a period of 400 years, while the March equinox, the June solstice and the September equinox occur 1, 8 and 5 hours earlier in the year, respectively, the winter solstice takes place 2 hours *later*.

The reason for these differences is that after a period of 400 years the eccentricity and the longitude of the perihelion of the Earth's orbit have changed sensibly, causing a change of the duration of *each* of the four seasons (though their sum remains equal to that of the tropical year).

Near A.D. 2000, the duration of the astronomical spring is 92.76 days, that of the summer 93.65 days. By the year 3000, the length of the spring will have decreased to 91.97 days, but that of the summer will have increased to 93.92 days.

Due to the variation of the lengths of the seasons, the times of the equinoxes and solstices cannot 'keep still' in the year. These instants must inevitably oscillate — with a period of many centuries — even in a calendar which would be perfectly adjusted to the tropical year. Near the year 2000 we have the following mean intervals:

between two successive

March equinoxes:	365 days 5 hours 49 minutes 01 second
June solstices:	365 days 5 hours 47 minutes 56 seconds
September equinoxes:	365 days 5 hours 48 minutes 30 seconds
December solstices:	365 days 5 hours 49 minutes 33 seconds

However, the mean of these *four* time intervals is exactly equal to the length of the tropical year, 365 days 5 hours 48 minutes 45 seconds.

Irregularity

From the last column of Table 57.A it appears that even during a short period the intervals between two successive March equinoxes are not equal. There are similar irregularities for the beginning of the other three astronomical seasons. These variations are due to planetary perturbations and the nutation.

The motion of the Earth in its orbit around the Sun is somewhat perturbed by the attraction of the planets (mainly Venus, Mars and Jupiter) and by that of the Moon. Due to these perturbations the Earth is now somewhat behind, now somewhat ahead of its 'normal', unperturbed position. The total effect due to the Moon and the planets can amount to 30" in heliocentric longitude. The Earth needs 12 minutes to travel over this distance. Hence, due to this effect an equinox or a solstice can occur up to 12 minutes earlier or later than normally, and the effect will of course be ever different from one year to the next.

The axis of the Earth oscillates around its mean position, a variation called the nutation. Due to the nutation, the true vernal equinox oscillates back and forth with respect to the mean equinox. The principal term of this variation has an amplitude of 17.2 arcseconds and a period of 18.6 years. There are smaller periodic terms, among which one with an amplitude of 1".3 and a period of six months. So, due to the nutation the true vernal equinox can be as much as 19" behind or ahead of the mean vernal point. This causes the seasons to begin up to 8 minutes earlier or later than normally.

Due to the two effects (planetary perturbations and nutation) the equinoxes and the solstices can occur up to 20 minutes earlier or later than normally would be the case.

TABLE 57.B

Times of the March equinoxes, 1971 to 2000. Dynamical Time is used.

Date	True	Mean	Difference (seconds)
	h m s	h m s	
1971 March 21	6 38 51	6 44 42	−351
1972 March 20	12 22 08	12 33 42	−694
1973 March 20	18 13 12	18 22 43	−571
1974 March 21	0 07 24	0 11 44	−260
1975 March 21	5 57 27	6 00 45	−198
1976 March 20	11 50 24	11 49 46	+ 38
1977 March 20	17 43 05	17 38 47	+258
1978 March 20	23 34 23	23 27 48	+395
1979 March 21	5 22 47	5 16 49	+358
1980 March 20	11 10 33	11 05 50	+283
1981 March 20	17 03 43	16 54 51	+532
1982 March 20	22 56 44	22 43 52	+772
1983 March 21	4 39 39	4 32 52	+407
1984 March 20	10 25 15	10 21 53	+202
1985 March 20	16 14 39	16 10 54	+225
1986 March 20	22 03 38	21 59 55	+223
1987 March 21	3 52 54	3 48 56	+238
1988 March 20	9 39 33	9 37 57	+ 96
1989 March 20	15 29 12	15 26 58	+134
1990 March 20	21 20 14	21 15 59	+255
1991 March 21	3 02 55	3 05 00	−125
1992 March 20	8 49 02	8 54 02	−300
1993 March 20	14 41 38	14 43 03	− 85
1994 March 20	20 29 01	20 32 04	−183
1995 March 21	2 15 27	2 21 05	−338
1996 March 20	8 04 07	8 10 06	−359
1997 March 20	13 55 42	13 59 07	−205
1998 March 20	19 55 35	19 48 08	+447
1999 March 21	1 46 53	1 37 09	+584
2000 March 20	7 36 19	7 26 10	+609

Table 57.B gives, for the years 1971 to 2000, the time of the true March equinox, and that of the mean equinox. The latter has been calculated by neglecting the planetary perturbations and the nutation. Contrary to Table 57.A, the uniform Dynamical Time is used here. The difference between the two instants is mentioned

in the last column; a positive (or negative) sign indicates that the equinox occurs later (or earlier) than normally.

The effect of the 18.6-year period of the nutation is clearly seen. From 1971 to 1975 the March equinox took place 'too early', from 1976 to 1990 it occurred too late, and so on.

For the years 1960 to 2040 the differences are graphically shown in Figure 57.*b*. Here also the nutation period is clearly visible, while the 'noise' is due to the ever variable effect of the perturbations by the Moon and the planets. The largest positive difference is +996 seconds for the March equinox of 2001; the largest negative is −709 seconds, in 2032.

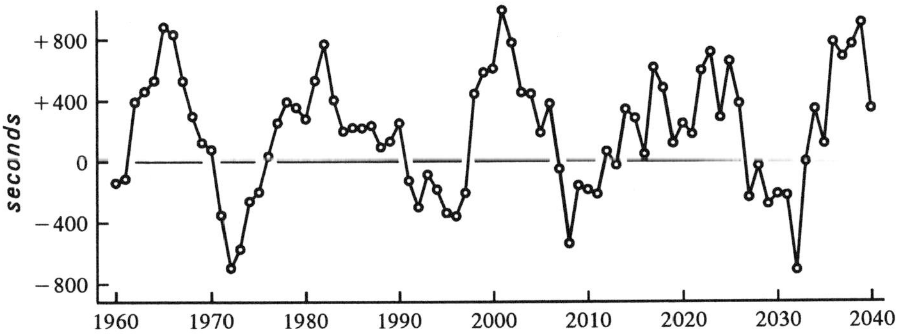

Fig. 57.b : *Difference, in seconds of time, between the true
and the mean March equinoxes, from 1960 to 2040.*

Some particular dates

Accurate times of the equinoxes and solstices for the years 1 to 3000 are given on pages 101-175 of my *Astronomical Tables of the Sun, Moon and Planets* (2nd ed., Willmann-Bell; 1995). These instants are expressed in Dynamical Time. If we use dates measured in Universal Time, the following facts are found.

During the second half of the 21st century and the end of the 22nd, spring equinox sometimes falls on March 19. For the first time since 1796, this will occur in A.D. 2044. In 2007, the equinox will take place on March 21, but then not again on this date before A.D. 2102.

In 1975 the June solstice occurred on the 22nd. Not before 2203 will it again fall on June 22. In 2008, summer will begin on June 20, the first time since 1896.

In 1968, autumn equinox took place on September 22, the first time since 1897. In the beginning of the 20th century, the equinox sometimes occurred on

September 24; this took place the last time in 1931. Not before A.D. 2303 will the astronomical autumn again begin on September 24. But, unless in the year 2307 the difference ΔT between Dynamical Time and Universal Time will be smaller than 8 minutes 25 seconds, that will be the last equinox on September 24 for many more centuries.

In 2080, for the first time since 1697, winter solstice will occur on December 20. Since the Gregorian calendar reform in 1582, the winter solstice occurred only once, in 1903, on December 23. Not before 2303 will it again take place on this date.

58. *The weekday of Christmas Day*

Does Christmas Day, December 25, occur with the same frequency on the different days of the week? The answer is *yes and no*! It depends whether a short or a long period is considered. Let us first examine the weekdays on which Christmas Day falls during a period of 28 years:

1970	Fri.	1977	Sun.	1984	Tue.	1991	Wed.
1971	Sat.	1978	Mon.	1985	Wed.	1992	Fri.
1972	Mon.	1979	Tue.	1986	Thu.	1993	Sat.
1973	Tue.	1980	Thu.	1987	Fri.	1994	Sun.
1974	Wed.	1981	Fri.	1988	Sun.	1995	Mon.
1975	Thu.	1982	Sat.	1989	Mon.	1996	Wed.
1976	Sat.	1983	Sun.	1990	Tue.	1997	Thu.

We note that after one year Christmas Day falls one day later in the week, except when there is a bissextile (leap) day between the two dates, and in this case the jump is 2 weekdays.

In this period of 28 years, Christmas occurs four times on a Sunday, four times on a Monday, etc. In other words, it falls 4 times on *each* of the seven days of the week.

We know that after a cycle of 28 years the dates of the year fall again on the same days of the week. For example, Christmas Day occurred on a Sunday in 1904, as well as in the years 1932, 1960 and 1988. The reason is that 28 is a multiple of 4 (the cycle of bissextile years), and that the number of days in 28 years, namely 10227, is a multiple of 7.

From this it results that Christmas Day, or any other date of the year, occurs *with the same frequency* on each of the seven days of the week. *But . . .* this holds only as long as the above-mentioned cycle of 28 years continues without interruption. Hence, the rule is valid:

— in the Julian calendar (that is, before A.D. 1582);
— in the Gregorian calendar only as long as there is no century year which is not a bissextile year, such as 1700, 1800, 1900, 2100.

Consequently, the above-mentioned 'rule' (that the Christmas weekdays all occur with the same frequency) holds in our present Gregorian calendar *for a short term only*, as long as no no-bissextile-century-year interferes. It is precisely these century years (1700, 1800, 1900; 2100, 2200, 2300; 2500, . . .) which perturb the game because they interrupt the periodicity of 28 years. For example, Christmas occurs on a Tuesday in the years 1979, 2007, 2035, 2063, and 2091; but 28 years later, in 2119, it will be a Monday.

In the Gregorian calendar there is a cycle of 400 years, or 146097 days, which is a multiple of 7, after which the dates of the year repeat again on the same weekdays. For example, December 25 occurs on a Sunday in 1588 as well as in 1988, 2388, 2788, and so on.

Hence, if one wishes to know *on the long run* the frequencies of the different Christmas weekdays in the Gregorian calendar, we need to investigate the problem on the base of the 400-year cycle instead of that of 28 years. Because 400 is not a multiple of 7, it is at once evident that, on the long run, the Christmas weekdays *cannot* all occur with the same frequency.

To find the frequency of the different Christmas weekdays, it is sufficient to search, for any arbitrary period of 400 successive years in the Gregorian calendar, how many times Christmas falls on a Sunday, on a Monday, etc.

In this count, much time is saved by using the cycle of 28 years as long as this periodicity remains valid. Let us, for instance, start from the year 1900. Note that in 1900 Christmas occurs *after* the non-bissextile February of that year. Counting seven times 28 years, we arrive at the year 2095. So the period 1900-2095 no longer needs to be investigated: we already know that during these 196 years Christmas falls the same number of times (28) on each of the seven days of the week. The same holds for the periods 2100-2183 and 2200-2283 (12 times for each of them). It only remains to calculate the weekday of Christmas for the remaining years, namely 2096-2099, 2184-2199, and 2284-2299.

As final result, we find that during the period 1900-2299 Christmas Day falls

58 times on a Sunday,
56 times on a Monday,
58 times on a Tuesday,
57 times on a Wednesday,
57 times on a Thursday,
58 times on a Friday,
56 times on a Saturday.

Hence, *on the long run*, in the Gregorian calendar Christmas Day occurs most frequently on a Sunday, a Tuesday, and a Friday (each of them 58 times in 400 years). The least frequent weekdays for Christmas are Monday and Saturday (56 times each per 400 years).

Similar results can be obtained for other dates of the year. For instance, the 4th of July occurs most frequently on a Monday, a Wednesday, or a Saturday.

Of course, all this is merely of theoretical interest only. For *practical* applications the differences we found can be neglected. It is an instructive problem, however, and a good lesson in reasoning.

As a final remark, let us note that if a period is considered that is not an integer multiple of 400 years, incorrect results are obtained. Because in the Gregorian calendar the periodicity of the weekdays is *exactly* 400 years, one would yield incorrect values for the frequencies if the investigation is based on a period which is not a multiple of 400 years, even if it is very long, for instance 1000 or 5000 years.

59. *The distribution of Easter Sundays*

In the *Julian* calendar, the Christian Easter date has a periodicity of 532 years. This means that after a period of 532 years the Easter Sundays repeat in the same order on the same dates of the year. So for example:

485 April 21	1017 April 21	1549 April 21
486 April 6	1018 April 6	1550 April 6
487 March 29	1019 March 29	1551 March 29
488 April 17	1020 April 17	1552 April 17
489 April 2	1021 April 2	1553 April 2
490 March 25	1022 March 25	1554 March 25
491 April 14	1023 April 14	1555 April 14
and so on.		

Table 59.A gives the frequency of the different dates of Easter in the 532-year period. We note that in the Julian calendar the two extreme possible dates, March 22 and April 25, occur four times each in a period of 532 years (second column), so this gives a mean frequency of one case every 133 years (third column).

In a period of 532 years, Easter Sunday occurs 8 times on each of the dates March 23 and 24, and April 23 and 24. Three dates (March 25, April 21 and 22) take place 12 times in a 532-year period. All other Easter dates, from March 26 to April 20 inclusively, occur either 16 or 20 times.

Of course, the sum of the numbers in the second column of the table is equal to 532, the Easter period in the Julian calendar.

TABLE 59.A

The distribution of Easter Sunday in the Julian calendar

Easter date	Number in a period of 532 years	Mean time interval in years	Easter date	Number in a period of 532 years	Mean time interval in years
March 22	4	133	April 8	20	26.6
23	8	66.5	9	16	33.25
24	8	66.5	10	16	33.25
25	12	44.33	11	20	26.6
26	16	33.25	12	16	33.25
27	16	33.25	13	16	33.25
28	20	26.6	14	20	26.6
29	16	33.25	15	16	33.25
30	16	33.25	16	20	26.6
31	20	26.6	17	16	33.25
			18	16	33.25
April 1	16	33.25	19	20	26.6
2	16	33.25	20	16	33.25
3	20	26.6	21	12	44.33
4	16	33.25	22	12	44.33
5	20	26.6	23	8	66.5
6	20	26.6	24	8	66.5
7	16	33.25	25	4	133

In what follows we shall consider the distribution of the date of Easter in the *Gregorian* calendar. Here things are more complicated. When one looks at the frequency of the different calendar dates on which Easter Sunday occurs, some remarkable facts are discovered.

Figure 59.*a* shows the frequencies as a function of the date for the 1000 successive Easter dates of the years 2000 to 2999. Note the regular occurrence of the maxima. During the third millennium, Easter occurs more often on March 31, and on April 5, 10, 16 and 21, than for instance on April 2, 7, 13 and 18. May we expect that the frequency curve will have the regular shape of the dashed line if much more years are taken into consideration?

To investigate this, let us consider several more millennia — see Table 59.B. In Figure 59.*b* the distribution curve for the years 2000-2999 is given again, together with that for the next 1000 years. It is clearly visible that in this latter period the maxima take place somewhat earlier in the year than during the period 2000-2999. The maximum of April 5, for instance, now occurs on April 4. After another 1000 years, the displacement again amounts to one day, as is seen in Table 59.B — see the numbers printed in bold.

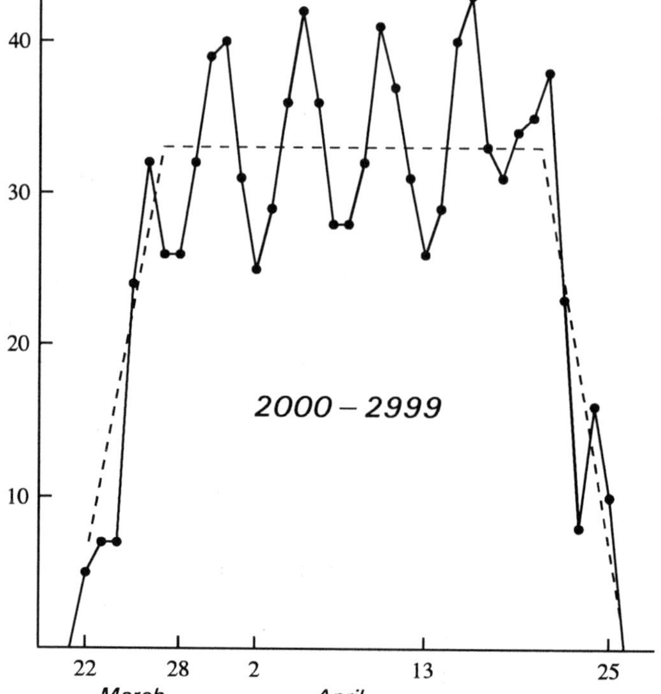

Fig. 59.a :

Distribution of the Easter dates in the years 2000 to 2999.

After approximately 6000 years the maxima will fall again on the initial dates, so here we have a periodicity of about 6000 years for what concerns the position of the maxima in the distribution.

The last column of Table 59.B gives the distribution of the dates of Easter in the period 2000-7999. This is represented in Figure 59.*c*. For the dates March 28 to April 17, we now have indeed a curve that is more or less flat, with a mean frequency of 200 (per 6000 cases). The frequencies for March 22, 23, ... 27 are equal to 1/7, 2/7, ... 6/7 of this amount, respectively.

The peak with value 231 on April 19 is disturbing, however. During the period 2000-7999, Easter occurs more frequently on April 19 than on any other date. (Surprisingly, this is not the case during the shorter interval 2000-2999, however! See the second column of Table 59.B).

Even if a much longer period is considered, that remarkable peak persists on April 19. Certainly it is not interesting for us to know on what date Easter will fall in that distant future. But our curiosity is a mathematical one only, and to perform a statistical investigation about the distribution of Easter Sunday as calculated by the current ecclesiastical rules, a large number of years must be taken into consideration.

TABLE 59.B

*Distribution of Easter Sunday in the Gregorian calendar
per millennium, years 2000 to 7999*

Date	2000 to 2999	3000 to 3999	4000 to 4999	5000 to 5999	6000 to 6999	7000 to 7999	Total
March 22	5	6	1	2	9	6	29
23	7	1	9	12	14	**13**	56
24	7	13	21	**18**	**17**	8	84
25	24	24	**26**	**18**	10	15	117
26	**32**	**26**	17	15	21	32	143
27	26	19	23	29	**39**	**36**	172
28	26	26	38	**43**	34	30	197
29	32	37	**41**	36	25	33	204
30	39	**42**	32	31	28	28	200
31	**40**	33	26	30	31	**41**	201
April 1	31	29	27	31	**42**	40	200
2	25	26	32	**41**	40	28	192
3	29	34	**43**	40	31	27	204
4	36	**41**	35	28	28	28	196
5	**42**	38	30	27	31	35	203
6	36	29	27	28	37	**44**	201
7	28	27	31	37	**41**	34	198
8	28	30	**40**	**43**	33	31	205
9	32	41	39	31	25	30	198
10	**41**	**42**	31	27	28	32	201
11	37	32	27	27	32	**41**	196
12	31	28	29	34	**43**	38	203
13	26	27	37	**41**	36	29	196
14	29	35	**44**	36	27	26	197
15	40	**42**	35	29	28	29	203
16	**43**	33	28	27	30	36	197
17	33	30	27	31	39	**44**	204
18	31	30	32	41	**42**	32	208
19	34	37	**43**	47	36	34	**231**
20	35	**43**	38	30	28	28	202
21	**38**	35	20	22	26	34	175
22	23	19	18	23	33	29	145
23	8	14	24	26	24	14	110
24	16	20	21	18	8	6	89
25	10	11	8	1	4	9	43

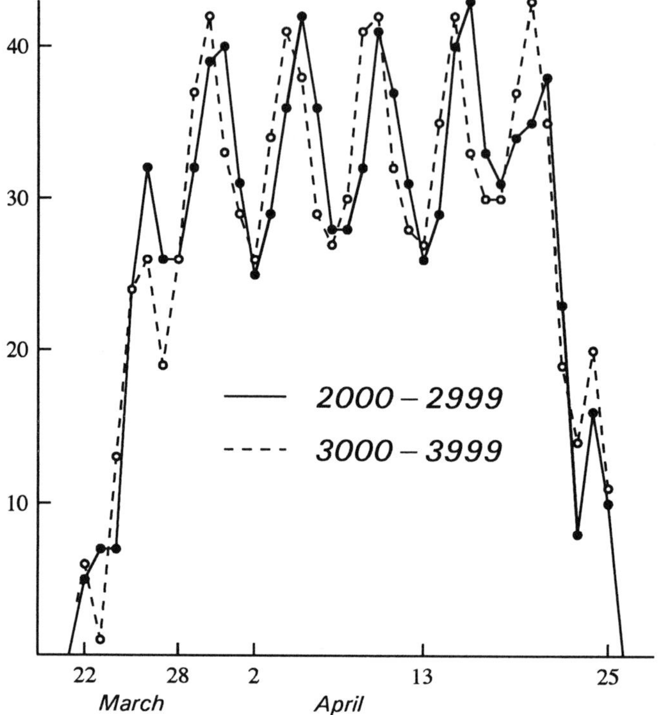

Fig. 59.b :

Distribution of the Easter dates in the years 2000 to 2999 (solid curve) and 3000 to 3999 (dashed curve).

Other unexpected facts appear from Table 59.B. For instance, considered over a period of 6000 years, Easter Sunday occurs somewhat more frequently on April 1 than on April 2, on April 5 than on April 4, on April 8 than on April 9, and on April 12 than on April 11. And calculation shows that this holds for successive 6000-year periods, for instance for the years 8000 to 13999, 14000 to 19999, and so on.

I don't know who was the first to discover that peak on April 19. When I found it in 1972, I published the above-mentioned details in a Dutch journal [1]. Some years later, Prof. Manfred Oswalden, of Klosterneuburg, Austria, provided additional data and explanations [2]. In the Gregorian calendar a period of 5700000 years is required for the cyclical recurrence of the Easter dates. Of course, *in practice* this very long period is meaningless, because anyway the correspondence of our present calendar with the motions of Sun and Moon can last for a few thousand years only. Long before those 5.7 million years, an adjustment of the calendar will be needed. Therefore, the period of 5700000 years is of theoretical interest only.

In the second column of Table 59.C, due to Oswalden, the distribution of the epacts in a period of 210 years is given. The *epact* of a year is the age of the 'ecclesiastical' Moon at the beginning of that year.

TABLE 59.C

Date of Easter	Number of epacts in a period of 210 years E	Number of Sundays in a period of 400 years S	Number in a period of 5 700 000 years $N = 475 \times E \times S$	Mean time interval in years
March 22	1	58	27 550	206.90
23	2	57	54 150	105.26
24	3	57	81 225	70.18
25	4	58	110 200	51.72
26	5	56	133 000	42.86
27	6	58	165 300	34.48
28	7	56	186 200	30.61
29	7	58	192 850	29.56
30	7	57	189 525	30.08
31	7	57	189 525	30.08
April 1	7	58	192 850	29.56
2	7	56	186 200	30.61
3	7	58	192 850	29.56
4	7	56	186 200	30.61
5	7	58	192 850	29.56
6	7	57	189 525	30.08
7	7	57	189 525	30.08
8	7	58	192 850	29.56
9	7	56	186 200	30.61
10	7	58	192 850	29.56
11	7	56	186 200	30.61
12	7	58	192 850	29.56
13	7	57	189 525	30.08
14	7	57	189 525	30.08
15	7	58	192 850	29.56
16	7	56	186 200	30.61
17	7	58	192 850	29.56
18	**7.42**	56	197 400	28.88
19	**8**	58	220 400	25.86
20	7	57	189 525	30.08
21	6	57	162 450	35.09
22	5	58	137 750	41.38
23	4	56	106 400	53.57
24	3	58	82 650	68.97
25	1.58	56	42 000	135.71

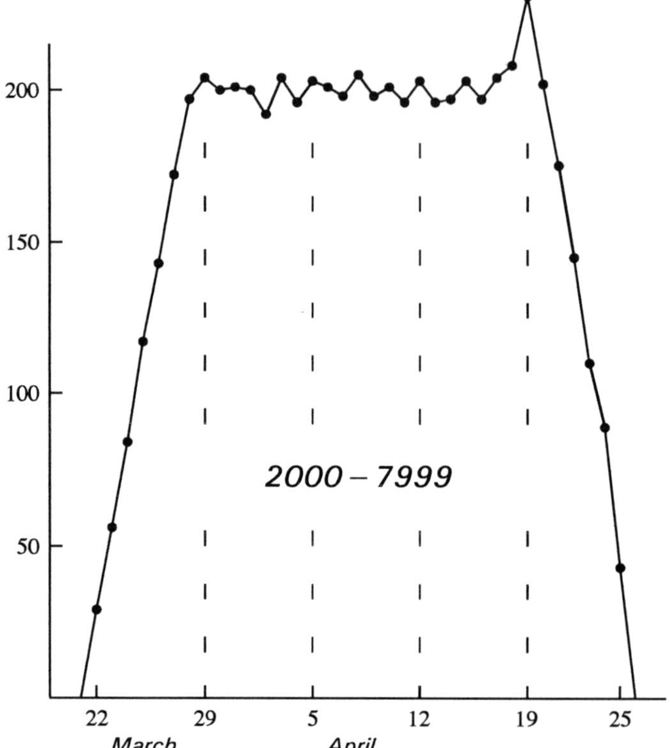

Fig. 59.c :
Distribution of the Easter dates in the years 2000 to 7999. Notice the peak at April 19.

To each year of the Gregorian calendar an epact is assigned according to a well-defined ecclesiastical rule. There are 30 different possible epacts (0 to 29), and each of them yields a date for the 'ecclesiastical Easter Full Moon', the Paschal Full Moon, in the period March 21 to April 18. Each of these dates, then, gives seven possible dates for Easter, namely the next Sunday. If, for instance, the Paschal Full Moon falls on April 2, the possible Easter Sundays are April 3 to 9.

The period of 210 years is obtained by multiplying 7 (the number of different weekdays) by 30 (the number of epacts). Of course, the sum of the numbers E in the second column of the table is 210. Notice the larger values of E for April 18 and particularly for April 19. These numbers are obtained by strictly following the ecclesiastical rule for the calculation of the epacts.

Indeed, there are only 29 possible dates for the Paschal Full Moon, namely March 21 to April 18 inclusively, while there are 30 possible epacts (0 to 29). Epact 25 plays a double role: Paschal Full Moon partly on April 17, and partly on April 18. Besides, April 17 is also assigned to epact 26, and April 18 to epact 24.

The numbers 7.42 and 1.58 in the second column of the table are given to two decimals only. Their exact values are $7^8/_{19}$ and $1^{11}/_{19}$, respectively.

The third column of Table 59.C indicates how many times in a period of 400 years (the periodicity of the weekdays in the Gregorian calendar) the date falls on a Sunday. This number S is equal to 56, 57 or 58 — see also Chapter 58.

The fourth column indicates how many times the date is Easter Sunday in the period of 5700000 years. This value is given by the formula $N = 475 \times E \times S$. The number 475 results from the division of 5700000 by 30×400.

Note the greater frequency of April 19 as Easter Sunday. But April 18 also exceeds all other dates. A graphical representation of the numbers N yields a curve very similar to that of Figure 59.c.

I have calculated the dates of Easter Sunday for 5700000 successive years, and could confirm the values of N (fourth column of Table 59.C) found by Oswalden.

And now we also understand why Easter occurs somewhat more frequently on April 1 than on April 2, etc.: April 1, 3, 5, 8, 10, 12, 15 and 17 fall more often on a Sunday than do April 2, 4, 9, 11 and 16. Reread what we said in the previous chapter about the weekday of Christmas.

Methods for calculating the dates of Easter Sunday in the Julian and in the Gregorian calendar are published in Chapter 8 of my *Astronomical Algorithms* (Willmann-Bell, ed.; 1991).

References

1. J. Meeus: "De frequentie van de Paasdata", *Hemel en Dampkring* (Netherlands), Vol. 71, No. 4, pages 132-135 (April 1973).
2. M. Oswalden: *Der Sternenbote* (Austria), Vol. 23, No. 5, pages 74-77 (May 1980).

60. The date of Easter — some interesting data

Series of April Easter dates

In the sequence of the successive dates of Easter, dates in March are always isolated. That is, two successive Easter Sundays cannot both occur in March. An Easter date in March is always preceded (one year earlier) and followed (one year later) by an Easter in April. For example:

<pre>
1996 April 7
1997 March 30
1998 April 12
</pre>

Sometimes there also occurs an isolated April Easter. The last ones occurred in 1770, 1781, 1838 and 1990. The next ones will take place in 2085 and 2142.

On the other hand, in the Gregorian calendar up to 10 successive Easter Sundays can fall in April. (In the Julian calendar the maximum number is 7). Such a series of ten successive April Easter dates did *not yet* take place. It will happen for the first time in the middle of the 29th century. Here is the list of these series between the years 1583 and 8000. Notice the very irregular distribution: there will be only one case during the third millennium, but no less than six during the next one, after which there will be a large gap until the next case:

$$2856 - 2865$$
$$3228 - 3237$$
$$3380 - 3389$$
$$3448 - 3457$$
$$3600 - 3609$$
$$3820 - 3829$$
$$3972 - 3981$$
$$7893 - 7902$$

Series of seven successive April Easters are more frequent. There are 47 such cases in the period 1583-3000. Here also the distribution is irregular: five series start between the years 2100 and 2200, but only one between 2300 and 2400, and one between 2500 and 2600:

1607–1613	1884–1890	2047–2053	2256–2262	2571–2577	2780–2786
1645–1651	1895–1901	2074–2080	2267–2273	2609–2615	2791–2797
1664–1670	1922–1928	2104–2110	2294–2300	2628–2634	2818–2824
1675–1681	1941–1947	2131–2137	2324–2330	2639–2645	2875–2881
1732–1738	1952–1958	2169–2175	2408–2414	2666–2672	2886–2892
1743–1749	1979–1985	2188–2194	2419–2425	2696–2702	2943–2949
1800–1806	2017–2023	2199–2205	2446–2452	2723–2729	2981–2987
1827–1833	2036–2042	2237–2243	2484–2490	2761–2767	

Recurrences

Recurrence of Easter after 11 years. — Sometimes, after 11 years Easter Sunday repeats on the same date. This can yield groups of at most four members taking place at intervals of 11 years, and then the last case always occurs in a leap year. Twelve such groups start between the years 1900 and 2100:

1903 –	1914 –	1925 –	1936	April 12
1923 –	1934 –	1945 –	1956	April 1
1927 –	1938 –	1949 –	1960	April 17
1947 –	1958 –	1969 –	1980	April 6
1971 –	1982 –	1993 –	2004	April 11

```
1991 - 2002 - 2013 - 2024        March 31
1995 - 2006 - 2017 - 2028        April 16
2015 - 2026 - 2037 - 2048        April  5
2019 - 2030 - 2041 - 2052        April 21
2039 - 2050 - 2061 - 2072        April 10
2059 - 2070 - 2081 - 2092        March 30
2063 - 2074 - 2085 - 2096        April 15
```

Repetition of a group of Easter dates after 152 years. — Years ago, a Mr. Franz Servais, of Echternach, Luxemburg, noticed that all hundred Easter dates of the years 1948 to 2047 will repeat in the same order 152 years later, from 2100 to 2199. This was mentioned in one short sentence on page 170 of the French journal *L'Astronomie* of April 1951. So we have

```
1948 March 28      →     2100 March 28
1949 April  17           2101 April  17
1950 April   9           2102 April   9
        etc.                     etc.
2047 April  14           2199 April  14
```

In 1980, Prof. Manfred Oswalden — already cited in the previous chapter — told me that in the Gregorian calendar such repetitions of whole groups of Easter dates after 152 years are rather frequent, that they occur during the whole Easter period of 5 700 000 years, and that such groups contain either 48, or 52, or 100 years. So, for instance:

Group of Easter dates	*Repetition after 152 years*	*Length of the group (years)*
1700 - 1747	1852 - 1899	48
1948 - 2047	2100 - 2199	100
2248 - 2299	2400 - 2451	52
2348 - 2399	2500 - 2551	52
2600 - 2699	2752 - 2851	100
2900 - 2947	3052 - 3099	48
3148 - 3247	3300 - 3399	100
3448 - 3499	3600 - 3651	52
3548 - 3599	3700 - 3751	52
3800 - 3899	3952 - 4051	100
4200 - 4299	4352 - 4451	100
4500 - 4547	4652 - 4699	48
4748 - 4847	4900 - 4999	100

Repetition of a whole century. — Prof. Oswalden also found that in the Gregorian calendar *every* century repeats *within* the Easter period of 5 700 000 years:

Century	First repetition of the Easter dates
1583 – 1599	85 183 – 85 199
1600 – 1699	343 600 – 343 699
1700 – 1799	343 700 – 343 799
1800 – 1899	85 400 – 85 499
1900 – 1999	343 900 – 343 999
2000 – 2099	344 000 – 344 099
2100 – 2199	85 700 – 85 799
2200 – 2299	427 800 – 427 899
2300 – 2399	344 300 – 344 399
2400 – 2499	86 000 – 86 099
etc.	

Ecclesiastical and astronomical Easter

Most amateurs know the rule: Easter falls on the Sunday which follows the first Full Moon occurring on or after the day of spring equinox. In fact, according to the ecclesiastical rules, the vernal equinox is fixed at March 21, not by the actual motion of the Sun. Moreover, the date of the Full Moon that occurs on or next after the vernal equinox — the Paschal Full Moon — is taken from ecclesiastical tables, not from astronomical ephemerides. Hence, the 'ecclesiastical Moon' is not strictly identical to the real Moon.

From the adopted rules, it results that the earliest possible date for Easter is March 22, and the latest possible is April 25. Hence, there are 35 dates on which Easter can fall.

Because spring equinox does not always occur on March 21 (see Chapter 57), and because the Paschal Full Moon does not exactly coincide with the true Full Moon, unavoidably there are years when Easter falls on an incorrect date, astronomically speaking.

For example, in 1974, Full Moon took place on Saturday April 6 at 21^h00^m Universal Time, so that normally Easter should fall the next day, Sunday April 7. Actually, Easter was on April 14.

In 1981, Full Moon occurred on Sunday April 19 at 7^h59^m UT, so Easter should have fallen on the *next* Sunday, April 26. Actually, it occurred on April 19.

In 2038, the spring equinox will occur on March 20 at 12^h40^m UT, and Full Moon on March 21 at 2^h09^m, so Easter should fall on the next Sunday, March 28. Actually, the date will be April 25.

Steven Verhezen has investigated for the years 1583 to 2582 which Easter dates, calculated according to the ecclesiastical rules, do not coincide with the date which would be obtained by using exact astronomical data. For each of these years, he searched the first Full Moon taking place after the March equinox (this sometimes happens on the same date, as in 1962), and the *next* Sunday he considered to be the 'astronomical' date of Easter.

Verhezen found that in no less than 78 years (out of the 1000 investigated ones) the date of the astronomical Easter differs from that of the ecclesiastical Easter. His results, which were published in the Dutch journal *Hemel en Dampkring*, Vol. 71, No. 4, page 131 (April 1973), are reproduced in Table 60.A.

TABLE 60.A

Years with different ecclesiastical and astronomical Easter dates, 1583 to 2582.   M = March, A = April.

Year	Eccles. Easter	Astron. Easter	Year	Eccles. Easter	Astron. Easter	Year	Eccles. Easter	Astron. Easter
1590	A 22	M 25	1962	A 22	M 25	2299	A 16	A 23
1598	M 22	M 29	1967	M 26	A 2	2316	A 16	A 9
1609	A 19	A 26	1974	A 14	A 7	2336	A 5	M 29
1622	M 27	A 3	1981	A 19	A 26	2339	M 26	A 2
1629	A 15	A 8	2038	A 25	M 28	2353	M 22	A 26
1666	A 25	M 21	2049	A 18	A 25	2372	M 26	A 23
1685	A 22	M 25	2069	A 14	A 7	2390	A 8	A 1
1693	M 22	M 29	2076	A 19	M 22	2394	A 17	A 24
1700	A 11	A 4	2089	A 3	M 27	2410	A 25	M 28
1724	A 16	A 9	2095	A 24	M 27	2417	A 2	A 9
1744	A 5	M 29	2096	A 15	A 8	2421	A 18	A 25
1778	A 19	A 12	2106	A 18	A 25	2429	A 22	M 25
1798	A 8	A 1	2114	A 22	M 25	2437	M 22	M 29
1802	A 18	A 25	2119	M 26	A 2	2448	A 19	M 22
1818	M 22	M 29	2133	A 19	M 22	2451	A 16	A 23
1825	A 3	A 10	2147	A 16	A 23	2467	A 24	M 27
1829	A 19	A 26	2150	A 12	A 19	2468	A 15	A 8
1845	M 23	M 30	2170	A 1	A 8	2471	A 5	A 12
1876	A 16	A 9	2171	A 21	M 24	2486	A 21	M 24
1900	A 15	A 22	2174	A 17	A 24	2488	A 4	M 28
1903	A 12	A 19	2190	A 25	M 28	2491	M 25	A 1
1923	A 1	A 8	2201	A 19	A 26	2492	A 13	A 20
1924	A 20	M 23	2221	A 15	A 8	2495	A 10	A 17
1927	A 17	A 24	2245	A 13	A 20	2515	M 31	A 7
1943	A 25	M 28	2277	A 22	M 25	2519	A 16	A 23
1954	A 18	A 25	2296	A 19	M 22	2546	A 17	A 24

We see that, while the extreme possible dates for the ecclesiastical Easter are March 22 and April 25, the astronomical Easter can occur on March 21 (as in 1666) and on April 26 (1609, 1829, 1981, 2201 and 2353). Even March 20 is a possible date, for instance when the equinox occurs on March 19, and Full Moon later on that same date. This does not happen, however, in the 1000-year period considered here.

Christian Easter and Jewish Pesach

The Jewish Pesach (15 Nisan) generally does not fall on the same date as the Christian Easter. In some years, however, the two dates do coincide. J. E. Weiland found that this occurs 32 times during the period 1583–3000. Here is the list of these cases, as published in the Austrian journal *Der Sternenbote*, Vol. 24, No. 4, page 55 (April 1981). Notice the large gaps from 1609 to 1805, 1981 to 2123, and 2566 to 2715.

1609 April 19	2170 April 1	2715 April 11
1805 April 14	2201 April 19	2718 April 7
1825 April 3	2299 April 16	2735 March 31
1903 April 12	2319 April 6	2742 April 12
1923 April 1	2417 April 2	2762 April 1
1927 April 17	2448 April 19	2793 April 18
1954 April 18	2471 April 5	2813 April 7
1981 April 19	2495 April 10	2820 April 19
2123 April 11	2515 March 31	2891 April 15
2143 March 31	2546 April 17	2911 April 5
2150 April 12	2566 April 6	

61. Rounding numbers

Results of measures or of calculations should be rounded correctly and meaningfully, where it is needed.

Rounding should be made to the *nearest* value. For instance, 15.88 is to be rounded to 15.9, or to 16, not to 15. However, calendar dates and years are exceptions. For example, March 15.88 denotes an instant belonging to March 15: it means 0.88 day after March 15, 0^h. Hence, if we read that an event occurs on March 15.88, it takes place on March 15, not on March 16. Similarly, 1984.69 denotes an instant belonging to the year 1984, not 1985.

Only meaningful digits should be retained. Suppose that a rectangular field has a length of 45 meters and a width of 23. What is the ratio F of these two distances? Shall I write that the answer is $F = 1.956521739130435$ under the pretext that this is actually the value my computer gives when I divide 45 by 23?

In no way. Common sense and some understanding of the obtained accuracy are indispensable here. In our example it is evident that the length of the field is not *exactly* 45 meters. When such a length is mentioned, it generally is rounded to the nearest integer number. So we may assume that the field's length actually lies between 44.5 and 45.5 meters. The same holds for the width, which in our example is somewhere between 22.5 and 23.5 meters. If, for instance, the sizes are actually 45.32 and 22.76 meters, the ratio is equal to 1.99, so we see that in that case even the second decimal of the above-mentioned value of F is incorrect! Even if the lengths were 45 and 23 meters 'exactly' to the nearest millimeter, the value of the ratio would be somewhere between 1.956457 and 1.956586.

Here is another example. Calculate the visual magnitude m of Jupiter on 1995 March 20 at 0^h Universal Time, using Müller's formula

$$m = -8.93 + 5 \log r\Delta$$

where r is Jupiter's distance to the Sun, and Δ its distance to the Earth, both in astronomical units. In the formula, the logarithm is to the base 10. It is *not* the LOG function found in most programming languages and which is to base $e = 2.71828\ldots$. From an accurate ephemeris of Jupiter, the distances at the above-mentioned instant are found to be $r = 5.3601791$ and $\Delta = 5.0350549$. Müller's formula then gives $m = -1.77408244$. But giving all these decimals would be ridiculous and it would give the reader a false impression of high accuracy.

Since the constant -8.93 in Müller's formula is given to 0.01 magnitude only, no higher accuracy can be expected in the result of the calculation. And, in any case, the meteorological phenomena in the atmosphere of Jupiter are such that the magnitude of that giant planet cannot be predicted with an accuracy better than 0.01 or even 0.1. Besides, Müller's formula is not 'exact' because it does not take into account the phase effect which, although being rather small, is not entirely negligible.

As another example, John Mosley [1] mentions a commercially available program giving rising and setting times of heavenly bodies to the nearest 0.1 second, which is impossibly precise.

Some 'feeling' and sufficient astronomical knowledge are necessary here. For instance, it would be completely irrelevant to give the illuminated fraction of the Moon's disk accurate to 0.000000001.

The rounding should be performed *after* the whole calculation has been made, not before the start or before the input of the data into the computer. For example, if $1.4 + 1.4$ should be calculated to the nearest integer, and if we first round the given numbers, we would obtain $1 + 1 = 2$. In fact, $1.4 + 1.4 = 2.8$, which is rounded to 3.

A similar error occurs when distances, already rounded to the nearest mile, are converted to kilometers. In this case, the value of 17 km, for instance, will never be reached because

10 miles will give 16.09 km, which is rounded to 16 km,
11 miles will give 17.70 km, which is rounded to 18 km.

Paulos [2] writes: "Consider a precise number that is well known to generations of parents and doctors: the normal human body temperature of 98°.6 Fahrenheit. Recent investigations involving millions of measurements have revealed that this number is wrong: normal human body temperature is actually 98°.2 Fahrenheit." The error is due to the fact that the original measurements were averaged and then rounded to the nearest degree centigrade: 37° Celsius. When this temperature was converted to Fahrenheit, however, *the rounding was forgotten*, and 98.6 was taken to be accurate to the nearest tenth of a degree. [The formula for converting from centigrade to Fahrenheit is $F = 32 + 1.8C$].

When in 1949 Neptune's second satellite, Nereid, was discovered, a note was published in the French journal *L'Astronomie* [3]. The author was Gabrielle Camille Flammarion, the wife of the famous Camille Flammarion. There it was stated that the diameter of Nereid is 322 kilometers. How could such a precise value be given for that remote, small satellite of magnitude 19, which even in the largest telescopes appears as a tiny dot? The answer is obvious: the good lady had read in an American message that the diameter of Nereid was *200 miles*. To obtain the value in kilometers, she multiplied the number by 1.61, the factor for converting miles into kilometers, and she obtained 'exactly' 322 kilometers! Apparently, she did not understand that the announced diameter of 200 miles was only approximate, and that the true value could have been somewhere between, say, 170 and 230 miles.

The reverse of what we said at the beginning of this chapter is also true: published, rounded values should be considered with a thorough knowledge of the matter. A Belgian almanac gave the following times for sunset at Uccle, near Brussels, in 1995:

July	8	19^{h}56^m UT
	9	19 55
	10	19 55
	11	19 54

The following question has been posed: why does the Sun set one minute earlier on July 9 than on the previous day, but sets at the same instant on July 10 as on July 9? Then the next day there is again a difference of one minute.

The person who states such a question overlooks that the published times *are rounded to the nearest integer minute*. Indeed, it is not interesting to give the times of sunrise and sunset to the nearest second. Besides, that would not make much sense: the atmospheric refraction at the horizon, which must be taken into account in the calculation, is somewhat variable as a function of the pressure and the temperature of the air, which are not predictable. So the exact value of the refraction at the horizon cannot be predicted, and this excludes the possibility to predict the times of rising and setting of the heavenly bodies with an accuracy of, say, one second.

Of course, adopting a *mean* value for the atmospheric refraction at the horizon, we *can* calculate the times of sunset to the nearest second. When this is done for the four mentioned dates at Uccle, we obtain

$$
\begin{array}{lll}
\text{1995 July} & 8 & 19^h 56^m 01^s \text{ Universal Time} \\
& 9 & 19\ 55\ 22 \\
& 10 & 19\ 54\ 40 \\
& 11 & 19\ 53\ 55
\end{array}
$$

The differences between the successive times are now 39, 42, and 45 seconds, respectively — quite regular! And now we see that the instants for July 9 and 10 are *not* equal, although both are *rounded* to the same value, $19^h 55^m$. Hence, on July 10 the Sun didn't set at the same time as on the previous day!

As a final remark, let us mention that trailing zeros can be important. For instance, 18.0 is not the same as 18. The first value means that the actual number lies between 17.95 and 18.05, while the second value has been rounded to the nearest integer and can actually be equal to any number between 17.5 and 18.5. For this reason, trailing zeros *must* be given in the result to indicate the accuracy: a star of magnitude 7 is not the same as a star of magnitude 7.00.

Of course, this rule doesn't hold for values which by definition are integers. An hexagon has 6 sides, and January has 31 days — exactly 6 and exactly 31, in these cases!

References

1. John Mosley, *Sky and Telescope*, Vol. 78, No. 3, page 300 (September 1989).
2. John Allen Paulos, *Skeptical Inquirer*, Vol. 20, No. 1, page 44 (January-February 1996).
3. G. C. F., *L'Astronomie*, Vol. 63, page 252 (September-October 1949).

62. *Predicting sunspot activity*

The text which follows was first published in the *Journal of the British Astronomical Association*, Vol. 101, No. 2, pages 115-116 (April 1991). It is reproduced here under a slightly modified form.

In the astronomical literature we find, now and then, a prediction about the next maximum or minimum of sunspot activity, but rarely do we read whether such forecasts have proven correct. So, let me give some examples.

In 1984, the British scientist D. R. Whitehouse [1] predicted that the next minimum of the solar cycle would occur in 1988.2, and the subsequent maximum at the epoch 1991.8. The uncertainty of both estimates was stated to be as small as 0.2 year. The actual times were 1986.7 and 1989.8, so that Whitehouse's predicted times were in error by 1.5 and 2.0 years, respectively! In the present state of our knowledge, nobody can predict the time of the next sunspot maximum or minimum with an accuracy of 0.2 year. — *Note:* the decimal form 1988.2 means 0.2 year after the beginning of the year 1988, etc.

Dong Shi-lun [2] predicted that the sunspot cycle No. 22 would reach a maximum in 1992.1 $\pm$ 1.5, with a maximum smoothed mean of 78 $\pm$ 11. The correct time was 1989.8 (2.3 years earlier than predicted), and the maximum value of the smoothed monthly means was 162, or more than twice the value expected by Dong Shi-lun.

In 1955 Schove [3] issued predicted times for the sunspot maxima of the second half of the 20th century. His predictions, stated to be accurate to ± 2 years, compare as follows with the actual values:

Predicted time	Actual time	Difference (years)
1958.5	1957.9	−0.6
1972.5	1969.1	−3.4
1984.5	1979.9	−4.6
1994.5	1989.8	−4.7

Enter the planets

Now and then, an author states that the variations of sunspot activity, and in particular the 11-year cycle, can be explained by tidal or other effects due to the planets. The number of those papers published during the last 100 years might suggest that something must be true in these 'planetary theories'. However, each such theory differs completely from the others, and a close examination shows that each study contains gross errors, misinterpretations, arbitrary statements or arbitrary manipulations of the planetary motions. What a nice thing if really the solar phenomena were determined by the positions of the planets! For, in that case, sunspot activity could be predicted many years in advance. Indeed, some authors tried to predict sunspot maxima using their planetary 'theory' — but without success.

In 1900, for instance, E. W. Brown (yes, *the* Ernest W. Brown who a few years later would become famous for his theory of the motion of the Moon) gave a 'possible explanation' of the sunspot cycle, by considering the tidal effects of the two giant planets Jupiter and Saturn [4].

After having made erroneous reasonings and arbitrary assumptions, Brown wrote: "If the theory developed here is correct, the actual minimum should not be reached until 1901 or 1902, if it is attained as early. The next maximum should be a period of *great activity*, (. . .) and it should be attained in 1908."

Actually, the sunspot maximum took place in October 1905, with a secondary maximum in February 1907, and it was the *lowest* sunspot maximum of the whole period 1820–1990.

The mean length of the sunspot cycle, from the maximum of March 1750 to that of October 1989 (22 cycles) is 10.89 years; from the maximum of 1917 to that of 1989 the mean length is even shorter: 10.3 years. The sidereal revolution period of Jupiter, 11.86 years, is sensibly longer than that. Hence, the two variations, if started in phase, become in opposite phase after a few tens of years. Consequently, the variable distance of Jupiter to the Sun cannot be the cause of the 11-year sunspot cycle. Jupiter was near the *perihelion* of its orbit at the time of the sunspot maximum of 1928, but it was near the *aphelion* at the sunspot maximum of 1957. And the synodic period Jupiter–Saturn cannot explain things either: its length, 19.86 years, is sensibly less than twice the sunspot cycle.

Due to the presence of the planets, the barycenter of the solar system does not coincide with the center of the Sun. It can even wander outside the Sun's globe during several years — see Chapter 26.

The late French scientist Alexandre Dauvillier believed, against all evidence, the height of a sunspot maximum to be proportional to the distance of the solar system's barycenter to the Sun's center at that epoch. On this basis, he predicted [5] that the sunspot maximum of 1979 would be high, with a Wolf value of 160. This prediction proved correct: the smoothed monthly means reached a maximum of 167 in November 1979.

However, one single prediction which comes true proves nothing, as it could so easily be a lucky chance. But *one* prediction which fails proves that the 'theory' on which it is based is in error. And, indeed, for the next solar maximum Dauvillier failed lamentably.

In April 1990 the planets' positions were such that the barycenter of the solar system passed close to the center of the Sun. Therefore, Dauvillier [5] predicted that the maximum of the sunspot cycle No. 22 would be very shallow: "Le maximum d'activité de 1990 *sera le plus faible du siècle.*" Alas, cycle 22 actually was one of the most powerful on record; the sunspot maximum of 1989 was as high as those of 1778 and 1947, and since 1750 only the maxima of 1957 and 1979 were stronger.

In memoriam the 80-year period

Figure 62.*a* shows the variation of the smoothed monthly means of the Zurich (Wolf) sunspot numbers from 1750 to 1995. On the horizontal axis the years of the sunspot maxima are mentioned. We see that the high maxima of 1769, 1778 and 1787 were followed by three low maxima (1804, 1816, 1829). Then there were high maxima again (1837 to 1870), followed by lower ones (1884, 1893, 1905), after which the maxima were stronger again.

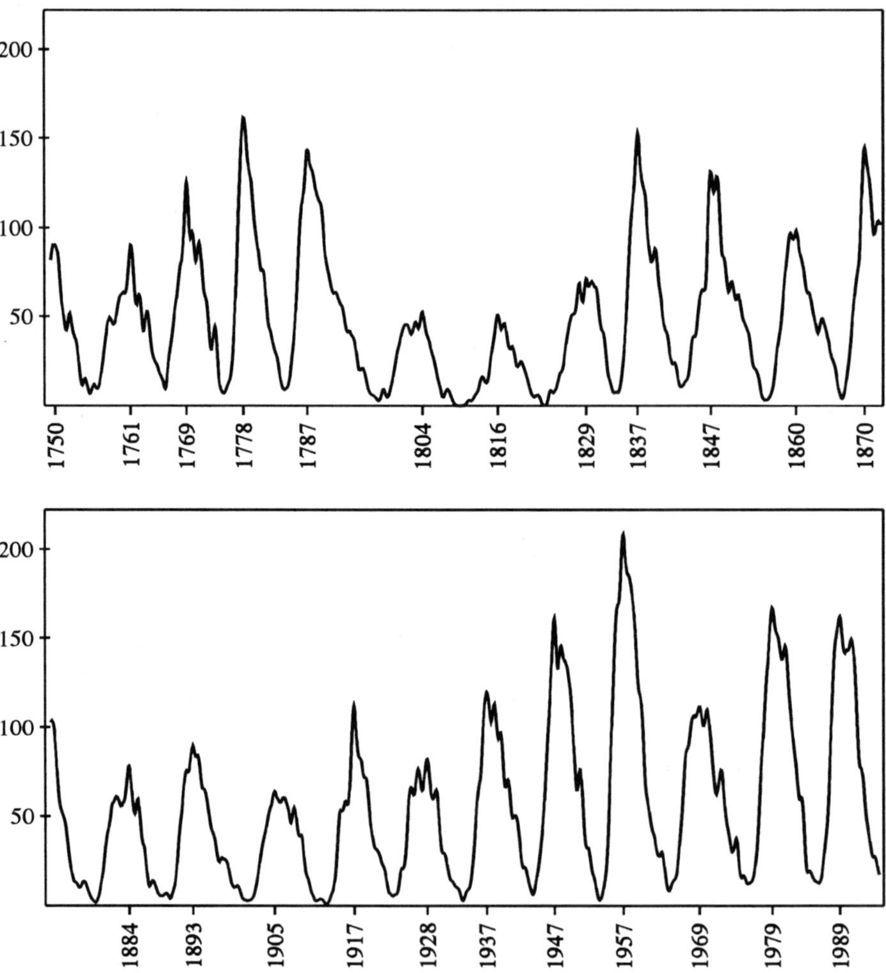

Fig. 62.a : *Variation of the smoothed monthly means of the Zurich (Wolf) sunspot numbers, 1750 to 1995.*

From this material, Gleissberg [6] deduced the existence of a period of about 80 years in the height of the sunspot maxima. The last maximum Gleissberg could dispose of was the high one of 1947. If the 80-year cycle really existed, the next maxima would again be low ones. Surprisingly, the next maximum (1957) was the highest on record. And, after the rather moderate maximum of 1969, there came two more high ones (1979, 1989)!

So we may conclude that the 80-year period, whose existence seemed well-established in the middle of the 20th century, very probably doesn't exist.

Bray and Loughhead [7] write: "It is evident that the basic period of the sunspot cycle, defined as the time between successive maxima or minima, is approximately 11 years. However, the great variations in the height of the maxima have led numerous authors to seek secondary cycles which, when superimposed on an 11-year cycle, correctly reproduce the observed curves. It is easy to represent the past course of the sunspot number variation by such means, but the real test lies in a prediction of the *future* variation: in this test such techniques have not been successful. It must be emphasized, therefore, that the only reliably established periodicity of the sunspot cycle is the well-known 11-year period."

As an example, let us consider the case of Cohen and Lintz [8] who, after applying the 'maximum entropy spectral analysis' to past sunspot cycles, and after stating that "the data suggest that our predictions for the smoothed numbers at sunspot maxima may be in error by up to $\pm 25\,\%$", made the following predictions. The remarks between brackets are mine.

"Cycle 21 will have a rather broad peak [the peak was rather sharp], possibly reaching its maximum value late in 1982 [the maximum took place late in 1979]. Mean sunspot numbers for this peak may not, however, exceed 50 [they reached 167]. The sunspot minimum following cycle 21 should occur around 1988 [it took place in September 1986], and could exhibit smoothed sunspot numbers as low as 2 [the minimum of the smoothed monthly means was as high as 12.8]. Twelve month running mean sunspot numbers greater than 100 will not be observed again until approximately 2015 [they occurred from 1979 to 1981, and from mid-1988 to early 1992]."

So, in spite of the use of 'maximum entropy spectral analysis', the prediction by Cohen and Lintz was completely wrong. Certainly, at this moment there is no known method for making reliable long-term forecasts of the sunspot activity.

Note. — The smoothed monthly means, used for the preparation of this text, are based on the definitive monthly Zurich sunspot numbers, but they have been calculated with the formula which I advocated in 1958 [9]. This formula, mentioned on page 316 of this book, gives a greater weight to the central months and yields a better smoothed curve than the 'classical' Zurich procedure.

References

1. D. R. Whitehouse: *Astronomy and Astrophysics*, Vol. 145, No. 2, pages 451-453 (1985). — Cited in *Sky and Telescope*, Vol. 70, page 308 (October 1985).

2. Dong Shi-lun: *Acta Astronomica Sinica*, Vol. 27, No. 1, page 53 (1986).

3. D. J. Schove: *Journal of Geophysical Research*, Vol. 60, No. 2, page 136 (June 1955).

4. E. W. Brown: *Monthly Notices Royal Astronomical Society*, Vol. 60, No. 10, pages 599-606 (Suppl. Number; 1900).

5. A. Dauvillier: *Ciel et Terre* (Belgium), Vol. 93, No. 4, pages 204-214 (July-August 1977).

6. W. Gleissberg: "Die Häufigkeit der Sonnenflecken", *Scientia Astronomica*, Vol. 2, Berlin, Akademie-Verlag (1952).

7. R. J. Bray and R. E. Loughhead: *Sunspots*, page 239 (London, 1964).

8. T. J. Cohen and P. R. Lintz: *Nature*, Vol. 250, pages 398-400 (1974).

9. J. Meeus: *Ciel et Terre*, Vol. 74, No. 6, pages 445-449 (November-December 1958).

Index